Scientists Confront

CREATIONISM

CH4 methane
NH3 Ammonia
H2 Hydrogen
H2O water
HCN Hydrogen Cyanide

Scientists Confront CREATIONISM

Intelligent Design and Beyond

EDITED BY

Andrew J. Petto and

Laurie R. Godfrey

W. W. NORTON & COMPANY

New York • London

"Darwinism, Creationism, and 'Intelligent Design'" is adapted and reprinted by permission of the publisher from *Darwinism Comes to America* by Ronald L. Numbers, pp. 1–23, Cambridge, MA: Harvard University Press, copyright ©1998 by the President and Fellows of Harvard College. All rights reserved.

Kevin Padian and Kenneth D. Angielczyk are particularly grateful to the Paleontological Society for allowing them to revise, update, and reprint their contribution to *The Paleontological Society Papers,* Volume 5 (1999), edited by P. H. Kelley, J. R. Bryan, and T. A. Hansen.

Originally published under the title *Scientists Confront Intelligent Design and Creationism*

Printed in the United States of America

For information about special discounts for bulk purchased, please contact W. W. Norton Special Sales at specialsales@wwnorton.com or 800-233-4830

Manufacturing by Courier Westford
Book design by Chris Welch
Production manager: Devon Zahn

Library of Congress Cataloging-in-Publication Data

Scientists confront intelligent design and creationism.
Scientists confront creationism : intelligent design and beyond / edited by Andrew J. Petto and Laurie R. Godfrey.
p. cm.
Originally published as: Scientists confront intelligent design and creationism. 2007.
Includes bibliographical references and index.
ISBN 978-0-393-33073-1 (pbk.)
1. Evolution (Biology) 2. Intelligent design (Teleology) 3. Creationism. I. Petto, Andrew J. II. Godfrey, Laurie R. III. Title
QH366.2.S43 2007a
576.8—dc22
2007039482

W. W. Norton & Company, Inc.
500 Fifth Avenue, New York, N.Y. 10110
www.wwnorton.com

W. W. Norton & Company Ltd.
Castle House, 75/76 Wells Street, London W1T 3QT

2 3 4 5 6 7 8 9 0

To the memory of
Dr. Clarence Churchill Sherwood —
for his lifelong dedication
to science and to his family.

CONTENTS

CONTENTS

ACKNOWLEDGMENTS

SIR ISAAC NEWTON WROTE TO Robert Hooke that his success was due to his standing on the "shoulders of giants"—an acknowledgment that we achieve most when others have laid a suitable foundation. For this reworked version of *Scientists Confront Creationism*, it is fitting to point out that the yeoman's work of promoting evolution and preserving scientific literacy against pseudoscientific adulteration is done by ordinary folk. It is because we hear questions and concerns from parents, teachers, journalists, clergy, and school administrators that we have such an extensive knowledge of anti-evolutionary pseudoscience. Without them, most of us in the sciences and related fields would spend our time on our own narrow specialties. Because of these people and the challenges that are being raised against evolution in their own communities, we knew we had to bring the original 1983 book—*Scientists Confront Creationism*—up-to-date, and it is for them that we knew we had to succeed.

In an edited volume such as this one, each of the contributors depends on a network of supporters—some of them disciplinary colleagues but most of them nonspecialists who ask questions that our colleagues would never think of. Because of these questions, we have a better book, and, even though we can never list all those people, none of our contributions would be as useful without them. The added value of their comments and questions is an integral part of the book, and we would be remiss not to mention it.

Because of the range and breadth of opposition to evolution among the general public, we are aware that literally hundreds of colleagues are actively involved in this issue throughout the world. Each one of those would have made a valuable contribution and added to the book significantly, but we were forced to make difficult choices about what to include and what to leave out. Many of the potential contributors have their own books on these and related issues, and we commend those to our readers who want to know more.

Unlike the academic publications that many of us write, a book on anti-evolutionism runs the risk of being outdated even before it is printed. When we began the book, there were several challenges to evolution in state boards of education and local school boards. As legal and curricular challenges developed, the situation changed several times over the course of the production of this book, and we know that things are not now—and may never be—"settled." We appreciate the patience of our editors at W. W. Norton as our scrambling to revise the chapters to reflect the changing anti-evolution landscape resulted in several delays in publication. We think—and our editors at Norton seemed to accept—that the result is a more coherent and up-to-date book. And we are also sure that the major setbacks to "intelligent design" in Ohio and in Dover, Pennsylvania, will result in a significant repackaging of anti-evolutionary ideas—but not to their elimination from our culture—in the next few years.

Finally, readers will note that we set "intelligent design" in quotation marks but do not mark other pseudoscientific concepts in this way. We decided on this convention primarily because "intelligent design" is a technical term in general use in the fields of engineering and industrial design. Of course, in those fields, it refers to the work of human designers who consider the practical aspects of their work from the point of view of the people who will have to use them. The anti-evolutionary concept promoted by the Seattle-based Discovery Institute as "intelligent design theory" is a neologism that misappropriates the term and applies an entirely unrelated meaning to the phrase. We want to be sure that readers understand that we object only to this misuse of the term by anti-evolutionists and that we are distinguishing this use from the legitimate use by engineers and industrial designers.

AJP & LRG

Milwaukee, Wisconsin,

and

Pelham, Massachusetts

May 2006

PREFACE

Why We Did It Again

Laurie R. Godfrey
& Andrew J. Petto

WHEN *SCIENTISTS CONFRONT CREATIONISM* WAS being produced in the early 1980s, evolution supporters were riding the crest of a heady wave. State and federal courts had recently struck down laws restricting the teaching of evolution in public schools or requiring equal time for "creation science" (Matsumura 2001). Many supporters of evolution were relieved by these legal victories, often assuming that the matter was settled. Indeed, subsequent court cases that were decided in favor of evolution seemed to bolster that view, particularly the landmark *McLean v. Arkansas Board of Education* (1982; 529 F. Supp. 1255).

On the other hand, one could see the stream of challenges to evolution—and in particular to evolution education—as evidence of the persistent commitment of and engagement by anti-evolutionists to challenging the central role of evolution in science education. In retrospect, the apparent quiescent period following *McLean v. Arkansas* represented in part another "reformation" or re-creation in the anti-evolutionists' quest for a "scientific alternative" to evolution. A new strategy was evolving among those who had openly opposed evolution on religious grounds in their search for a framework that appeared more secular and "scientific." In part, this strategy responded to language in Justice Antonin Scalia's opinion in *Edwards v. Aguillard*: "The people of

Louisiana, including those who are Christian Fundamentalists, are quite entitled, as a secular matter, to have whatever scientific evidence there may be against evolution presented in their schools . . ." (Scalia 1987). More recently, the "evidence against evolution" slogan has emerged in opposition to science education standards that promote evolution education (Petto 2000; Meikle 2002).

Anti-evolutionism is back, and it has new aliases. But, as many chapters in this book will show, these aliases are little more than new labels for the same tired arguments—sometimes with new voices and faces. The substance of the argument has changed little in the years since this book's first edition in 1983. Many people, however, are not aware of the history of anti-evolutionism and its intellectual bankruptcy, and thus unknowingly support what they erroneously assume is a scientific research program. In doing so, they do not recognize that this "new scientific alternative" is merely the old-style "creation science" repackaged and relabeled for secular consumption.

We seek here to address the newer "alternatives" and approaches, such as "intelligent design theory," and (1) to show explicitly their links to the "creation science" that motivated the first edition; and (2) to provide a critique of more recent anti-evolutionist materials and formulations. The anti-evolutionist initiative in the past twenty years has been multidimensional and has crossed many disciplines. In this volume, specialists from the natural and social sciences and the humanities provide a perspective for interested readers about this persistent cultural phenomenon. The chapters included here, however, are little more than the threshold to a vast and growing literature.

We compiled this second edition with the expectation that the need to support evolution in the public domain will be ongoing—particularly in public education—despite its firm position in the sciences. We wish for two outcomes: (1) that it will communicate to a public audience the strength of scientific support for evolution across professional disciplines; and (2) that it will encourage our colleagues in the sciences and the humanities to write, speak out, and become active in their own communities in ways that directly address anti-evolutionist activity.

Anti-evolutionism is not a passing phenomenon, nor is it a matter of logic or intellectual integrity. It is, however, a phenomenon deeply embedded in cultural history, trends, and institutions. We offer this revised edition as an overview of the complex world of anti-evolutionism at the beginning of the twenty-first century. We invite our readers to use these chapters as a springboard for their own professional and civic engagement in support of evolution and evolution education.

References

Matsumura, M. 2001. Ten significant court decisions. Available at www.ncseweb.org/resources/articles/5690_10_significant_court_decisi_2_15_2001.asp last accessed August 27, 2007).

Meikle, E. 2002. Ohio reflections. *Reports of the National Center for Science Education* 22 (6): 4–5.

Petto, A. J. 2000. Creeping creationism in PA's science education standards. *Reports of the National Center for Science Education* 20 (4): 13–15.

Scalia, A. 1987. Justice Scalia, with whom the Chief Justice joins, dissenting. *Edwards v. Aguillard* 482 U.S. 578.

Introduction: The Problems with Creationism

THIS BOOK IS ENTITLED *Scientists Confront Creationism*, yet it is clear from looking at the contents that this isn't just about science. Indeed, the three major sections of the book are a good summary of the issues: creationism has deep historical and cultural roots in the United States (part I), it is of course an alleged challenge to a scientific theory (part II), and it involves complex questions of philosophy and science education (part III). In other words, creationism is not a single problem but rather a panoply of loosely connected threads, all of which continue to pose a threat not only to science funding but, more important, to the quality of education and ultimately even to free speech in modern liberal societies.

Historically (part I of this book), of course, creationism has preceded evolutionary theory (despite some interesting speculations from materialist philosophers in ancient Greece). Most important, natural theology was the dominant paradigm when Charles Darwin published *The Origin of Species* in 1859. In fact, Darwin spent a considerable amount of time directly answering the arguments of those such as William Paley (1802), who attributed features of the natural world to the actions of a purposeful designer. Paley was a first-rate naturalist (unlike, one cannot help adding, modern proponents of "intelligent design" such as Michael Behe, William Dembski, and Phillip Johnson), and Darwin took him very seriously.

Contrary to what modern creationists like to claim, Darwinism was also vigorously challenged within the scientific community after its alleged rise to the status of biological "dogma." Bowler (1983) recounts in detail the story of how, for example, neo-Lamarckism[1] was very much alive and well in scientific circles, both in Europe and the United States, well into the twentieth century. Many professional scientists at the time saw Darwin's theory in decline and the future of biology uncertain. But the rediscovery of the laws of genetics—and their incorporation into the theories of natural selection and descent with modification—led to the so-called neo-Darwinian synthesis of the middle part of the twentieth century (Mayr and Provine 1980). Though modern creationists refuse to take note of this remarkable scientific fusion, the theory of evolution is now as much in doubt in biology as quantum mechanics is in physics (which, of course, doesn't guarantee that either of them is "true," since science can by its nature only provide provisional answers).

Ever since the defeat of scientific neo-Lamarckism at the turn of the twentieth century, therefore, creationists and "intelligent design" (ID) proponents (better thought of as neo-creationists, as Eugenie Scott points out in this volume) have fought a battle that at best was based on a barely viable philosophical position and at worst has simply been a stubborn exercise in attempting to impose one's parochial religious ideology on the rest of the world. Furthermore, with the exception of the Scopes "monkey trial," this battle has been set back throughout the twentieth cen-

1. It was Jean-Baptiste Lamarck's view that environmental change created "needs" in organisms that responded by differential usage of organs. In response, the organs grew or shrunk, and evolution occurred as this intragenerational modification experienced by the parents is inherited by descendants. In contrast, Darwinian models are based on differential survival and reproduction of those who possess favored variations, and evolution occurs as the environment "tests" these variations, allowing some to leave descendants that also have these features and producing modifications in the proportions of certain variations among future generations.

tury by an endless stream of legal defeats. At the local level, however, creationists have been doing much better, shamelessly bypassing laws and regulations and successfully intimidating school boards and teachers into avoiding or watering down "the issue." This is part of the reason books like the one you are about to read are much needed. (Another reason is that if one does an online search for books using the keyword *evolution,* one finds mostly creationist propaganda, much of it directed to children.)

Scientifically (part II), as I have noted, creationists and ID supporters don't have a leg to stand on, but they nevertheless manage to make a lot of noise. One of the recurring tricks I have experienced when talking with creationists is that they like to portray the debate as one concerning the "science of origins." Don't waste time looking up any "Origins Science" Department at the local college or university; like much else associated with creationism, it is a pure rhetorical device. The "origins" (plural) to which creationists refer are three: the origin of humans (sometimes extended to the origin of any species, though an increasing number of creationists allow for some "micro-evolution" within vaguely defined "kinds" of living organisms), the origin of life, and the origin of the universe.

Even a superficial acquaintance with science reveals that these three subjects actually belong to three very distinct disciplines (evolutionary biology, biophysics, and cosmology), and that only the first one has anything to do with the theory of evolution proper. Darwinian mechanisms cannot get going until after life originates on a planet, and, last time I checked, planets and galaxies were not making babies that could be subjected to natural selection. In other words, to put it in terms of "origins" is misleading at best, which once again clearly reveals the thin veneer of science thrown over the creationist Trojan horse.

This is not to say that the current version of the theory of evolution does not face serious unanswered questions from a genuinely scientific perspective. Of course it does. So do other dominant theories in science. For example, even though quantum mechanics is by all accounts the single most successful scientific theory ever produced (because it accounts for so much, and with such a high degree of accuracy in its predictions), physicists still do not know how to reconcile it with Einstein's theory of relativity. The two theories make conflicting predictions in certain cases (for example, the behavior of a black hole), yet they both seem to be correct (Greene 1999). Clearly, something is amiss, but this has *stimulated* physicists to look for new and broader solutions, such as "superstring theory," not to give up and say, "Well, it must be that God did it, let's leave it at that."

In the same way, evolutionary biologists are still working to determine how new patterns of development arise through the history of life on earth (for recent discussions of some of the relevant issues, see Schlichting and Pigliucci 1998; Wilkins 2002; West-Eberhard 2003). There will be questions that perhaps will never be answered, simply because it is unlikely that we will ever uncover enough evidence—the great diversification of invertebrate life at the beginning of the Cambrian period, more than 500 million years ago, being one possible example. But to claim from this that something is profoundly wrong with the theory of evolution is to fundamentally misunderstand, or willfully ignore, the very nature of scientific inquiry—which brings us to the third component of the problem, philosophy (part III).

As much as scientists, by and large, don't particularly enjoy musing about philosophy, science itself is based on fundamental philosophical assumptions, and creationists have tried to make the most of this alleged "weakness." For example, in order for science to work, one has to assume (as a methodological tool) "naturalism"—that is, that all we need in order to comprehend nature is a solid understanding of the laws and processes that we can observe and test in the natural world. The supernatural cannot be called forth, simply because it cannot provide us with any "explanation" in the sense of a useful and testable set of statements. But then again, all of us—including creationists—make the same assumption in most of our everyday dealings. While it is true that some people hold to such an extremely religious view of life that they think that God is directly responsible when they find a parking place at the mall, we can't actually function as human beings in modern technological societies if we go around attributing everything to the supernatural.

For that matter, both scientists and creationists make even more radical—and philosophically questionable—assumptions, such as "realism," the idea that there is a physical world out there, and that everything is not just in our (or God's) mind. Plenty of idealist and skeptical philosophers have repeatedly pointed out that there is not and cannot be any *proof* of realism: one has to accept it "on faith." Then again, those same philosophers apparently had no trouble setting aside their qualms about reality when they went to the grocery store. Bertrand Russell once said that sometimes one wishes that philosophers who question reality would get into a car and drive into a wall at a speed proportional to their disbelief in the existence of said wall.

Philosophy of science is no laughing matter, and scientists would do well to read up on it, especially if they wish to counter creationism in the public arena (and they ought to do so, as professional researchers and educators). Alas, it turns out that to try to understand what the scientific method is and how it works is not trivial (Chalmers 1999). Scores of philosophers have managed to give partial, more or less correct accounts of it. Beginning in the seventeenth century with discussions of induction by Francis Bacon (2000) and John Stuart Mill (1851), through the concepts of falsification of Karl Popper (1968) and the influence of "paradigm shifts" examined by Thomas Kuhn (1970) in the twentieth century, philosophers of science have tried to formulate the general principles of the scientific method based on the ways in which science is practiced by scientists.

Why not ask scientists themselves, then? After all, they are the ones *doing* the job. But that is like asking a football player how he gets a touchdown, or a painter what her latest canvas *really* means. In all of these cases, you will hear a lot of blabbering nonsense and very little useful information. That is because all these activities (science, sports, and arts) share an interesting commonality that is seldom explicitly acknowledged by scientists: they require long years of apprenticeship. One does not become a professional player by reading manuals on how to handle the ball. Rather, one has to start with some degree of natural talent and then practice constantly while learning from watching and interacting with people who know how to play well.

It is no different in science, which is why we have graduate schools where we *train* PhDs. It is hard to convey to a creationist the idea that getting a PhD by working in a laboratory for years is not "brainwashing"; it simply is the best way we know to produce the new crop of scientists. Of course individual scientists are going to be "biased"—in the general sense that they will have a point of view when they approach any particular scientific problem, and such point of view is likely to have been influenced by their PhD advisers (though it is common to see students take positions diametrically opposite to those of their mentors).

That being the case, where does the alleged scientific "objectivity" come from? Philosopher Helen Longino (1990) has convincingly argued that there are two sources of (long-term) objectivity in science: (1) the fact that the scientific community is made up of a culturally diverse group of people, with both men and women, members of different racial and ethnic groups, and individuals who espouse all sorts of ideological and religious positions; (2) the fact that there is a real world with which any scientific theory, sooner or later, has to come to terms. It is entirely possible—and it has happened several times in the history of science—that scientists will temporarily fall for the wrong explanation of a class of natural phenomena (for example, phrenology—the idea that one's character can be surmised by the shape of one's head). But eventually the facts will simply stubbornly refuse to comply with our preconceptions, and we will be forced to move on or give up our quest to understand nature.

How does creationism measure up to Longino's two criteria? On the one hand, creationists tend to represent a very narrow sample of human backgrounds and opinions: most of them are religious fundamentalists (often Christians or Muslims, but including Orthodox Jews and others). More important, the few testable statements that have come out of creationism (for example, the occurrence of a worldwide flood about four thousand years ago) have been tested, and clearly rejected! Yet, as in the case of other pseudosciences such as astrology, its followers refuse to bow to the judgment imposed on their theories by the real world. It is philosophy of science that condemns creationism and ID as inviable options, for the basic reasons that—contrary to science—they are not rational enterprises, and they don't "work."

I will close with a note about the fourth problem with creationism, though it is treated also in part III, together with philosophy. It concerns science education, which is where the front line of the evolution–creation controversy lies. This is a battle over the minds of our children, and too many educators simply have not been doing their job, partly because—ironically—they don't *know* how to do it (Pigliucci 2002).

The problem is too complex to attempt even a superficial summary here, but let me at least provide a couple of obvious pointers. In the best Catholic tradition, I can start with a *mea culpa* and admit that too many scientists don't know how to teach and largely don't care about it. We don't know how to do it because we get no training in it. Or, rather, the only exposure we have to teaching is usually of the worst kind. During graduate school, we are thrown into a lab classroom with very little idea of what to tell the students and with the mandate to "get it done" and return as soon as possible to our *real* jobs, doing research. That trend continues when we are professional scientists, working in departments that put a high premium (understandably) on publication and, especially, grant writing. Teaching? Well, that's something you can *buy yourself out of* if you have a grant, or are encouraged to do in the most "efficient" way possible: little preparation, no research into alternative teaching techniques, and as little contact with the undergraduates as possible.

Our pre-college teachers fare little better, unfortunately.

Many of them don't know enough about the science they are supposed to teach, because too often they didn't receive anything more than superficial training in it. Those who do know the science are overwhelmed by the amount of material they must cover (which keeps increasing steadily, since science is a cumulative enterprise, unlike—say—creationism), or they are too busy simply keeping their students' attention focused longer than the duration of an MTV sound bite. Is it any wonder that most of our high school and undergraduate students are ignorant of the intricacies of the theory of evolution and do not understand how science itself works?

The outlook, however, is not as bleak as I have hinted at so far. More scientists, philosophers, and educators are becoming involved in the evolution–creation controversy and are striving to promote better teaching of science and development of critical thinking skills. This is a major positive shift. Lately, even several professional scientific societies have timidly started to come out of the ivory tower and become involved in sponsoring symposia on teaching, as well as a variety of public outreach activities. This change will take years, probably decades, to yield measurable effects, but, ironically, it is a positive trend to which creationism has unwittingly contributed.

Take my own case, for example. Before 1996, I had never written a word about creationism, nor was I particularly interested in teaching methods or in talking to the public. But then I accepted a job at the University of Tennessee and immediately witnessed an attempt by the Tennessee legislature to pass an anti-evolution law. This shock was a wake-up call. It helped channel the energy of some of my colleagues and graduate students toward doing something about the problem. We felt we had to give back to the community simply because, after all (as creationists never cease to remind us), the public pays for our salaries (if one works at a public university) and for most of our research money (even if one works at a private institution). Engaging in public outreach and fighting against irrational nonsense is not merely an option to fill our spare time, it is also our moral obligation as members of a free society.

Massimo Pigliucci
Stony Brook, New York

Dr. Massimo Pigliucci is professor of ecology and evolution and philosophy at Stony Brook University. He has research interests in philosophy of science—in particular, in conceptual and critical analyses of fundamental ideas of evolutionary theory—and he is the author of *Denying Evolution: Creationism, Scientism, and the Nature of Science.*

References

Bacon, F. [1620] 2000. *Novum organum.* Cambridge: Cambridge University Press.

Bowler, P. J. 1983. *The eclipse of Darwinism.* Baltimore: Johns Hopkins University Press.

Chalmers, A. F. 1999. *What is this thing called science?* rev. ed. Brisbane, Australia: University of Queensland Press.

Darwin, C. [1859] 1910. *On the origin of species by means of natural selection, or the preservation of favored races in the struggle for life.* New York: A. L. Burt.

Greene, B. 1999. *The elegant universe: Superstrings, hidden dimensions, and the quest for the ultimate theory.* New York: W. W. Norton.

Kuhn, T. 1970. *The structure of scientific revolutions.* Chicago: University of Chicago Press.

Longino, H. E. 1990. *Science as social knowledge: Values and objectivity in scientific inquiry.* Princeton, NJ: Princeton University Press.

Mayr, E., and W. B. Provine. 1980. *The evolutionary synthesis. Perspectives on the unification of biology.* Cambridge, MA: Harvard University Press.

Mill, J. S. 1851. *A system of logic, ratiocinative and inductive, being a connected view of the principles of evidence, and the methods of scientific investigation.* London: J. W. Parker.

Paley, W. 1802. *Natural theology: or, Evidences of the existence and attributes of the Deity, collected from the appearances of nature.* London: Faulder.

Pigliucci, M. 2002. *Denying evolution: Creationism, scientism and the nature of science.* Sunderland, MA: Sinauer Associates.

Popper, K. R. 1968. *Conjectures and refutations: The growth of scientific knowledge.* New York: Harper and Row.

Schlichting, C. D., and M. Pigliucci. 1998. *Phenotypic evolution, A reaction norm perspective.* Sunderland, MA: Sinauer Associates.

West-Eberhard, M. J. 2003. *Developmental plasticity and evolution.* Oxford: Oxford University Press.

Wilkins, A. S. 2002. *The evolution of developmental pathways.* Sunderland, MA: Sinauer Associates.

Scientists Confront
CREATIONISM

Nothing in biology makes sense
except in the light of evolution.

—THEODOSIUS DOBZHANSKY

PART ONE

Creationism and "Intelligent Design"

Darwinism, Creationism, and "Intelligent Design"[1]

Ronald L. Numbers

THE PUBLICATION IN 1859 OF Charles Darwin's epoch-making book *On the Origin of Species* touched off a national debate that continues to divide American society. Scarcely a week passes without some evolution-related story appearing in the news—religious leaders declaring the scientific legitimacy of biological evolution, politicians expressing their belief in divine creation, local school boards wrangling over the teaching of origins, state legislatures debating whether to fire teachers who present evolution as a fact, biology textbooks carrying disclaimers denying the factual basis of evolution, scientists claiming that they have discovered evidence of "intelligent design" in the natural world, and public-opinion polls showing that nearly half of all Americans believe in the recent special creation of the first humans.

Early Reactions

At first Americans reacted coolly to Darwinism—a term commonly used as a synonym for organic evolution. American naturalists embraced biological evolution gingerly, and those who did

1. This chapter is adapted and reprinted by permission of the publisher from *Darwinism Comes to America* by Ronald L. Numbers, pp. 1–23, Cambridge, MA: Harvard University Press,

accept it tended to downplay the importance of Darwin's preferred mechanism of natural selection operating on random variations (Numbers 1998). As long as the scientific community remained skeptical about the merits of Darwinism, theologians could remain on the sidelines, confident that speculations about monkeys becoming men would never be taken seriously as science. By the mid-1870s, however, most American naturalists who expressed themselves on the subject were speaking out positively, and by the close of the decade only a handful of prominent scientists continued to regard Darwinism as a false theory.

Darwin's success in convincing fellow naturalists of the truth of evolution prompted more and more religious leaders, such as James McCosh, the president of Princeton College, to take a public stand. Theological liberals in the Protestant camp fairly quickly adapted their reading of Scripture and doctrinal beliefs to accommodate biological evolution, but most theological conservatives, representing the majority of Americans, viewed Darwinism, especially when applied to humans, as erroneous, if not downright dangerous. They feared that the notion of "might makes right" would undermine Christian morality and that tracing human genealogy back to apes would invalidate the concept of humans' being created in the image of God. With few exceptions, however, even the most literalistic Bible-believers accepted the antiquity of life on earth as revealed in the paleontological record. They typically did so either by interpreting the days of Genesis 1 as vast geological ages (the day–age theory) or by inserting a series of catastrophes and re-creations or ruins and restorations into an imagined gap between the first two verses of the Bible (the gap theory). By the close of the nineteenth century, virtually the only Christians writing in defense of the recent appearance of life on earth and attributing the fossil record to the action of Noah's Flood were Seventh-day Adventists, a fundamentalist group numbering fewer than 100,000 members. Until the 1970s this uncompromisingly literal reading of Genesis, developed and popularized by the Adventist "geologist" George McCready Price, generally went by the name of "Flood geology" (Roberts 1988; Numbers 1992).

Like Protestants, Catholics split along the progressive–conservative divide, with most prelates and priests remaining on the latter side. In the 1890s Father John Zahm, a priest-scientist at the University of Notre Dame, took the lead in trying to harmonize Catholicism with a theistic version of Darwinism, but the Vatican effectively silenced him in 1897. The next year, his book *Evolution and Dogma* appeared on the Index of Forbidden Books, casting a theological shadow over evolution among American Catholics for years to come (Appleby 1999). American Jews overwhelmingly rejected evolution till the mid-1870s, when some Reform rabbis began to urge its acceptance. By the early 1890s, evolution had established itself in the Reform community, though traditional Jews often expressed skepticism (Swetlitz 1999).

Evolution and Anti-Evolution before Sputnik

Little organized opposition to evolution appeared before the early 1920s, when fundamentalist Christians, led by the Presbyterian layman and three-time Democratic presidential candidate William Jennings Bryan, launched a state-by-state crusade to outlaw the teaching of human evolution in public schools. By the end of the decade, they had succeeded in only three states—Tennessee, Mississippi, and Arkansas—but the ruckus they raised retarded the dissemination of evolutionary ideas in American classrooms for over three decades. The anti-evolution campaign of the 1920s attracted enthusiasts around the country, in the North as well as in the South, in urban as well as rural areas, but it garnered little support among America's scientific elite. Its leading scientific authorities were Arthur I. Brown, an obscure Canadian surgeon whose handbills touted him as "one of the best informed scientists on the American continent"; S. James Bole, a science teacher at the fundamentalist Wheaton College who had earned a master's degree in education with a thesis on penmanship in an Illinois school district; Harry Rimmer, a Presbyterian preacher and self-styled "research scientist" who had briefly attended a homeopathic medical school; and the self-trained George McCready Price, whom the journal *Science* identified as "the principal scientific authority of the Fundamentalists" (Numbers 1992).

Anti-evolutionists in the 1920s may have agreed on the evils of Darwinism, but they disagreed spiritedly over the correct interpretation of Genesis 1. As one frustrated creationist observed in the mid-1930s, fundamentalists were "all mixed up between geological ages, Flood geology and ruin, believing all at once, endorsing all at once" (Numbers 1992, 102). As long as they remained split over the meaning of Genesis, he reasoned, they could scarcely expect to convert the world to their creationist way of thinking. The 1930s and 1940s witnessed multiple attempts to create a united fundamentalist front against evolution, but each one failed because of the intransigence of the various partisans, especially the Flood geologists, who refused to compromise on the recent appearance of life on earth and the geological significance of Noah's Flood. In 1941, at the invitation of the president of the Moody Bible Institute, a group of five evangelical Christian scientists met in Chicago to establish the American Scientific Affiliation (ASA), which initially opposed evolution but soon came to accept organic development over time, punctuated by divine interventions, especially for the creation of matter, life, and humans. More liberal Christians increasingly identified evolution as simply God's method of creation and ignored the problem of reconciling science and Scripture.

Meanwhile, during the same years, biologists, after decades of disagreeing over the mechanism of evolution to the point of fostering reports of Darwinism lying on its "deathbed," began to forge a common explanation of evolution, which came to be known, perhaps misleadingly, as the modern or neo-Darwinian synthesis. Geneticists, taxonomists, and paleontologists, who had long worked in virtual isolation from one another, finally began interacting—and agreeing on the centrality of natural selection in the evolutionary process. In doing so, they repudiated other evolutionary explanations, particularly ones that gave evolution the appearance of having a purpose (Mitman and Numbers 1996). This created, in the words of the historian-biologist William B.

Provine, an "evolutionary constriction" that squeezed any talk of supernatural design out of biological discourse. "The evolutionary constriction," he asserts, "ended all rational hope of purpose in evolution," thus making belief in Darwinism the functional equivalent of atheism (Provine 1992). Many evolutionists remained devout Christians and Jews, but it became increasingly difficult to do so on the basis of the scientific evidence for evolution.

The Creationist Revival

The evolutionary constriction scarcely influenced the content of high-school biology textbooks until after 1957, when the Soviet Union successfully launched Sputnik into space, greatly embarrassing the American scientific establishment. Politicians and science-policy experts quickly fingered the inferior scientific education of Americans as the underlying cause of the country's slide to second place in the space race. To remedy the situation, the federal government began pouring large amounts of money into improving science textbooks for high-school students. In biology, where leading practitioners were complaining that "one hundred years without Darwinism are enough," the funds went to the Biological Sciences Curriculum Study (BSCS), which produced a series of texts featuring evolution as the centerpiece of modern biology. When these unabashedly pro-evolution texts descended on American classrooms in the early 1960s, they produced howls of protest from conservative Christians, who regarded the BSCS books as an ungodly "attempt to ram evolution down the throats of our children"(Numbers 1992).

Just as the BSCS controversy was breaking, two fundamentalists, John C. Whitcomb, Jr., an Old Testament scholar, and Henry M. Morris, a hydraulic engineer, brought out a book called *The Genesis Flood* (1961), which presented Price's Flood geology as the only acceptable interpretation of the first books of Genesis. Their insistence on beginning with a literal reading of the Bible and then trying to fit science into that context, rather than constantly accommodating the Bible to the findings of science, struck a responsive chord with many concerned Christians. In substantial, though undetermined, numbers they abandoned the once-favored day–age and gap theories, which allowed for the antiquity of life on earth, in exchange for the strict creationism of Flood geology, which limited the history of life to no more than 10,000 years and affirmed creation in six twenty-four-hour days.

Two years after the appearance of *The Genesis Flood*, Morris and nine other like-minded creationists banded together to form the Creation Research Society (CRS). The members of the CRS, all possessing some scientific or technical training at the graduate level, not only attacked evolution, as their intellectual forebears had done in the 1920s, but any compromise with theories of ancient life. Many members, including Morris himself, insisted that God had created the entire universe, not just earthly life, within the past 6,000 years or so. In contrast to the anti-evolutionists of the 1920s, who could claim no well-trained scientists of their own, the ten founders of the CRS included five biologists with earned doctorates from major universities, a biochemist and an engineer with PhD degrees, two biologists with master's degrees, and a tenth member who pretended to have an MA in geology.

These young-earth creationists, as they came to be called, proved highly effective in promoting Price's Flood geology among conservative Christians. The San Diego–based Institute for Creation Research, which Morris had established in 1972, served as unofficial headquarters. To help gain a platform for their views in public-school classrooms—and to endow them with a measure of scientific respectability—the proponents of Flood geology around 1970 renamed Price's model of earth history "scientific creationism" or "creation science." The sequence and timing of key events, such as a recent special creation and subsequent worldwide Flood, remained the same, but all direct references to biblical characters and places, such as Adam and Eve, the Garden of Eden, and Noah and his ark, disappeared from the stripped-down narrative. Within a decade or two, the tireless proselytizers for scientific creationism had virtually co-opted the generic creationist label for their hyperliteralist views, which only a half-century earlier had languished on the margins of American fundamentalism. People who called themselves creationists during the last quarter of the twentieth century typically assumed that most listeners would identify them as believers in a young earth.

The explanation for this dramatic shift in creationist thinking is difficult to nail down. Developments such as the evolution offensive launched by the BSCS help to explain the robust revival of creationism in the post-Sputnik period, but they scarcely account for the dramatic shift among Bible-believing Christians from old-earth to young-earth interpretations of Genesis 1. Facile generalizations about educational deprivation and cultural alienation simply will not suffice. Highly educated citizens may have been more likely than their less-well-trained neighbors to subscribe to evolution, but a quarter of those Americans who professed belief in the recent special creation of the first humans had graduated from college. Within the evangelical community of Christians, fundamentalists have displayed greater enthusiasm for scientific creationism than their Pentecostal brethren and sisters; yet few students of American religion would argue that the former are more socially alienated or economically depressed than the latter.

Whatever the reasons for the efflorescence of anti-evolutionism in the late twentieth century, the prodigious popularity of scientific creationism among conservative Christians almost certainly related more to theological than social impulses. Many converts were attracted by the creation scientists' insistence on giving the Bible priority over science. As believers who took the Bible as literally as possible, they found the young-earth creationists' nonfigurative reading of the days of creation, the genealogies of the Old Testament, and the universal deluge of Noah to be especially appealing. No longer did they have to *assume* (as day–age advocates did) that Moses meant "ages" when he wrote "days" in Genesis 1, nor did they have to *assume* (as gap theorists did) that Moses, without explanation or comment, skipped over the longest period of earth history—between the creation "in the beginning" and the far later Edenic creation—simply to accommodate Scripture to science.

The theological factors that encouraged adoption of creation science varied among and within denominations. Independent Baptists, Missouri Synod Lutherans, and Seventh-day Adventists, to name three of the religious groups most receptive to creation science, each possessed distinctive motivations for embracing it. Premillennial Baptists and Adventists (but not amillennial Lutherans), who interpreted the prophecies of Revelation, the last book of the Bible, as indicating an apocalyptic end of the world associated with the Second Coming of Christ, tended to see an intimate link between the beginning described in the first book of the Bible and end-time events. As the Baptist Morris once observed, "If you take Genesis literally, you're more inclined to take Revelation literally." The Adventists (but not Baptists and Lutherans) possessed an extra-biblical endorsement of creation science in the divinely inspired writings of their prophetess, Ellen G. White. Though all three traditions read the Bible through literalistic lenses, the Missouri Lutherans (but not Baptists or Adventists) well into the twentieth century went so far as to defend Ptolemaic astronomy, which placed the earth rather than the sun in the middle of the solar system.

Balanced Treatment

In the early 1980s, state legislatures across the United States debated a creationist-inspired model bill that called for the balanced treatment of "evolution-science" and "creation-science" in public schools. Two states, Arkansas and Louisiana, enacted this proposed legislation into law, but the new statutes quickly encountered judicial opposition. In 1982, after a trial that brought more attention to the creation-evolution controversy than any event since Clarence Darrow confronted William Jennings Bryan in the Scopes trial of 1925, a federal judge in Little Rock declared the Arkansas law to be an unconstitutional breach of the wall separating church and state. Five years later, the U.S. Supreme Court, after hearing a case from Louisiana, upheld a lower-court decision that creation science served a religious, not scientific, purpose. However, one justice, writing for the majority, left the schoolhouse door open a crack for creationism to slip through. "Teaching a variety of scientific theories about the origins of humankind to schoolchildren," he wrote, "might be validly done with the clear secular intent of enhancing the effectiveness of science instruction" (Larson 1989).

Since the 1920s the scientific establishment had paid little attention to the snipings of creationist critics, but the creationist successes (and near successes) of the early 1980s finally aroused them from their apathy. Organizations from the National Academy of Sciences to little-known local science societies, fearing the loss of public patronage and cultural authority, denounced those who challenged their conclusions and sought to adulterate science with religion. Just how much influence their fulminations had on the voting of state lawmakers is hard to assess. In at least one state, creationist educators rather than evolutionary biologists brought about the defeat of the model bill. A scientist at the University of Oklahoma told of witnessing how a legislative committee in that state reached its decision to oppose the balanced-treatment act. At a public hearing in early 1981, a joke-cracking, down-home school superintendent from a rural district begged the legislators to "leave us alone—we know what we're doing. We're not teaching evolution—we're teaching biblical creation." The bill under consideration required that creation and evolution be given equal time if either were taught; because most Oklahoma schools were teaching only creation, the bill, seen as another "example of big government telling the local school boards what to do," would force them to expose students to evolution. Needless to say, the bill (and evolution) went down to defeat (Sonleitner 1981).

Creationism in the 1990s

The Supreme Court's decision effectively ended efforts to mandate the inclusion of creationism in public-school curricula, but it did little to slow down creationist initiatives to undermine evolution. Instead of agitating for balanced-treatment acts at the state level, creationists refocused much of their energy on individual schools and school districts, where in many instances considerable support for creationism already existed. In the early 1990s the National Center for Science Education (NCSE), which monitored creationist endeavors throughout the country, warned that people unfamiliar with pre-collegiate education would "be surprised at the amount of official anti-evolutionism that is found there, especially among administrators." In the fall of 1992 the center drew attention to "a sharp surge upwards" in creationist attacks on evolution. These often took the form of calling for downgrading the status of evolution from "fact" to "theory" or for presenting students with "evidence against evolution," a notion the director of the center, Eugenie C. Scott, dismissed as "merely 'scientific' creationism in sheep's clothing" (Scott 1992, 1993a)

Some educators employed novel solutions to solve the recurring evolution problem. In response to complaints about the inclusion of evolutionary cosmology in elementary-school textbooks, the superintendent of schools in Marshall County, Kentucky, ordered that the offending two pages be glued together. The Cobb County school district in suburban Atlanta, Georgia, went directly to the publisher of a troublesome fourth-grade text and asked that a chapter entitled "The Birth of Earth" be deleted. Modern electronic publishing allowed Macmillan/McGraw-Hill, the publisher, to excise seventeen pages, thereby producing a custom-made text exclusively for the students of Cobb County (Schmidt 1996; Scott 1996a).

The Alabama state school board in 1995 voted six-to-one in favor of inserting the following disclaimer in all biology textbooks used in the state (Anonymous 1995a):

A MESSAGE FROM THE ALABAMA STATE BOARD OF EDUCATION

This textbook discusses evolution, a controversial theory some scientists present as a scientific explanation for the origin of living things, such as plants, animals and humans.

No one was present when life first appeared on earth. Therefore, any statement about life's origins should be considered as theory, not fact.

The word "evolution" may refer to many types of change. Evolution describes changes that occur within a species. (White moths, for example, may "evolve" into gray moths.) This process is microevolution, which can be observed and described as fact. Evolution may also refer to the change of one living thing to another, such as reptiles into birds. This process, called macroevolution, has never been observed and should be considered a theory. Evolution also refers to the unproven belief that random, undirected forces produced a world of living things.

There are many unanswered questions about the origin of life which are not mentioned in your textbook, including:

- Why did the major groups of animals suddenly appear in the fossil record (known as the "Cambrian Explosion")?
- Why have no new major groups of living things appeared in the fossil record for a long time?
- Why do major groups of plants and animals have no transitional forms in the fossil record?
- How did you and all living things come to possess such a complete and complex set of "Instructions" for building a living body?

Study hard and keep an open mind. Someday, you may contribute to the theories of how living things appeared on earth.

Biology textbooks in Alabama subsequently began arriving from the publishers with the above message pasted into the front. The Republican governor, Fob James, who presided over the board, strongly backed the disclaimer, saying that he personally believed the biblical account of the origin of life to be true (Scott 1995).

In the mid-1990s, controversies over creationism erupted not only in Georgia, Kentucky, and Alabama but also in Virginia, Pennsylvania, New Hampshire, Ohio, Indiana, Michigan, Wisconsin, New Mexico, California, and Washington (Matsumura 1995a; Matsumura 1996; Matsumura and Petto 1996; Matsumura 1997; Nelson 1997). Tennessee legislators defeated a bill, at first expected to "blast through the House Education Committee like a rocket," that would have allowed the firing of any teacher who presented evolution as fact rather than theory (Matsumura 1995b; Cheek 1996). Such activity prompted one frustrated anti-creationist to exclaim that "Creationism is like a vampire, and every time you think the thing is finally dead, someone pulls the damned stake out again" (Scott 1996b).

As a Republican candidate for the presidency of the United States in 1980, Ronald Reagan had insisted that "if evolution is taught in public schools, creation also should be taught" (Numbers 1992, 300). In 1995 the Republican presidential candidate Pat Buchanan adamantly denied any kinship with simian ancestors: "I don't believe it is demonstrably true that we have descended from apes. I don't believe it. I do not believe all that." He *did* believe that parents had "a right to insist that Godless evolution not be taught to their children" (Anonymous 1995b). During the 1990s various state Republican parties added creationist planks to their platforms. And in all regions of the country—North, South, East, and West—creationists stood for election to local school boards. They often won (Matsumura 1994; Lemonick 1996).

Support for creationism ran deep in North American society. Despite the nearly unanimous endorsement of evolution by leading biologists, a Gallup poll in 1993 showed that 47 percent of Americans continued to believe that "God created man pretty much in his present form at one time within the last 10,000 years," and an additional 35 percent thought that the process of evolution had been divinely guided. Only 11 percent subscribed to purely naturalistic evolution. (Seven percent expressed no opinion.) Fifty-eight percent of the public favored teaching creationism in the schools. In Canada, which had experienced comparatively little controversy over origins, 53 percent of adults rejected evolution (Anonymous 1993; Scott 1993b; Cole 1996).

In 1986, during a visit to New Zealand, the American paleontologist and anti-creationist Stephen Jay Gould assured his hosts that scientific creationism was so "peculiarly American" it stood little chance of "catching on overseas" (Anonymous 1996). His colleague Richard C. Lewontin seemed to agree. "Creationism is an American institution," he declared, "and it is not only American but specifically southern and southwestern" (Lewontin 1983, xxv). So it may have seemed at the time, but scientific creationism was already traveling far beyond the borders of the United States, enjoying growing popularity in Europe, Asia, and the South Pacific (Numbers 1992). In 1980 Australian antievolutionists established the Creation Science Foundation (CSF) in Queensland, where for a period in the 1980s creation appeared in the state syllabus for secondary schools (Numbers 2002; see also Numbers and Stenhouse 2000). Within a short time, the CSF became the world's second leading center for the propagation of scientific creationism (after Morris's Institute for Creation Research). In the mid-1990s, its star speaker, Kenneth A. Ham, opened an international creationist organization in Florence, Kentucky (near Cincinnati), as "an outreach of the CSF." The Korea Association for Creation Research, also founded in 1980, expanded so rapidly that it, too, established branches in the United States.

Even in Islamic countries such as Turkey, creationism made extensive inroads. In the 1980s, the ministry of education translated several creation-science books into Turkish and distributed them to teachers throughout the country (Edis 1999; Sayin and Kence 1999; Shapiro 1999). In a report on evolution sent to Turkish educators, the minister of education dismissed Darwinism as a handmaiden of materialism based on "nothing but some interpretations and guesswork." He recommended that biology textbooks "provide all of the evidence in favor of and against the theory of evolution," so that Turkish youth could "gain the habit of objective and scientific thinking" (Edis 1994).

Varieties of Evolutionism and Creationism

Although popular accounts of the creation-evolution controversies of the 1990s often characterized them as a bipolar debate between naturalistic evolutionists and supernaturalistic creationists, opinions on origins actually spanned a wide intellectual spectrum. On one end were the *naturalistic evolutionists,* often atheists or agnostics (such as the Oxford biologist Richard Dawkins and the Tufts philosopher Daniel C. Dennett), who saw no evidence of, or need for, a Creator God. Nearer the center were the *theistic evolutionists,* often devout Christians (such as the physicist Howard J. Van Till of Calvin College and many members of the evangelical American Scientific Affiliation), who saw little or no evidence of God in nature but who, for theological reasons, believed that God had created the world by means of evolution. On the other side of the center were the *"intelligent-design" theorists* (such as Australian biochemist-physician Michael Denton, Berkeley law professor Phillip E. Johnson, Lehigh biochemist Michael J. Behe, and the editors of the journal *Origins & Design*), who rejected naturalistic evolution and claimed to see evidence of an Intelligent Designer in the complexity of nature, but who often accepted the antiquity of life on earth. At the opposite end of the spectrum from the naturalistic evolutionists, beyond a mixed group of *old-earth creationists* (such as the California-based astronomer Hugh Ross) were the *scientific creationists,* typically fundamentalist Christians (such as Henry M. Morris and most members of the Creation Research Society), who compressed the entire history of the universe into little more than 6,000 years and postulated a divine creation in six literal days.

The "point man" for naturalistic evolution in the 1990s was Dawkins, author of such books as *The Blind Watchmaker* (1986), which an admirer described on the dust jacket as perhaps "the most important book on evolution since Darwin." The title, a Paleyesque reference to the role of natural selection in creating organized complexity, left little doubt of Dawkins's position:

> Natural selection, the blind, unconscious, automatic process which Darwin discovered, and which we now know is the explanation for the existence and apparently purposeful form of all life, has no purpose in mind. It has no mind and no mind's eye. It does not plan for the future. It has no vision, no foresight, no sight at all. If it can be said to play the role of watchmaker in nature, it is the blind watchmaker. (Dawkins 1986, 5–6)

In an oft-quoted statement, Dawkins praised Darwin for making "it possible to be an intellectually fulfilled atheist," and he repeatedly went out of his way to bait creationists, all of whom he believed to be "ignorant, stupid or insane" (Dawkins 1989, 34). He dismissed the first chapters of Genesis as just another creation myth "that happened to have been adopted by one particular tribe of Middle Eastern herders" and theistic evolution as a superfluous attempt to "smuggle God in by the back door." No wonder one of Dawkins's patrons, Charles Simonyi, a rich Microsoft executive who endowed a special professorship for Dawkins at Oxford, fondly called his beneficiary "Darwin's Rottweiler," a reference to the nineteenth-century evolutionist and agnostic Thomas H. Huxley, famous as "Darwin's bulldog" (Downey 1996).

If Dawkins played the role of point man for late-twentieth-century naturalistic evolutionists, Tufts University philosopher Daniel C. Dennett gladly served as their hatchet man. In a book called *Darwin's Dangerous Idea* (1995), which Dawkins warmly endorsed, Dennett portrayed Darwinism as "a universal solvent, capable of cutting right to the heart of everything in sight"—and particularly effective in dissolving religious beliefs (Dennett 1995, 515). The most ardent creationist could not have said it with more conviction, but Dennett's agreement with them ended there. He despised creationists, arguing that "there are no forces on this planet more dangerous to us all than the fanaticisms of fundamentalism" (Dennett 1995, 516). Displaying a degree of intolerance more characteristic of a fundamentalist fanatic than an academic philosopher, he called for "caging" those who would deliberately misinform children about the natural world, just as one would cage a threatening wild animal. "The message is clear," he wrote: "those who will not accommodate, who will not temper, who insist on keeping only the purest and wildest strain of their heritage alive, we will be obliged, reluctantly, to cage or disarm, and we will do our best to disable the memes [traditions] they fight for" (Dennett 1995, 519–20). With the bravado of a man unmindful that only 11 percent of the public shared his enthusiasm for naturalistic evolution, he warned parents that if they insisted on teaching their children "falsehoods—that the earth is flat, that 'Man' is not a product of evolution by natural selection—then you must expect, at the very least, that those of us who have freedom of speech will feel free to describe your teachings as the spreading of falsehoods, and will attempt to demonstrate this to your children at our earliest opportunity" (Dennett 1997). Those who resisted conversion to Dennett's scientific fundamentalism would be subject to "quarantine."

Evolutionary ideologues such as Dawkins and Dennett made headlines for their hard-nosed views, but not all naturalistic evolutionists took such a draconian line. For example, Stephen Jay Gould, a self-described agnostic, expressed dismay at such tough rhetoric and welcomed any signs of rapprochement between naturalistic evolutionists and theists. He celebrated Pope John Paul II's recognition in 1996 of the theory of evolution being "more than a hypothesis," even though the pontiff rejected materialistic evolution and insisted on the special creation of the soul (John Paul II, 1997). Gould knew that the Dawkinses and Dennetts of the world would say, "C'mon, be honest; you know that religion is addlepated, superstitious, old-fashioned b.s.; you're only making those welcoming noises because religion is so powerful, and we [that is, evolutionists] need to be diplomatic in order to assure public support and funding for science," but he refused to countenance their intolerance. Scientists who claimed that Darwinism had disproved the existence of God should have "their knuckles rapped for it"—as should those who claimed that evolution was God's method of creation (Gould 1992). In the late 1980s, he even trained a young-earth creationist and doctoral candidate, Kurt P. Wise, in his laboratory and, though bemused by his beliefs, always treated him with respect, even after he joined the faculty of the fundamentalist William Jennings Bryan College in Dayton, Tennessee (Numbers 1992).

Theistic evolutionists, representing 40 percent of the American population, occupied a spectral range from borderline naturalistic evolutionism to progressive creationism, but the most prominent Protestant advocates in the 1990s leaned more toward the former than the latter. Ever since its creation in the early 1940s, the American Scientific Affiliation (ASA) has served as the primary forum for evangelical Christian scientists interested in issues related to creation and evolution. During its early years the ASA leaned toward creationism, but that began to change in the late 1950s. During the Darwin centennial of 1959, at a time when even the most progressive evangelical scientists were still limiting the scope of evolution, Walter R. Hearn, a biochemist in the society, pushed for uncompromising evolution, which he regarded as God's method of creation. Though a fervent theist, he denied that scientists could discover direct evidence of God's activity by looking through a microscope. By the 1990s views such as his had come to dominate ASA thinking (Numbers 1992).

Unfortunately for the ASA and theistic evolutionists generally, the din of debate between naturalistic evolutionists and scientific creationists often drowned out their irenic voices. The evangelical physicist Howard Van Till, one of the most vocal theistic evolutionists of the 1980s and 1990s, described evolution as "an ordinary natural process—a process that is not fundamentally different in character or status from other natural processes, such as a summer sunrise, a winter snowstorm, the blooming of a flower, or the birth of a child" (Van Till 1986). Though evolution did "not require the introduction of phenomena that go beyond the ordinary pattern of material behavior," it nevertheless occurred within "the Creator's domain of action" and thus was not "inherently naturalistic" (Van Till 1986, 252–53). This brand of theistic evolution appealed to many Protestants (as well as Catholics and Jews) who sought middle ground in the battle between creation and evolution, but it elicited only scorn from naturalistic evolutionists and scientific creationists, who agreed on few matters, noted one observer, except that "theistic evolution is woefully—even perniciously— confused." One side disliked it because it was theistic; the other, because it was evolution. But mostly it was ignored (Ratzsch 1996). After all, noted the anti-evolutionist Phillip Johnson condescendingly, theistic evolutionists occupied "just a few backwater positions" in the debate over origins (Belz 1996, 18).

Scientific (or young-earth) creationism pretty much dominated creationist discourse after the 1960s, but large pockets of evangelical Christianity remained loyal to (if comparatively silent about) the once-dominant old-earth models associated with the day–age and gap interpretations of Genesis 1. According to Henry M. Morris in 1995, "the most influential current scientist" then writing in defense of the day–age interpretation of Genesis 1 was Hugh Ross, a Toronto-trained astronomer who believed that

Morris's scheme was not only bad science but bad exegesis (Ross 1994; Morris 1995). Perhaps the most visible defender of the gap theory in late-twentieth-century America was the lusty charismatic TV evangelist Jimmy Swaggart, whose Pentecostal tradition had long detected two different creation events in the first chapter of Genesis (Numbers 1992, 308).

"Intelligent Design"

Just as scientific creationists had captured the fancy of journalists in the 1980s, "intelligent design" (ID) theorists grabbed headlines in the 1990s with their bold calls for rewriting the basic rules of science and their claims of having found indisputable evidence of God. The intellectual roots of the ID movement go back centuries, but its contemporary incarnation dates from the mid-1980s. In 1984 three Protestant scientists, Charles B. Thaxton, Walter L. Bradley, and Roger L. Olsen, brought out *The Mystery of Life's Origin*, in which they attributed the complex process of originating life to a divine Creator. The most striking feature of their book was not its text but its foreword, contributed by Dean H. Kenyon, a Roman Catholic professor of biology at San Francisco State University and the coauthor of a major text on the chemical origins of life (Thaxton et al. 1984). Confessing that he no longer held to naturalistic evolution, Kenyon joined the authors of the book in identifying "a fundamental flaw" in current theories about the origins of life (Kenyon 1984). "A major conclusion to be drawn from this work," he wrote, "is that the undirected flow of energy through a primordial atmosphere and ocean is at present a woefully inadequate explanation for the incredible complexity associated with even simple living systems, and is probably wrong."

Two years later Michael Denton, an Englishman living in Australia, wrote an iconoclastic book, *Evolution: Theory in Crisis* (1986), questioning the validity of neo-Darwinism and arguing for evidence of divine design in nature. Although he had grown up in a religiously conservative family, he no longer maintained ties with organized religion or harbored any sympathy for young-earth creationism. He did, however, see humans and other organisms as the products of God-ordained laws of nature (Denton 1998). Neither *The Mystery of Life's Origin* nor *Evolution: Theory in Crisis* attracted much attention among mainstream scientists, but they both helped to lay the intellectual foundation for the ID movement of the 1990s.

The first ID book to reach a wide audience—and one of the first explicitly to adopt the "intelligent design" slogan—appeared in 1989 as the first edition of *Of Pandas and People: The Central Question of Biological Origins.* The authors, Percival Davis and Dean Kenyon, designed this slim, illustrated volume as a supplement to high-school biology texts that were written from the Darwinian point of view. Using six case studies, they compared Darwinian and ID explanations to see which better matched the scientific data. Not surprisingly, "intelligent design"—defined as a frame of reference that "locates the origin of new organisms in an immaterial cause: in a blueprint, a plan, a pattern, devised by an intelligent agent"—always won (Davis and Kenyon 1993). An appended "Note to Teachers" (Hartwig and Meyer 1993) went out of its way to distinguish ID theory from religious fundamentalism and scientific creationism, but skeptical evolutionists regarded the theory as simply a "creationist alias." By 1996 at least two states, Alabama and Idaho, and a number of local school districts had evaluated the suitability of the book for adoption in public schools. Most, if not all, had recommended against its use (Scott 1996c).

In 1991 the infant ID movement received a big boost from an unlikely source: a University of California (Berkeley) law professor, Phillip E. Johnson. A few years earlier, the Presbyterian Johnson had stumbled across Dawkins's *The Blind Watchmaker* and discovered, as he put it, that the argument for evolution was more rhetorical than factual. Being a lawyer, he recognized the practice all too well. In a book titled *Darwin on Trial* (Johnson 1991), he sought to expose the soft underbelly of Darwinism by critically examining the evidence for the blind-watchmaker thesis. Both in that book and in a subsequent work, *Reason in the Balance: The*

Case Against NATURALISM in Science, Law & Education (1995), Johnson disclosed what he saw as the core problem with naturalistic evolution: its unwarranted assumption of naturalism as the only legitimate way of doing science. This bias, he argued, unfairly limited the range of possible explanations and ruled out, *a priori*, any consideration of theistic factors (Johnson 1997).

On the dust jacket of *Darwin on Trial*, Denton hailed it as "the best critique of Darwinism I have ever read," while Gould dismissed it as "scarcely more than an acrid little puff," unworthy of a serious response (Gould 1992, 118–20). In a scathing review in *Scientific American*, Gould insisted that "Science can work only with naturalistic explanations; it can neither affirm nor deny other types of actors (like God) in other spheres (the moral realm, for example)." When the editor of *Scientific American* denied Johnson's request for "equivalent space" to respond to Gould, the ID camp saw the editor's action as a confirmation of their suspicions of official discrimination against theistic views. When, about the same time, Kenyon's department at San Francisco State University ordered him to quit teaching "creationism" (he had introduced students to the "evidences" against Darwinism and for "intelligent design"), the ID camp acquired a martyr (Scott 1993c; Johnson 1995, 29–30).

Until the mid-1990s, no major academic or trade press had published a work supporting "intelligent design" or, indeed, creationism of any kind. That changed in 1996, when the Free Press in New York released Michael J. Behe's *Darwin's Black Box: The Biochemical Challenge to Evolution.* Behe, a Roman Catholic biochemist at Lehigh University, had first become aware of the alleged difficulties of Darwinism through reading Denton's book. Later he read Johnson's *Darwin on Trial*, which confirmed his growing doubts about the adequacy of naturalistic evolution to explain molecular life. When a reviewer of Johnson's book in the journal *Science* treated it harshly, Behe rushed to the lawyer's defense with a letter to the editor. He subsequently began exchanging correspondence with Johnson and drafting his own book-length reply to naturalistic evolutionists such as Dawkins, whom he regarded as "the best modern popularizer of Darwinism around." In *Darwin's Black Box* Behe argued that biochemistry had "pushed Darwin's theory to the limit . . . by opening the ultimate black box, the cell, thereby making possible our understanding of how life works." The "astonishing complexity of subcellular organic structure" led him to conclude—on the basis of scientific data, he asserted, "not from sacred books or sectarian beliefs"—that "intelligent design" had been at work. "The result is so unambiguous and so significant that it must be ranked as one of the greatest achievements in the history of science," he declared without a trace of false modesty. "The discovery [of "intelligent design"] rivals those of Newton and Einstein, Lavoisier and Schrödinger, Pasteur and Darwin" (Anonymous 1996).

As newspapers and magazines spread the news of Behe's discovery, he won recognition as a modern-day William Paley (the most famous natural theologian of the early nineteenth century). The influential evangelical magazine *Christianity Today* honored *Darwin's Black Box* with its Book of the Year award for 1996 (Anonymous 1997). Like so many other ID theorists, Behe distanced himself as far as possible from the scientifically disreputable young-earth creationists, going so far as to concede the possibility that the universe had been around for billions of years and that life on earth had developed from a common ancestor. But such disclaimers scarcely deterred critics from deriding his views as "thinly veiled creationism." The great nemesis of theistic science, Dawkins, chided Behe on television for lazily relying on "intelligent design" when he should have gone out looking for scientifically acceptable explanations of his data (Wheeler 1996; Woodward 1997).

By the mid-1990s the ID theorists, many of whom had been collaborating since the 1980s, were coalescing into an institutionalized movement: organizing conferences, establishing a center, and publishing a journal, *Origins & Design*. The journal's masthead listed the leading lights of ID theory, with Denton, Thaxton, Kenyon, Johnson, and Behe all serving on the editorial advisory board. A trio of young Christian philosophers of science—Paul A.

Nelson, Stephen C. Meyer, and William A. Dembski—headed up the editorial office. Many of these same persons were connected with the generously endowed Center for the Renewal of Science and Culture (renamed The Center for Science and Culture in 2002), affiliated with the Discovery Institute in Seattle. Although individually they espoused a wide range of views on origins (from Denton's and Behe's virtual theistic evolutionism to Nelson's young-earth creationism), collectively they staked out a position between theistic evolutionism, "American evangelicalism's ill-conceived accommodation to Darwinism," and scientific creationism, American fundamentalism's ill-conceived effort to base science on Scripture (Dembski 1995). Their goal was "an intellectual revolution" that would redraft the basic rules of science to include nonnaturalistic explanations of phenomena. Unlike some midcentury creationists who felt the need to form a united front on creation before challenging evolution, the ID theorists set aside the construction of a specific creation model while they tried to mount a unified attack against Darwinism. "When the Goliath [of naturalistic evolution] has been tumbled," they reasoned, "there will be time to work out more details of how creation really did occur" (Belz 1996).

"In so pluralistic a society as ours," Dembski once asked rhetorically, "why don't alternative views about life's origin and development have a legitimate place in academic discourse?" He already knew the answer: the scientific establishment's alleged bias toward atheistic materialism, a bias acknowledged by at least a few members of America's scientific elite (Dembski 1995; see also Dembski 1998). The distinguished evolutionary biologist Richard Lewontin, for example, worried in the 1990s that the public might actually believe what Dawkins and other careless popularizers told them about evolution, which often rested on "unsubstantiated assertions or counter-factual claims." In an unvarnished statement that conformed precisely to what the ID theorists had been claiming, he described the workings of the modern scientific mind: "We take the side of science in spite of the patent absurdity of some of its constructs, *in spite* of its failure to fulfill many of its extravagant promises of health and life, *in spite* of the tolerance of the scientific community for unsubstantiated just-so stories, because we have a prior commitment, a commitment to materialism" (Lewontin 1997, 31).

As they no doubt anticipated, the "intelligent design" crowd took a beating from all sides: scientific creationists, theistic evolutionists, and, of course, naturalistic evolutionists. Although a few young-earth creationists, such as Nelson and Wise, applauded the effort to discover evidence of God in nature (Frair 1997; Swanson 1997), the leaders of creation science, despite believing in divine design, never warmed up to ID theory. One scientist on the staff of the Institute for Creation Research faulted the advocates of ID for their "lack of reliance on the literal statements of Scripture and the construction of alternative models of origin, which involve long periods of years." Henry M. Morris, the grand old man of scientific creationism, admired the efforts of ID theorists to refute Darwinism but lamented their apparent lack of concern for theological niceties (Morris 1997). In embracing the paleontological record of life and death on earth, as many did, they seemed "indifferent to the fact that this means accepting a billion years of a suffering, dying biosphere before Adam's fall brought sin and death into the world." Morris predicted that, despite having compromised on the plain meaning of the Bible, the proselytizers for ID theory would find no more favor with naturalistic evolutionists than he himself had.

The theistic evolutionists in the ASA, who also believed in a divinely designed world, remained skeptical of ID theory for other reasons. In an editorial on "intelligent design" in the ASA journal, *Perspectives on Science and Christian Faith,* Gordon College chemist J. W. Haas, Jr., surmised that "most evangelical observers—especially working scientists—are deeply skeptical" (Haas 1997). Though supportive of theistic worldviews, they balked at being "asked to add 'divine agency' to their list of scientific working tools." To rely on "intelligent design" to explain complex biological organisms was, said Haas quoting Dawkins, "a pathetic cop-out of [one's] responsibilities as a scientist." Besides, Haas noted, ID theorists rarely applied their methods to disciplines outside of biology, leaving "the rest of us as physicists, chemists, mathematicians, or geologists . . . to go our 'godless' ways in spite of the complexities we face at the quantum level or with the weather."

Reactions to ID theory among naturalistic evolutionists were overwhelmingly negative (Pennock 2001; Forest and Gross 2004). One annoyed critic no doubt captured the feelings of many when he described it as "the same old creationist bullshit dressed up in new clothes" (Webb 1996). When the Jewish magazine *Commentary* in 1996 published a version of ID theory by the mathematician and novelist David Berlinski (1996), letters of protest poured across the editor's desk. Dennett ridiculed Berlinski's stylish essay as "another hilarious demonstration that you can publish bull———t at will—just so long as you say what an editorial board wants to hear in a style it favors" (Dennett 1996). Another reader characterized Berlinski's "intuitions about the Design of the World as neither more nor less reliable than those of flat-earthers, goat-entrail readers, or believers in the Oedipus complex" (Wessel 1996).

Historians are notoriously unreliable prophets, but it seems safe to predict that America will continue to witness spirited, indeed acrimonious, debates over the scientific, theological, and political consequences of evolution for the foreseeable future (see, e.g., Giberson and Yerxa 2002). As much as some people—scientists no less than fundamentalists—might like to dictate what their fellow citizens should believe in regard to origins, it appears unlikely that they will succeed in a constitutional, democratic, and divided republic, where only about one in ten adults subscribes to naturalistic evolution and a majority believes the Book of Genesis to be inerrant (Gallup 1990; Wills 1990; Larson and Witham 1997). As long as the Bible remains the most trusted and widely read text in America and scientists maintain their considerable cultural authority, consensus seems unlikely, even if desirable.

References

Anonymous, 1986. Creationism in NZ "Unlikely." *New Zealand Herald* (July 3): 14.

———. 1993. God is alive. *Maclean's* (April 12): 53.

———. 1995a. Alabama school board votes to put evolution message in biology texts. Associated Press news release, November 10.

———. 1995b. Pat Buchanan takes on Darwin. *NCSE Reports* 15 (Winter): 3–4.

———. 1996. The evolution of a skeptic: An interview with Dr. Michael Behe, biochemist and author of recent best-seller, *Darwin's Black Box. The Real Issue* 15 (November/December): 1, 6–8.

———. 1997. CT 97 book awards. *Christianity Today* (April 28): 12.

Appleby, S. 1999. Exposing Darwin's "hidden agenda": Roman Catholic responses to evolution, 1875–1925. In *Disseminating Darwinism: The role of place, race, religion, and gender*, ed. R. L. Numbers and J. Stenhouse, 173–208. New York: Cambridge University Press.

Behe, M. J. 1996. *Darwin's black box: The biochemical challenge to evolution.* New York: Free Press.

Belz, J. 1996. Witness for the prosecution. *World* (November 30–December 7): 18.

Berlinski, D. 1996. The deniable Darwin. *Commentary* (June): 19–29.

Cheek, D. 1996. Bill may evolve into law. Nashville *Tennessean* (February 27): 1A–2A.

Cole, J. R. 1996. Gallup Poll again shows confusion. *NCSE Reports* 16 (Spring): 9.

Davis, P., and D. H. Kenyon. 1993. *Of pandas and people: The central question of biological origins*, 2nd ed. Dallas: Haughton Publishing Co.

Dawkins, R. 1986. *The blind watchmaker.* New York: W. W. Norton.

———. 1989. Review of *Blueprints: Solving the mystery of evolution*, by Maitland A. Edey and Donald C Johanson. *New York Times* (April 9), sec. 7: 34.

Dembski, W. A. 1995. What every theologian should know about creation, evolution, and design. *Transactions* 3 (May–June): 1–8.

———. 1998. *The design inference: Eliminating chance through small probabilities.* Cambridge: Cambridge University Press.

Dennett, D. C. 1995. *Darwin's dangerous idea: Evolution and the meaning of life.* New York: Simon and Schuster.

———. 1996. Denying Darwin: David Berlinski and critics [letter]. *Commentary* (September): 6.

———. 1997. Appraising grace: What evolutionary good is God? *The Sciences* 37 (January/February): 39–44.

Denton, M. 1986. *Evolution: Theory in crisis.* Bethesda, MD: Adler and Adler.

———. 1998. *Nature's destiny: How the laws of biology reveal purpose in the universe.* New York: Free Press.

Downey, R. 1996. Darwin's watchdog, *Eastsideweek* (December 11).

Edis, T. 1994. Islamic creationism in Turkey. *Creation/Evolution* 14 (Summer): 3–12.

———. 1999. Cloning creationism in Turkey. *Reports of the National Center for Science Education* 19 (November/December): 30–35.

Forest, B., P. R. Gross. 2004. *Creationism's Trojan horse: The wedge of intelligent design.* New York: Oxford University Press.

Frair, W. 1997. Review of *Darwin's black box. Creation Research Society Quarterly* 34: 113.

Gallup, G. H., Jr. 1996. *Religion in America 1990.* Princeton, NJ: Princeton Religion Research Center.

Giberson, K. W., and D. A. Yerxa. 2002. *Species of origins: America's search for a creation story.* Lanham, MD: Rowman and Littlefield.

Gould, S. J. 1992. Impeaching a self-appointed judge. *Scientific American* (July): 118–20.

Haas, J. W., Jr. 1997. On intelligent design, irreducible complexity, and theistic science. *Perspectives on Science and Christian Faith* 49 (March): 1.

Hartwig, M., and S. C. Meyer. 1993. Note to teachers. In *Of pandas and people: The central question of biological origins,* 2nd ed., ed. P. Davis and D. H. Kenyon. Dallas: Haughton Publishing Co.

John Paul II. 1997. Magisterium is concerned with question of evolution for it involves conception of man. Address to the Pontifical Academy of Sciences, October 22, 1996. Reprinted in *Quarterly Review of Biology* 72 (4): 381–83.

Johnson, P. E. 1991. *Darwin on Trial.* Washington, DC: Regnery Gateway.

———. 1995. *Reason in the balance: The Case against naturalism in science, law & education.* Downers Grove, IL: InterVarsity Press.

———. 1997. *Defeating Darwinism by opening minds.* Downers Grove, IL: InterVarsity Press.

Kenyon, D. H. 1984. Foreword. In *The mystery of life's origin: Reassessing current theories,* ed. C. B. Thaxton, W. L. Bradley, and R. L. Olsen. New York: Philosophical Library.

Larson, E. J. 1989. *Trial and error: The American controversy over creation and evolution,* updated edition. New York: Oxford University Press.

Larson, E. J., and L. Witham. 1997. Scientists are still keeping the faith. *Nature* 386: 435–36.

Lemonick, M. D. 1996. Dumping on Darwin. *Time* (March 18): 81.

Lewontin, R. C. 1983. Introduction. In *Scientists confront creationism,* ed. L. R. Godfrey, xxv. New York: W. W. Norton.

———. 1997. Billions and billions of demons. *New York Review of Books* (January 9): 28–32.

Matsumura, M. 1994. Evolution in an election year. *NCSE Reports* 14 (Fall): 3, 10.

———. 1995a. Georgia: Creationism pushed at state and local levels. *NCSE Reports* 15 (Winter): 8–9.

———. 1995b. Tennessee upset: "Monkey bill" law defeated. *NCSE Reports* 15 (Winter): 6–7.

———. 1996. Textbook evolution disclaimer in Fairfax County, VA. *NCSE Reports* 16 (Fall): 16.

———. 1997. New Mexico: State legislature joins the fray. *Reports of the National Center for Science Education* 17 (January/February): 4.

Matsumura, M., and A. J. Petto. 1996. New anti-evolution strategy rejected by New Hampshire legislature. *NCSE Reports* 16 (Spring): 20.

Meyer, S. C. 1993. A Scopes trial for the '90s. *Wall Street Journal* (December 6).

Mitman, G. A., and R. L. Numbers. 1996. Evolutionary theory. In *Encyclopedia of the United States in the twentieth century*, vol. 2, ed. S. I. Kutler, 859–76. New York: Scribner's.

Morris, H. M. 1995. ICR and progressive creationism. *Acts & Facts* 24 (February): 2–4.

———. 1997. Defending the faith. *Back to Genesis*, no. 97 (January), a–c, insert in *Acts & Facts* 26 (January 1997).

Nelson, J. 1997. Creationism: The debate is still evolving. *USA Weekend* (April 18–20): 12.

Numbers, R. L. 1992. *The creationists.* New York: Alfred A. Knopf.

———. 1998. *Darwinism comes to America.* Cambridge, MA: Harvard University Press.

———. 2002. Creationists and their critics in Australia: An autonomous culture or the USA with kangaroos? *Historical Record of Australian Science* 14 (June): 1–12.

Numbers, R. L., and J. Stenhouse. 2000. Anti-evolutionism in the antipodes: From protesting evolution to promoting creationism in New Zealand. *British Journal for the History of Science* 33 (118): 335–50.

Pennock, R. C., ed. 2001. *Intelligent design creationism and its critics: Philosophical, theological, and scientific perspectives.* Cambridge, MA: MIT Press.

Provine, W. B., 1992. Progress in evolution and meaning in life. In *Julian Huxley: Biologist and statesman of science*, ed. C. K. Waters and A. Van Helden, 165–80. Houston: Rice University Press.

Ratzsch, D. 1996. *The battle of beginnings: Why neither side is winning the creation-evolution debate.* Downers Grove, IL: InterVarsity Press.

Roberts, J. H. 1988. *Darwinism and the divine in America: Protestant intellectuals and organic evolution, 1859–1900.* Madison: University of Wisconsin Press.

Ross, H. 1994. *Creation and time: A biblical and scientific perspective on the creation-date controversy.* Colorado Springs: NavPress.

Sayin, Ü., and A. Kence. 1999. Islamic scientific creationism: A new challenge in Turkey. *Reports of the National Center for Science Education* 19 (November/December): 18–20, 25–29.

Schmidt, K. 1996. The battle of the books. *Science* 273: 421.

Scott, E. C. 1992. Creationist cases blooming, *NCSE Reports* 12 (Summer): 1, 3, 5.

———. 1993a. In the trenches. *NCSE Reports* 13 (Summer): 6.

———. 1993b. Gallup reports high level of belief in creationism. *NCSE Reports* 13 (Fall): 9.

———. 1993c. Dean Kenyon and "intelligent design theory" at San Francisco State U. *NCSE Reports* 13 (Winter): 1, 5, 13.

———. 1995. State of Alabama distorts science, evolution. *NCSE Reports* 15 (Winter): 10–11.

———. 1996a. Big bang glue-on in Kentucky. *NCSE Reports* 16 (Summer): 1, 9.

———. 1996b. Close Ohio vote scuttles "evidence against evolution" bill. *NCSE Reports* 16 (Spring): 18.

———. 1996c. Monkey business. *The Sciences* 36 (January/February): 20–25.

Shapiro, A. 1999. Fundamentalist bedfellows: Political creationism in Turkey. *Reports of the National Center for Science Education* 19 (November/December): 15–17.

Sonleitner, F. J. 1981. Creationists embarrassed in Oklahoma. *Creation/Evolution* (Spring): 22–27.

Swanson, S. 1997. Debunking Darwin? "Intelligent-design" movement gathers strength. *Christianity Today* (January 6): 64–65.

Swetlitz, M. 1999. American Jewish responses to Darwin and evolutionary theory, 1860–1890. In *Disseminating Darwinism: The role of place, race, religion, and gender*, ed. R. L. Numbers and J. Stenhouse, 209–46. New York: Cambridge University Press.

Thaxton, C. B., W. L. Bradley, and R. L. Olsen. 1984. *The mystery of life's origin: Reassessing current theories.* New York: Philosophical Library.

Van Till, H. 1986. *The fourth day: What the Bible and the heavens are telling us about the creation.* Grand Rapids, MI: William B. Eerdmans.

Webb, D. K. 1996. Letter to the editor, *Origins & Design* 17 (Spring): 5.

Wessel, K. F. 1996. Denying Darwin: David Berlinski and critics [letter]. *Commentary* (September): 11.

Wheeler, D. L. 1996. A biochemist urges Darwinists to acknowledge the role played by an "intelligent designer." *Chronicle of Higher Education* (November): A13.

Whitcomb, J. C., Jr., and H. M. Morris. 1961. *The Genesis Flood.* Philadelphia: Presbyterian and Reformed Publishing Co.

Wills, G. 1990. *Under God: Religion and American politics.* New York: Simon and Schuster.

Woodward, T. 1997. Meeting Darwin's wager. *Christianity Today* (April 28): 14–21.

Creation Science Lite: "Intelligent Design" as the New Anti-Evolutionism

Eugenie C. Scott

"Intelligent design" creationism (ID) arose in the last decades of the twentieth century. Although claimed to be a qualitatively different set of ideas from creation science—the earlier and, arguably, most important form of twentieth-century anti-evolutionism—ID is a subset of creation science. It ignores many creation-science contentions such as the age of the earth or the reality of Noah's Flood, but it does not present any unique ideas not previously found in its ancestor. Despite proponents' efforts to distinguish ID from creation science, ID's roots in creation science go deep.

The late Henry M. Morris was the father of the movement known as "creation science." Morris, a Baptist, was trained as a hydraulic engineer. He began his career as a creationist in 1946 with the publication of his first book, *That You Might Believe,* while he was still in graduate school. The book and its successor, *The Bible and Modern Science* (1951), proclaimed that the universe was recently created in six twenty-four-hour days, and the earth's geology and life forms were shaped by a literal, historical Flood. The books were based not only on the Bible but, as their titles suggest, also on the facts and theories of science—as Morris understood them, at any rate. The modern creation-science movement crystallized in 1961 with the publication of Morris's book *The Genesis Flood*, coauthored with theologian John C. Whitcomb.

In *The Genesis Flood*, Whitcomb and Morris argued for "Flood geology"—the thesis that most modern geological features could be explained by Noah's Flood. The book's mix of theology and science is characteristic of creation science, and it was widely read in evangelical and fundamentalist circles. The authors proposed that there is scientific evidence that the earth is less than 10,000 years old, and that evolution is therefore impossible. This view became known as "young-earth creationism" (YEC), and it swiftly became popular among fundamentalists seeking scientific support for their religious views.

To promote scientific research supporting the young age of the earth and of the universe, the special creation of all living things, and the Flood of Noah, Henry Morris worked with a group of conservative Christian scientists to found, in 1963, the Creation Research Society (CRS), which began publishing *The Creation Research Society Quarterly* (CRSQ) in 1964. Although in the early days, only some of the CRS board members were Flood-geology proponents, the society founded by Morris soon evolved into an exclusively YEC organization.

Creation-science proponents endorse special creationism, a view of *how* God created. In this conservative Christian view, God created the universe, the solar system, earth, plants and animals, and humans in their present form; the universe we see today is essentially as it was created. The most popular form of special creationism holds that creation took place over six twenty-four-hour days. Some old-earth creationists, however, believe that God created the natural universe over a longer period of time: the essential idea of special creationism is creation of phenomena *in their present form*. Regarding biological organisms, special creationism holds that God created the "kinds" of living things with limited genetic variability. Thus it is possible to have evolution (or variation) within kinds, but one "kind" cannot change into another. Special creationism thus reflects biblical literalism, where God creates organisms and instructs them to "reproduce according to their kind."

Creation-science proponents contend that there are only two views of "origins": (special) creationism and evolution; thus, arguments against evolution are therefore arguments in favor of creationism. Literature supporting creation science thus centers on alleged examples of "evidence against evolution," which are considered to constitute positive evidence for creationism.

"Intelligent design" (ID) creationism can be dated from the publication of *The Mystery of Life's Origin* (Thaxton et al. 1984). The three authors proposed that the origin of life could not be explained without reference to an outside, intelligent cause. It was "fundamentally implausible that unassisted matter and energy organized themselves into living systems" (Kenyon 1984, viii).

The inspiration for *Mystery* was Jon Buell, a former campus minister who became president of the Dallas-based conservative Christian organization, The Foundation for Thought and Ethics (FTE). A former employee of Campus Crusade for Christ, Buell formed Probe Ministries in 1972 to promote Christian theism (Thomas 1990). Probe published booklets and articles by academic Christians (Witham 2002). Buell also recruited historian and chemist Charles Thaxton, engineer Walter Bradley, and geochemist Roger Olsen to write a document on scientific difficulties concerning the origin of life, which grew into *Mystery*. In an attempt to give the book credibility, Buell sought a secular publisher rather than a Christian one. According to Witham, Buell approached no fewer than 176 secular publishers before achieving success with Philosophical Library (Witham 2002). Nevertheless, *Mystery* was not widely reviewed in the scientific press.

Buell, Thaxton, Bradley, Olsen, and others associated with the FTE proposed a new form of creationism that did not rely directly on the Bible: there were no references to a universal Flood, to the special creation of Adam and Eve or any other creature, or to a young earth. But, echoing creation science, they emphasized supposed scientific problems of evolution. *Mystery* used the language of science, with only brief references in an epilogue to the necessity for intelligence to be involved in the origin of life—and even here, it was claimed that this intelligence need not be transcendent. The suggestion of Hoyle and Wickramasinghe (1979) that life on earth was produced by extraterrestrials of high intelligence is offered as one example of nondivine creation, although the authors express their preference for creation by God.

To jump-start the movement, Thaxton organized two conferences: "Going Beyond the Naturalistic Mindset: Origin of Life Studies" in 1984 and "Sources of Information Content in DNA" in 1988 (Nelson 2001; Witham 2002). Only a few creation-science proponents were invited.[1] Instead, most of the presenters were other anti-evolutionists who, feeling uncomfortable with traditional creation science, eagerly embraced the new approach as being both more intellectually rigorous and more likely to be accepted by the general public as a *scientific* alternative to evolutionary theory.

The second FTE book was actually begun before the publication of *Mystery*; reference to an "unbiased biology textbook" appeared in a 1981 Students for Origins Research publication. The book would be "sensitively written to 'present both evolution and creation while limiting discussion to scientific data.'" (Anonymous 1981). Early manuscripts had titles that clearly reflected the creation science orientation of FTE: *Creation Biology*; *Biology and Creation*; *Biology and Origins*.[2] As *Biology and Origins*, the book was shopped to secular publishers for more than two years before one was found—a small Texas press that specialized in seed catalogues (Scott 1989). By this time, the nascent movement had settled upon "intelligent design" as its referent, and this term appeared in *Of Pandas and People* (Davis and Kenyon 1989).

1. Participants in the 1988 conference included creation-science proponents Wayne Frair, Pattle Pun, and Paul Nelson, old-earth creationist Hugh Ross, and many theistic evolutionists associated with the American Scientific Affiliation, such as Gordon Mills, Howard Van Till, and Peter Rust. "Generic" anti-Darwinists such as Robert Augros, George Stanciu, and Michael Denton also participated. Future leaders of the ID movement such as William Dembski and Steve Meyer also presented papers.

2. These early manuscripts were obtained by plaintiffs' attorneys in the federal district court case *Kitzmiller v. Dover*, a 2005 trial over the legality of a school-board policy requiring the teaching of "intelligent design." The manuscripts are discussed later in this chapter.

Although *Pandas* was proposed for adoption as an approved textbook (and thus eligible for purchase using state funds) in at least two states (Idaho and Alabama) and in several school districts, its supporters were unsuccessful in achieving their goal of wide usage in public schools. A noisy controversy erupted in Alabama over whether *Pandas* should be on the approved list. Normally, publishers place copies of proposed textbooks in twenty-two specified locations around the state so members of the public can examine them; *Pandas* was unavailable at the public-examination centers until one day before the textbook committee's vote (Brande 1989). Under much public pressure from the science and education communities, the textbook committee voted to reject *Pandas*, and the publisher withdrew the book before the board of education could vote on it—probably to avoid the publicity from a negative vote (Brande 1990). But even as *Pandas* was floundering in Alabama, the ID movement was about to receive its biggest shot in the arm: the active involvement of a respectable, mainstream law professor from the esteemed University of California.

Phillip Johnson, who had been corresponding with Stephen Meyer and other ID proponents for years, published his landmark book, *Darwin on Trial*, in 1991. In March 1992, Johnson took a key role at a seminal conference at Southern Methodist University. "Darwinism: Scientific Inference or Philosophical Preference?" included presentations by Johnson and ID proponents Michael Behe, William Dembski, Stephen C. Meyer, and others, with countering presentations by mainstream biologists and philosophers such as Michael Ruse, Arthur M. Shapiro, and Frederick Grinnell. As is normal for scholarly conferences, a book of symposium proceedings subsequently was published (Buell and Hearn 1994). Other conferences followed, while the fledgling movement attempted to develop an anti-evolutionism independent of the creation-science movement. Still, religious themes were evident even in the titles of these conferences (for example, "The Death of Materialism and the Renewal of Culture" [1995]; "Mere Creation" [1996]; "Naturalism, Theism, and the

Scientific Enterprise" [1997]). Symbolic of the distancing of the ID movement from its sectarian roots, "Naturalism, Theism, and the Scientific Enterprise" in 1997 was the first ID conference held at a secular institution of higher learning—the University of Texas–Austin. The conference was hosted by an ID sympathizer in the philosophy department, Robert C. Koons.

Under Johnson's guidance—and taking advantage of his prominence and connections as a professor holding an endowed chair at a leading secular institution—the ID movement sought to find acceptance first and foremost from the secular academic community. At about this time, the rapidly expanding ID movement found a new institutional locus beyond the FTE at a conservative think tank in Seattle known as the Discovery Institute. Perhaps the proponents of ID reasoned that academics might respect the movement more if it were housed in a more neutral institution. The FTE had long been associated with evangelical Christianity and thus with creation science.

The Discovery Institute was founded in 1991 by a politician named Bruce Chapman to "promote ideas in the common sense tradition of representative government, the free market and individual liberty."[3] One program promoted evangelical Christian political activity ("Religion, Liberty and Civic Life"), though the majority of the programs echoed free-market and libertarian themes. In a 1996 press release, the Discovery Institute announced the creation of its Center for the Renewal of Science and Culture (CRSC), which was renamed the Center for Science and Culture in 2002. The CRSC rapidly replaced the Foundation for Thought and Ethics as the hub for ID activities. The FTE Web site, in fact, sounds almost wistful, describing itself as having "helped to inspire the robust and exciting international movement of 'intelligent design.'"[4]

3. Discovery Institute mission statement. See www.discovery.org/about.php (accessed September 5, 2005).

4. Foundation for Thought and Ethics mission statement. See www.fteonline.com/about.html (accessed September 5, 2005).

Tenets of "Intelligent Design"

Despite supporters' claims for scientific validity, "intelligent design" is a religious movement supported by conservative Christians concerned about the secularization of modern society. Beyond their desire to see an increase in personal piety, these theists seek a return of religion to public life. Bruce Chapman, president of the Discovery Institute, stated this view quite clearly in his description of the goals of the CRSC:

> To defeat scientific materialism and its destructive moral, cultural and political legacies. To replace materialistic explanations with the theistic understanding that nature and human beings are created by God. (CRSC Web site, October 1999)

and:

> Accordingly, our Center for the Renewal of Science and Culture seeks to show that science supports the concept of design and meaning in the universe—and that that design points to a knowable moral order. (Chapman 1998, 3)

Such labels as "scientific materialism" and "materialistic explanations" are hallmarks of the conservative Christian fear of modernism (Eve and Harrold 1991), a staple in creation-science literature. The museum at the Institute for Creation Research (ICR) features a "tree of evil," in which evolution is depicted as the source of a long list of dreadful "isms" ranging from communism to imperialism and including bestiality, infanticide, slavery, and child abuse. To the supporters of creation science, acceptance of evolution equates with atheism and a consequent degeneration of society. Similarly, Johnson's *Reason in the Balance* (1995) equates evolutionary naturalism with the modernist tolerance of homosexuality, pornography, abortion, genocide, and other "social evils"—real and imagined. Both creation-science and ID proponents posit a dichotomy—with God, creation, purpose, and goodness on one side, and evolution, meaninglessness, and social degeneration on the other (Pennock 1996).

From the beginning, the core of individuals who built the ID movement was concerned with the materialist focus of American society and of science, which they associate with materialism. For example, science explains events and observations through natural causes; the principle of *methodological* materialism rules out appeals to divine cause in science (Scott 1997; National Academy of Sciences 1998, 1999). Methodological materialism is distinct from *philosophical* materialism—the belief that matter, energy, and their interactions comprise the universe; no gods or supernatural powers exist. But ID proponents claim that methodological materialism is merely a front for philosophical materialism; they see a slippery slope between the former and the latter. In Phillip Johnson's words, "That naturalistic explanation of how life came to its present state of complexity and diversity is a major prop for naturalism in philosophy and for agnosticism in religion" (Johnson 1994a).

Although all of science is *methodologically* materialistic, it is evolution upon which the ID proponents focus. Evolution directly or by implication deals with existential issues: Where did people come from? How are we alike and different from other creatures? Are we special in God's eyes? Do people (and I as an individual) have a purpose? What is the meaning of life? Evolution thus makes a good foil for the attack upon the materialism of science, which in turn is a means to attack what ID proponents consider a perverse societal materialism. The "wedge" strategy (see Cole, in this volume) is based on precisely this premise: The edge of the wedge splits the solid log of evolution, which allows the abandonment of philosophical materialism in science and, in Chapman's words (1998), the "defeat [of] scientific materialism and its destructive moral, cultural and political legacies," allowing for the revival of "the theistic understanding that nature and human beings are created by God."

ID's antimaterialism leads its proponents to propose radical changes in how science is done. In 1984, the authors of *The Mystery of Life's Origin* distinguished between regular (or "operation science") and a supposedly different kind of science, "origin science," which requires or at least permits an alternate sort of scientific methodology. Like future ID proponents, the authors attributed historical and biological events to "intelligence," where the "intelligence" was understood as operating supernaturally. Origin science is defined as the science used to explain singular, unrepeatable events (the origin of life, for example), which supposedly are untestable and thus outside of science. Therefore, attribution of causality to God is acceptable in "origin science," but not in "operation science."

The abandonment of methodological materialism in science was also championed by creation-science advocates; it appeared only three years later in a book by two young-earth creationists, Norman L. Geisler and J. Kerby Anderson, with a foreword by Walter L. Bradley (Geisler and Anderson 1987). Geisler obliquely claimed precedent for the distinction between "operation science" and "origin science" (which he called "science of origin") in an obscure 1983 publication, but in general, both ID and creation-science proponents cite Thaxton and others as the source of the distinction. *Of Pandas and People*, in 1993, included a "Note to Teachers" by Mark Hartwig and Stephen Meyer in which they similarly distinguished "inductive sciences" and "historical sciences" and defended the idea of broadening science to include "intelligence" as a cause.

Other ID proponents have encouraged "theistic science" (Plantinga 1991) as a way of broadening science beyond methodological materialism. The argument is made that if we "arbitrarily" limit science to only natural cause, we may miss the true explanation—which is direct or indirect supernatural design. Moreland has proposed that the essence of theistic science is

> . . . a commitment to the belief that God, conceived of as a personal agent with great power and intelligence, has through direct, primary causation and indirect, secondary causation created and designed the world for a purpose. He has directly intervened in the course of its development at various points (for example, in directly creating the universe, first life, the basic

> kinds of life, and humans). And these kinds of ideas can enter into the very fabric of scientific practice. (Moreland 1993, 46)

As mentioned, the abandonment of methodological materialism in science is part of the strategy of reviving a theistic—in particular, a conservative Christian—understanding of the world and humanity's place in it. Because of the success of methodological materialism in explaining the natural world, however, scientists are unlikely to embrace theistic science.

"Intelligent design" creationism focuses on the key concept of creation science: God's design, which is viewed as analogous to human design. If humans have a need that can be met by a structure or artifact, we put together component parts to produce an artifact that meets the need. Nature also exhibits structures that are composed of parts that work together and function to get something done, much as do human artifacts. The leg of a deer has many component parts that, working together, allow a deer to run swiftly. The vertebrate eye has many individual parts that, combined, allow light collection, focusing, and transmission of images to the brain. Both as biologists and as lay people, we colloquially refer to this as "design," much as we refer to the result of human purposive action.

But do structures found in nature reflect intelligence in the same way that human artifactual design reflects intelligence? To William Paley in his 1802 *Natural Theology*, the parallels between design in nature and human design were so obvious that it seemed reasonable to him to infer intelligent cause for both: the structural complexity of biological phenomena such as the vertebrate eye surely could not have occurred through mere natural cause, hence such designed structures were the result of the divine hand. As will be discussed in the next section, a major tenet of ID is that there are some natural phenomena that have the distinctive attributes of intelligently designed phenomena; only a materialistic bias—claim the ID proponents—prevents scientists from recognizing that such phenomena indeed *are* produced by "an intelligence." If a structure *looks* designed, it *was* designed. Michael Behe has said, "Biologists routinely talk about machines in the cell, and they use the term literally not metaphorically" (Behe 2005).

But who or what is the designer (or Designer)? Here the ID proponents, mindful of the legal problems faced by creation science in being overtly religious, deny that the designer *necessarily* is God, although that is their preference. As proposed by Thaxton and others in *The Mystery of Life's Origin*, the creating "intelligence" could be material (as in extraterrestrial intelligence) or transcendent. But little green men are not really what ID proponents have in mind. On February 25, 2002, philosopher William Dembski was a guest on televangelist D. James Kennedy's show, "Truths That Transform." There he referred to his "design inference," saying, "It gets interesting when you apply these methods to the natural sciences where there is no human or extraterrestrial intelligence that could have been involved, but where, in fact, you're dealing with a design that is most likely transcendent."

The Two Foci of "Intelligent Design"

"Intelligent design" has both a scientific/scholarly focus and a "cultural renewal" focus—viewed by both its proponents and its critics as complementary. ID's scholarly focus posits that the universe (or at least components of it) has been designed by an "intelligence." As in creation science, ID recognizes that some adaptive complexity can be produced through natural processes (such as natural selection rearranging the limited genetic variation within a "kind"), but it holds that some extremely complex natural phenomena can only be explained through the action of "intelligence." So just as we attribute "John loves Mary," written in the sand on a beach, to an intelligent agent, when we see a complex, meaningful ("specified") pattern in nature—such as the DNA code—so also should we recognize that such a pattern had a designer (Davis and Kenyon 1989). ID elaborates on this position by claiming to provide a scientifically justifiable method to distinguish intelligently designed phenomena in nature from those phenomena that are designed by natural processes (Dembski 2005).

Thus, for example, biochemist Michael Behe contends that intelligence is required to produce irreducibly complex cellular structures (ones from which the removal of a single part would cause the structure to cease functioning) because such structures could not be produced by natural selection (Behe 1996). Similarly, Dembski holds that his "design inference" also can distinguish between apparent design that is a result of natural processes and that which must be attributed to the workings of intelligence (Dembski 1998).

Other contributors to this volume will discuss in greater detail why neither Behe's nor Dembski's arguments truly accomplish the goals their inventors have set for them; my purpose here is to compare ID with creation science. Although arguments against evolution based on probability have been a mainstay in creation science (Morris 1974a; Gish 1976; Perloff 1999), Dembski's design inference is, at least superficially, more impressive, couched as it is in a more sophisticated mathematical idiom. Behe's irreducible-complexity concept was anticipated in creation science; even his favorite example, the bacterial flagellum, was earlier discussed as a structure whose complexity called out for a "Designer" (Lumsden 1994).

Much as in Paley's conception, structures that are "too complex" to have occurred "by chance" (i.e., through natural cause, equated with chance) require special creation. Behe, following ID convention, does not mention God directly, but in essence his argument is that irreducibly complex structures are evidence for God's direct action. As such, ID becomes a sort of progressive creationism in which God intervenes at intervals to create irreducibly complex structures like DNA, the bacterial flagellum, the blood-clotting cascade, and any number of necessary biological functions. Dembski and others sometimes float a second option where all the irreducibly complex structures in the universe were created by God during the Big Bang, and they are merely unfolding like so many homunculi as time passes (Dembski 2001).

The second focus of ID is "cultural renewal," which consists of its efforts to promote a theistic sensibility to replace the alleged philosophical materialism of American society. Creation science also has as its goal to bring Americans to Christ; the leading creation-science organization, the Institute for Creation Research, states: "ICR's purpose is to advance the cause of true science and education, winning people to Christ, and strengthening the Christian witness by promoting creation thinking in science and Scripture" (Anonymous n.d.).

Until August 2002, this focus was reflected in the name of the main ID institution, Discovery Institute's Center for the Renewal of Science and Culture (CRSC) (Anonymous 2002). In that month, the word *renewal* was dropped from all Web pages, and the CRSC became the Center for Science and Culture (CSC). One may speculate that "cultural renewal" may have been too reminiscent of the goals of twentieth-century creation science, distracting attention from the effort to have ID considered science; scientific and other scholarly organizations do not typically have as their goal the "renewal" of culture.

Perhaps the most vocal proponent of the cultural-renewal focus of ID is Phillip Johnson. Although his first book, *Darwin on Trial,* made only a few references to the purported evils of materialism in American society, subsequent books have been much more evangelical in tone and have strongly and clearly promoted the ID vision—which Johnson helped craft—for a society with more theistic sensibilities. Conferences (such as the 1996 "Mere Creation" conference) have also promoted sectarian Christian views.

Although ID proclaims itself as a scholarly movement, its cultural-renewal focus is fundamentally incompatible with the openness and flexibility required in a true scientific theoretical perspective. When one has an ideological, political, or social goal, it is all too easy to misrepresent or ignore the empirical data when they do not support the goal; certainly creation science is infamous for doing so (Scott 1993). A few "intelligent design" proponents appear to be aware that the scholarly aspect of ID has taken a back seat to the political/ideological. Bruce Gordon has been especially eloquent on this issue, writing:

> . . . design-theoretic research has been hijacked as part of a larger cultural and political movement. In particular, the theory has been prematurely drawn into discussions of public science education, where it has no business making an appearance without broad recognition from the scientific community that it is making a worthwhile contribution to our understanding of the natural world. (Gordon 2001, 9)

and, "If design theory is to make a contribution in science, it must be worth pursing on the basis of its own merits, not as an exercise in Christian 'cultural renewal,' the weight of which it cannot bear" (Gordon 2001, 9).

Somc ID proponents (especially those such as Paul Nelson and Nancy Pearcey, who joined the ID camp from the creation-science side of anti-evolutionism) reject evolution altogether; others accept some common descent—at least within large if undefined groupings, paralleling the creation-science "kinds" of limited genetic variation (Dembski 1995). Common to all ID proponents, however, is the rejection of "Darwinism," which is used as an epithet to indicate an ideology ("ism") of philosophical materialism inextricably tied to evolution. The cultural-renewal aspect of ID—the movement seeking to replace philosophical materialism with Christian theism—is clearly seen in its exploitation of the term "Darwinism."

This fixation on "Darwinism/Darwinist" in ID literature is puzzling to scientists, who, after all, do not refer to physicists as Kelvinists or geologists as Lyellists. In evolutionary biology, "Darwinism" usually refers to the general idea of evolution by natural selection; it may specifically refer to the ideas held by Darwin in the nineteenth century. Usually the term is not used for modern evolutionary theory, which, because it goes well beyond Darwin to include subsequent discoveries and understandings, is more frequently referred to as "neo-Darwinism," or just "evolutionary theory." Evolutionary biologists hardly ever use "Darwinism" as a synonym for evolution, though some historians and philosophers of science do. In ID literature, however, "Darwinism" can mean evolution itself, natural selection, Darwin's ideas, or neo-Darwinism, but most commonly it refers to materialist ideology inspired by "Godless evolution."

The public, on the other hand, equates "Darwinism" with evolution (common descent), and this confusion is exploited by antievolutionists. For decades, creation-science proponents have cited the controversies among scientists over how evolution occurred (including the specific role of natural selection) in their attempts to persuade the public that evolution itself—the thesis of common ancestry—was not accepted by scientists, or at least was in dispute. Within the scientific community, of course, there are lively controversies over how much of evolution is explained by natural selection and how much by additional mechanisms such as those being discovered in evolutionary developmental biology ("evo-devo"). No one says natural selection is unimportant; no one says that additional mechanisms are categorically ruled out. But these technical arguments go well beyond the understanding of laypeople and are easily used to promote confusion over *whether* evolution occurred.

Note that none of the active proponents of either ID or creation science are contributing to the scientific discussion of these points: instead of debating evolutionary theory at professional scientific conferences or in journals, they do no research in evolutionary biology but merely report on the work of other scientists (Johnson 1991; Wells 2000), often distorting it severely in the process (Gould 1992; Scott and Sager 1992; Coyne 1996, 2001; Scott 2001a; Padian and Gishlick 2002). They write articles and books for the general reader rather than for specialists in science.

An example of confusing "Darwinism" with evolution is the "100 scientists" advertisement that was published by the Discovery

Institute in late 2001 and early 2002 in three national publications: the *New York Review of Books*, the *New Republic*, and the *Weekly Standard*.[5] Their full-page ad, entitled "A Scientific Dissent to Darwinism," consisted of two introductory paragraphs followed by a statement attested to by about 100 (the exact number varied) scientists and philosophers of science, whose names were listed. Judging from the periodicals' rate cards, the combined cost of the advertising campaign must have been in the neighborhood of $50,000.

The introductory paragraphs intentionally blur evolution and the Darwinian mechanism of natural selection, in an effort to leave a naïve reader with the impression that the scientists were objecting to evolution, rather than the undefined "Darwinism." The first paragraph claims: "The public has been assured, most recently by spokespersons for PBS's *Evolution* series, that 'all known scientific evidence supports [Darwinian] evolution' as does 'virtually every reputable scientist in the world.'" The quotation—minus the bracketed "[Darwinian]" insertion—came from an in-house PBS briefing book sent to television stations before the broadcast of the *Evolution* series in September 2001. The insertion of "Darwinian" changes the meaning of the sentence from a general (and accurate) statement of the acceptance of evolution (common descent) by the scientific community to an inaccurate statement of the universality of acceptance of natural selection (assuming that this is the meaning of "Darwinism" in this context) in the scientific community as the exclusive mechanism of evolution.

The statement the scientists were asked to sign also focuses on the mechanism of natural selection rather than common descent. It read:

> We are skeptical of the claims of random mutation and natural selection to account for the complexity of life. Careful examination of the evidence for Darwinian theory should be encouraged.

5. A PDF file of the advertisement can be linked from www.crsc.org, and an updated version of the number of signatories, titled "Over 500 Scientists Proclaim Their Doubts About Darwin's Theory," was available as of May 1, 2006, at www.discovery.org/scripts/viewDB/index.php?command=view&id=2732.

Setting aside the ambiguity of the phrases "complexity of life"[6] and "Darwinian theory," the statement directly misleads. Most biologists would agree that natural selection does not explain *everything* about evolution, and that science ought always to encourage careful examination of evidence. One scientist joked that even the famous evolutionary biologist Stephen Jay Gould could have signed the statement! Yet surely the intent of the advertisement was to cast doubt not only on the natural-selection mechanism but also on the idea of evolution itself, because the two are inextricably linked in the public mind and by many proponents of ID itself as "Darwinian theory." As Phillip Johnson himself has noted, "I will say for myself, however, that it is very difficult to separate the common ancestry theory from the 'blind watchmaker' mechanism espoused by Dawkins and other Darwinists" (Johnson 2001).

The ID proponents' objection to natural selection, like that of the creation scientists, ultimately is not scientific; it is difficult to imagine that mere disagreement with a scientific concept could produce an anti-evolutionist movement of such scale and vehemence. Animosity toward natural selection springs from the association of natural selection with existential issues of meaning and purpose, which are in turn associated with materialist (nontheistic or antitheistic) philosophy.

ID literature, like creation-science literature before it, abounds with references to natural selection as purposeless, undirected, or random. Johnson even defines evolution as a materialist enterprise: "The important claim of 'evolution' is that life developed gradually from nonliving matter to its present state of diverse complexity through purposeless natural mechanisms that are known to science" (Johnson 1990, 33). Similarly, in his *Defeating Darwinism by Opening Minds,* Johnson declares, "Modern science educators are

6. "Complexity of life" might refer to the complexity of the branching network of living things, the complexity of an individual organism, the complexity of components of an individual organism (such as organ systems or individual organs, or even complicated cellular or molecular structures), or the complex origin of life. Natural selection could indeed explain some of this complexity, though evolutionary biologists do not expect it to explain everything.

absolutely insistent that evolution is an *unguided* and mindless process, and that our existence is therefore a fluke rather than a planned outcome" (Johnson 1997, 15).

"Darwinism" supposedly takes God out of the picture, but the ID opposition to "Darwinism" goes even deeper than this. After all, many Christians believe that God works through natural selection to bring about the present diversity of life, but this approach is rejected by ID supporters. Another issue—common in creation-science literature but less frequently voiced in ID—is whether a benevolent, personal God would create through a painful, cruel, and inhumane process such as natural selection (Scott and Sager 1992). The apparent incompatibility of the image of a benevolent, personal creator with the harsh process of natural selection remains central to ID proponents' fixation on "Darwinism."

In summary, then, ID has three major tenets and two major goals. The major tenets are:

1. that some natural phenomena are the direct result of intelligent design;
2. that these intelligently designed natural phenomena can be scientifically identified;
3. an obsession with "Darwinism" that may take precedence even over opposition to evolution itself.

The major goals are:

1. the abandonment of evolution, followed by the abandonment of methodological materialism in science preliminary to the ultimate goal, which is . . .
2. the establishment of a society in which theism is triumphant over philosophical materialism.

"Intelligent Design" and Creation Science: Similarities and Differences

"Intelligent design" has many parallels to young-earth creation science, the dominant anti-evolution movement of the twentieth century. I will discuss similarities of and differences between the ID and creation-science movements in terms of six categories: Structure/Personnel, Motivations, Goals, Target Audience, Theological Orientation, and Content.

Structure/Personnel

Although there are many different creation-science organizations, all ultimately take their inspiration from the work of Henry M. Morris. Despite the efforts of Morris and associates to obtain academic respectability for creation science, the movement has never been taken seriously by faculty at most colleges and universities, whether secular or religious; only a handful of Bible colleges have creation-science proponents on their faculties or include creation science in their curricula. Having no credibility in the academic world, creation science is structured around a number of nonprofit organizations, including the flagship Institute for Creation Research (ICR), Answers in Genesis (AIG), Creation Moments (formerly the Bible-Science Association), and several freelance anti-evolution ministries, including those of Kent Hovind, Walter Brown, and Carl Baugh. Although excluded from the scholarly world, these institutions have been quite successful at the grassroots level, reaching large numbers of Americans through their publications, radio and television programs, lectures, seminars, workshops, and revivals. Some of these organizations have mailing lists with hundreds of thousands of names, as well as major magazine and book publishing and distribution operations. Internal Revenue Service documents reveal that revenue for AIG in the year 2000 was $5,677,620. Revenue for the ICR in 2001 was in excess of $3 million.

Institutionally, ID is also organized around nonprofit organizations, such as the Foundation for Thought and Ethics (FTE) and the Center for Science and Culture (CSC). Another Christian nonprofit organization, Access Research Network (ARN), evolved from an earlier creation-science collective, Students for Origins Research; ARN now has embraced ID as its major focus. ARN is almost entirely a Web-based organization; it does not hold conferences or have any discernible off-Web activities (www.arn.org). In general, the Internet has been a boon to anti-evolution organizations, providing great outreach and dissemination of message at low expense. The Web-based Leadership University also provides much ID information. William Dembski has founded an Internet-based professional society, the International Society for Complexity, Information, and Design, which holds online chats and "conferences," and lists an online quarterly journal, *Progress in Complexity, Information and Design* (www.iscid.org/pcid). Budgets for "intelligent design" organizations currently are smaller than those for creation-science organizations; official Washington state documents do not report the CSC's budget separately from the larger organization, but the Discovery Institute income in 1998 was $2,048,277, and FTE's income in 2000 was around $400,000. The smaller ARN had an income of $80,000 in 2000.

Whereas creation science is ignored at secular universities and many religious ones, ID has had much greater success attracting the attention of academics. What brought ID to broad public notice was not the 1984 publication of the first ID book, *The Mystery of Life's Origin*, but Johnson's *Darwin on Trial* (1991). Shortly after its publication, Johnson began to organize a number of younger scholars eager to build an anti-evolution movement with academic credibility. Although some of the ID supporters are recycled from the creation-science movement (that is, Paul Nelson, Nancy Pearcey, Siegfried Scherer), many are comparative newcomers to anti-evolutionism, including Robert C. Koons (University of Texas), Robert Kaita (Princeton), Scott Minnich (University of Idaho), and Henry Schaefer (University of Georgia).

In my experience, religious academics are initially drawn to ID because of the attraction of its promise to find an active role for God in nature; ID's lack of empirical specificity allows it to masquerade as a rather vague proclamation—"God did something, sometime, somehow, but He was involved!"—which religious academics and other religious Americans find attractive. I have met religious academics who believe that ID is a form of theistic evolution—which is ironic, considering that most ID proponents vociferously reject theistic evolution in any form. Once it becomes clear to them that ID is definitely *not* theistic evolution, but in fact has a specific sectarian and anti-evolution agenda, many academics withdraw their support. I have seen this occur among participants in the "Science and Religion" movement, which consists of theologians and scientists interested in the interplay of their disciplines. Although initially attracted to ID because of its religious orientation, these potential converts were eventually turned off by the thinness of the ID scholarship.

The conservative Christian media have adopted ID along with creation science. The young-earth creationist evangelist Hank Hanegraaff has hosted Phillip Johnson on his radio program, and has published articles by Paul Nelson and William Dembski in his journal (Dembski 2000; Nelson 2002). Televangelist D. James Kennedy continues to host creation-science proponents, but he has added Johnson and other ID supporters to his roster. Less extremely conservative Christian media have warmly embraced ID: *Christianity Today* has run frequent articles by ID proponents, and editorially supports the movement, and James Dobson's huge Focus on the Family organization regularly features the ID view.

As to the mainstream media, creation science generally appears only in the context of a news story, where, for example, a local school district might be disputing the lack of creationism in textbooks. "Intelligent design," by contrast, has been profiled as a significant cultural movement in such national publications as *U.S. News and World Report, Time,* the *Washington Post,* and the *New York Times.* To my knowledge, Henry Morris has never had an op-ed piece published in the Wall Street Journal, the *New York Times,* or the *Washington Post,* whereas ID proponents have appeared more than once in all three of these national publications.

Motivations

The motivations of creation-science supporters and the proponents of ID are quite similar. To creation-science supporters, evolution is the source of much evil in the world. In Henry Morris's words:

> Evolution is at the foundation of communism, Fascism, Freudianism, social Darwinism, behaviorism, Kinseyism, materialism, atheism, and in the religious world, modernism and Neo-orthodoxy.
>
> Jesus said, "A good tree cannot bring forth corrupt fruit" (Matthew 7:18). In view of the bitter fruit yielded by the evolutionary system over the past hundred years, a closer look at the nature of the tree itself is well warranted today. (Morris 1963, 24)

Evolution is viewed as an evil, Satan-inspired idea that leads people away from God and toward atheism and other philosophically materialist beliefs. Clearly, the stakes are very high for Morris: If a child studies evolution and loses his faith in God, the child will be lost to salvation. In addition, it is materialism—brought about by evolution—that generates the social evils expressed in the Morris "isms" quote above. Creation-science proponents, then, reject evolution notably because it contradicts the Bible but also because it leads to materialist philosophy.

ID proponents stress the latter criticism; opposition to philosophical materialism is, as mentioned earlier, a major motivator for ID anti-evolutionism. Like other conservative Christians, ID proponents are uncomfortable with materialistic explanations when they appear to supplant the direct hand of God. Because natural selection can explain structural complexity and the diversity of species without reference to divine action, ID proponents fear that it *removes* God from the process of creation (a view shared by creation science adherents.) If God is removed from direct action, He is a less personal God; the consequent loss of purpose and meaning is a large component of the motivation for the ID opposition to evolution. This view is expressed by creation science but articulated more fully in ID. In Phillip Johnson's words, "Is there any reason that a person who believes in a real, personal God should believe Darwinist claims that biological creation occurred through a fully naturalistic evolutionary process? The answer is clearly 'no'" (Johnson 1994b, 44).

Goals

The goals of the two movements are similar: to reduce the amount of secularism and philosophical materialism in American society and to revive a particular—usually conservative Christian—religious view by attacking the science of evolution. Because evolution is viewed as a pillar of materialist philosophy, ID proponents thus directly or indirectly discourage the teaching of evolution as valid science in the schools. If creation science and "intelligent design" enter into the curriculum, so much the better, but the primary goal is the disenfranchisement of evolution and/or "Darwinism" in the popular mind. An even more basic goal of both sides is to promote Christian evangelism, as shown by statements from the CSC and ICR Web sites.

Of course, protestations often come from both creation-science and ID proponents that they are not trying to remove evolution from the classroom. They claim that they merely want to give students "all the evidence" or "the complete range of scientific theories"—which, of course, in practice includes giving them creation science and/or ID along with evolution. But given their view that evolution is evil in and of itself, or at best a stalking horse for philosophical materialism, it is difficult to take such pronouncements at face value. I believe that creation-science and ID proponents deliberately avoid trying to ban evolution because they know such approaches are illegal (as in the 1968 Supreme Court case *Epperson v. Arkansas*). They also are savvy enough to realize that the public equates efforts to ban evolution with backwardness: the state or community that attempts to do so becomes a source of ridicule in editorial cartoons and late-night talk shows, as did Kansas in 1999 when its state board of education attempted to remove evolution from the state science education standards (Scott 1999). A strategy promoting inclusion of alternatives to evolution is far more publicly palatable.

Target Audience

Although creation-science proponents initially attempted to persuade the academic community that its scientific views were valid, they quickly abandoned this unsuccessful strategy in favor of concentrating on presenting this view to the general public (Numbers 1992). The ID movement has also tried to appeal to the intelligentsia, and its ideas have attracted far more attention—even if negative—in scholarly publications than have those of young-earth creationism. As Johnson proclaimed, "You have to have a joint popular movement and academic movement. You have to have both of them—the popular movement to back up the academics" (Foust 2001).

The creationism-versus-evolution controversy is largely a Christian issue; adherents of other religious views are hardly ever active anti-evolutionists (but see Edis 1999, Brass 2002, and Nussbaum 2002 for discussions of Muslim, Hindu, and Jewish versions of creationism). Different groups of Christians tend to be targeted by creation-science and ID proponents. Because of the structure of American Christianity, there is a different potential "market" for each group. According to several polls, upward of 85 percent of Americans describe themselves as Christian—a large pie to divide.

To find the potential audience for creation science, we need an estimate of the percentage of biblical-literalist Christians. This is not an easy task, however, since polls tend not to ask the question directly, and the terms used to describe conservative Christians (evangelical, fundamentalist, holiness, charismatic, pentecostal, etc.) are poorly defined, even by adherents. Even the classification of Christians into mainstream and conservative is not clear cut. In general, however, conservative Christians are those who believe in biblical inerrancy and prophecy, the historical resurrection of Christ, substitutionary atonement (i.e., Christ died for the sins of all), traditional views of heaven and hell, and a *personal* relationship with Jesus, which is directly tied to salvation (as opposed, for example, to salvation through good works). A higher percentage of conservative Christians than mainstream Christians regard the Bible as being literally true, according to a poll conducted by Barna Research Group.[7]

Barna first classifies respondents as "born agains" and then subdivides those into the more conservative "evangelical born agains" and less conservative "nonevangelical born agains." By Barna's fairly strict criteria, approximately 33 percent of Americans are born again, and 8 percent are evangelical. But how many Americans interpret the Bible literally, or largely literally? To obtain this estimate, consider that Barna describes his evangelicals as "by definition" believing "that the Bible is totally accurate in all that it teaches." He also finds that of the nonevangelical born agains, "only six out of ten people in this category strongly believe that the Bible is totally accurate in all that it teaches." Adding the 8 percent of the evangelicals to the 60 percent of the nonevangelical born agains gives us a total of approximately 23 percent of Americans who interpret the Bible literally.

These biblical-literalist Christians are the natural constituency of creation science, although they are a minority of American Christians. Still, the followers of Henry Morris can count on almost a quarter of the American public—nearly 75 million people—as potential supporters of their position. This is not an insubstantial number.

"Intelligent design," on the other hand, because it is theologically less limited, can appeal to the approximately 50 percent of Americans who are mainline Christians, as well as the conservative component of American Christianity. "Intelligent design" proponents would like to convince the mainstream Christians as well as the already-convinced conservatives that evolution is empirically unsupported science, propped up only by philosophical materialism. By keeping both theological and scientific details to a minimum, ID proponents seek to welcome mainstream Christians to the "big tent" while not offending conservatives (Scott 2001b).

7. www.barna.org/cgi-bin/PagePressRelease.asp?PressReleaseID=105&Reference=B (accessed January 1, 2002).

Phillip Johnson has actively promoted a closer relationship between ID and creation science, but creation-science proponents have not been uniformly supportive. Creation-science supporters have criticized ID for being insufficiently biblical. To them, it is not enough just to oppose evolution; one must also stress the authority of the Bible and a young earth. The criticism may be becoming stronger, however. In 1998, John Morris demurred:

> ICR very much appreciates the work of Johnson, Behe, and Berlinski, but we recognize that without biblical creationism they fall short of a God-pleasing mark. Any form of old-earth thinking, theistic evolution, or progressive creation is so similar to secular evolution that their defense is ultimately a waste of time. (Morris 1998, d)

The next year, Henry Morris said, "But this [design] approach, even if well-meaning and effectively articulated, will not work! . . . The evidence of design may impress the soul, but it will not save the soul!" (Morris 1999, a, b). Similarly, Carl Wieland of Answers in Genesis wrote a strong online criticism of ID in 2002, arguing that it lacks a "coherent philosophical framework" and ignores the "story of the past," concentrating only on the mechanism of naturalistic, materialistic natural selection. He also criticized the ID "refusal to identify the Designer with the biblical God," which he feels may result in ID's leading "to New-Age or Hindu-like notions of creation, as well as weird alien sci-fi notions." Recognizing the failings of the "big tent" strategy, Wieland wrote, "For tactical reasons, they have been urged (especially by their coolest and wisest head, Phil Johnson, who does not himself share that hostility [towards biblical creation]) not to publicly condemn their Genesis-believing fellow travelers, although this simmering opposition has burst forth from time to time."[8]

Yet Phillip Johnson claims that a union of creation-science and ID supporters has been achieved in Kansas. After the failure to rid the Kansas science-education standards of evolution in

8. Wieland, www.answersingenesis.org/docs2002/0830_idm.asp (accessed August 30, 2002).

1999, Johnson was interviewed in the local media. "He said that a coalition of creation science proponents and 'intelligent design' proponents was formed to support any future political movement opposing Darwinian evolution. Johnson said this was quite a feat, because both camps approach the creation debate from different perspectives" (Foust 2001). Similarly, during the Ohio state standards controversy, in 2001 and 2002, some young-earth creationists morphed into ID supporters, taking advantage of the broader appeal of the latter movement. It will be interesting to see if political expediency manages to override the deep theological differences between the two groups.

Theological Orientation

Both creation-science and ID proponents hold to the Christian doctrine of special creation. For strategic reasons, ID advocates ignore biblical literalism, which is essential to the YEC position. Still, many ID proponents, especially at the grassroots level, are biblical literalists who promote ID because it was thought to be more legally viable than creation science (but see "Over in Dover," below). But ID literature steers clear of promoting ideas such as the young earth, Noah's Flood, or the special creation of Adam and Eve.

Literature from both creation-science and ID proponents teems with references to the essentially religious nature of their respective movements. Creation science favors a biblical-literalist theology, and ID tends to be more theologically eclectic, encompassing almost the full spectrum of Christian religious views, with the exception of theistic evolution. The most important scientific concepts in ID are Behe's "irreducible complexity" and Dembski's "design inference"; taken at their word, they imply progressive evolution. After all, every time an IC structure is encountered, God had to have intervened, since natural processes cannot explain it. This makes God a serial intervenor, creating at intervals throughout time. ID logically entails progressive creationism, though in keeping with the big-tent strategy, such specificity is never underscored.

The cultural-renewal focus of ID has parallels in creation science. In the cover letters for donation appeals sent to their supporters, both the ICR's John Morris and AIG's Ken Ham refer to their organizations as "ministries"—to fight evolution is to evangelize and save souls. References to the "creation witness" are not uncommon in creation science, though the lower-key ID supporters avoid such language. Both creation-science and ID literature express deep concerns about social ills that are claimed to be rooted in the secularization of American culture (Morris 1974b; Johnson 1995; Morris and Morris 1996).

Both creation-science and ID supporters reject theistic evolution, for largely the same reasons. Both regard it as an unacceptable Christian compromise toward evolution. John Morris (1998, d) says, "Combining any form of long-age evolution with Christianity will not satisfy evolutionists, nor will it bring commendation from God or glory to Him." ARN lays it out pretty directly: "There are only two general views that aren't compatible with 'intelligent design': 1) a radical naturalism that denies the existence of any non-human intelligence, theistic or otherwise and 2) conventional theistic evolution" (www.arn.org/id_faq.htm, accessed June 26, 2001). William Dembski is equally abrupt: *"Design theorists are no friends of theistic evolution"* (1995, 3; emphasis in original).

It is peculiar, given the hostility of ID supporters toward theistic evolution, that one of the most prominent ID proponents, Michael Behe, is frequently identified as a theistic evolutionist. It appears as if perhaps Behe is changing his mind; a newspaper article said, "Behe says that the concept [of theistic evolution] is 'no threat to Christian beliefs' and he once agreed with it, but it isn't supported by the biological evidence" (Ostling 2002).

Content

Creation-science proponents and ID proponents agree on the basics: Materialist philosophy and evolution are to be opposed. Some (but not all) ID supporters even claim that "Darwinism" (the mechanism) rather than evolution itself is the issue, yet ID Web sites are full of "evidence" that calls evolution into question, and certainly the take-home message of Jonathan Wells's *Icons of Evolution* (2000) is that evolution did not occur. A major area of contrast, however, is the specificity with which empirical claims ("what happened") are made. Creation science is specific about the age of the earth, the lack of common ancestry of living things, the occurrence of a global Flood, the unreliability of radiometric dating, and so on. ID studiously avoids making empirical scientific claims to avoid conflict with creation-science supporters, viewed as valuable allies against the common foes—materialism and evolution.

When pressed, most—but not all—ID supporters will acknowledge that the earth is old, but they quickly brush aside the topic as unimportant. Phillip Johnson has counseled that all opponents of materialism should set aside their theological differences about such details as the age of the earth, the Flood, and so on, and work together to overcome the common materialist enemy. Johnson is quoted as saying, "I want to develop a challenge to materialistic evolution. Let's unite around the Creator. After that we can have a marvelous argument about the age of the earth" (Stafford 1997). But in most other respects, proponents of ID and creation science are presenting the same *content* to the public.

There are three themes that run through the creation-science and ID literatures; virtually every article, letter to the editor, debate topic, or other expression of creation science reflects one or more of them. These themes, described below, show yet another link between creation science and "intelligent design."

Pillars of Creationism

Evolution Is a "Theory in Crisis"

As expressed in the title of Michael Denton's 1985 book, *Evolution: A Theory in Crisis*, a strong theme of anti-evolutionism is that evolution is questionable science, not supported by evidence, and actually disproved by data that would be apparent to scientists if only they were not blinded by materialist presuppositions. Creation-science and ID proponents constantly repeat the canard that scientists today are questioning whether evolution actually occurred. Of course, the small number of scientists who reject evolution appears to be restricted to supporters of these two movements; the strong presence of evolution in scientific journals belies the claim that scientists are "giving up on evolution." Furthermore, the ubiquity of the teaching of evolution in secular universities as well as in such prestigious sectarian institutions as Brigham Young, Baylor, Notre Dame, Texas Christian, and the like, shows that evolution is in no danger of abandonment. But the contrary claim is repeatedly made to the general public, and it regularly appears in letters to the editor in newspapers across the country.

Among the topics presented as supposed "evidence against evolution" by proponents of either creation science or ID are the improbability of the origin of life from nonliving chemicals, the gaps in the fossil record, and an alleged limited genetic variation within a species (or "kind") that reportedly disallows the emergence of new "kinds" or body plans of organisms. Both creation-science and ID supporters also dispute the idea that natural selection is the "engine" of evolutionary change. Natural selection "works" only to move genes around within a "kind"; it supposedly is not powerful enough to produce new "kinds" or body plans (additional mechanisms are conveniently ignored). "Microevolution," or changes below the species level (changes within the created "kind"), are accepted, but not "macroevolution," which they define as equivalent to the inference of the common descent of living things. By contrast. in evolutionary biology, macroevolution involves such issues as rates of change, tempo and mode of change, the possibilities of species selection, and so forth.

These are old arguments, and there exist many references analyzing these and other alleged scientific claims made by anti-evolutionists (Futuyma 1982; Kitcher 1982; Godfrey 1983; Montagu 1984; Strahler 1987; Scott 2005), so I will not undertake their refutation here. The most recent salvo contending that evolution is a theory in crisis comes from ID proponent Jonathan Wells, who, in his *Icons of Evolution* (2000), maintains that high-school and college textbooks present erroneous, out-of-date, and even fraudulent data to students as "evidence" of evolution. Such "icons" as the peppered moth, Darwin's finches, and the Miller-Urey experiment are presented as evidence that scientists are dissembling or even lying to the public about the strength of the evidence in support of evolution. The book has been dismissed by scientists for misleading readers about the nature of evolution as well as the role of the "icons" as teaching tools (Coyne 2001; Pigliucci 2001; Scott 2001a; Padian and Gishlick 2002).

Both creation science and ID focus on what they perceive to be the "soft underbellies" of evolution. The examples they highlight are those for which they perceive a lack of scientific consensus on all the details: the Big Bang (rejected in creation science but promoted within ID through the anthropic principle), the origin of DNA, the origin of life, the Cambrian explosion, and the origin of humans. J. P. Moreland is upfront about the direct hand of God being involved in these events: "He has directly intervened in the course of its [the universe's] development at various points (for example, in directly creating the universe, first life, the basic kinds of life, and humans)" (Moreland 1993).

Not all the "arguments against evolution" proposed in creation science are echoed by the ID proponents, however. A staple of creation science is the idea that the second law of thermodynamics disallows evolution by prohibiting the increase in complexity seen in the record of the history of life; this is absent in ID except for certain subtle references to entropy in some of William Dembski's mathematical formulations (see Stenger, in this volume). ID proponents also do not repeat the critiques of radiometric dating and other age-of-the-earth arguments found in creation science. ID ignores issues such as the age of the earth, helping to maintain a big tent, including all anti-evolutionists, by papering over disagreements among them.

Although the rejection of "Darwinism" is universally shared by ID proponents, they present anything but a unified front when pressed for their own views. The amount of diversity within the ID ranks is striking. ID proponents include young-earth creationists such as Paul Nelson, Dean Kenyon, Siegfried Scherer, and Nancy Pearcey; old-earthers such as William Dembski and Michael Behe; and individuals reluctant to state their position, such as Phillip Johnson. With regard to evolution itself, supporters of ID include those who accept descent with modification, such as Behe and Jed Macosko; those who vehemently deny it, such as Paul Nelson and Jonathan Wells; and those who accept (as do most creation-science proponents) some limited descent with modification within as-yet-unspecified genetic boundaries, such as Phillip Johnson and William Dembski. The "big tent" results in the inability to speak with one voice on the topic of evolution, rendering incoherent ID's pronouncements on anything other than the untenability of evolution or "Darwinism."

Evolution and Religion Are Incompatible

The second "pillar of creationism" is the supposed incompatibility of evolution and religion. With creation science, the incompatibility of evolution with biblical literalism is most important; with ID, the incompatibility of evolution (and "Darwinism") with a sense of purpose and meaning of life is most important. Again, because the ID proponents are so religiously diverse—ranging from biblical literalists to a few lonely theistic evolutionists—agreement that "Darwinism" (materialism) disallows a sense of purpose and meaning represents the least common denominator of the movement.

Both creation science and ID equate evolution with atheism. Needless to say, this view is not universally shared among Christian theologians or scientists of faith, who find ways of accommodating various beliefs in God and various interpretations of Christianity with evolution (Matsumura 1995; Miller 1999; Russell et al. 1999; Haught 2000).

"Fairness," or "Equal Time"

When laws banning the teaching of evolution were struck down in 1987 by the Supreme Court decision *Epperson v. Arkansas*, anti-evolutionists proposed that it was only "fair" that if evolution is taught, creation science should be taught. Such laws were struck down in *Edwards v. Aguillard*, a Supreme Court decision making Louisiana's "equal time" law unconstitutional. Unsuccessful in the courts, the equity argument is nevertheless extraordinarily persuasive to the general public. Americans are a fair-minded people, and the appeal to "hearing both sides" is ingrained in our culture.

Edwards did not end efforts to "balance" the teaching of evolution, though it did change the terminology (Scott 1997). Instead of arguing for teaching both evolution and creation science, anti-evolutionists argue for teaching evolution as well as "evidence against evolution," or "scientific alternatives to evolution" (both of which on inspection consist of recycled creation-science arguments).

"Intelligent design" supporters also make the equity argument, as in 2001 and 2002 during the controversy over the Ohio state science standards (Evans 2002). Here a creationist group called Science Excellence for All Ohioans originally proposed including ID in the state science standards; when that failed, SEAO retreated to the equity position of "TEACH ORIGINS SCIENCE OBJECTIVELY. Allow students to be shown the scientific evidence that supports both viewpoints" which in their view would include ID.[9] Although the ID equity argument is not fundamentally different from the creation-science equal-time argument, the terminology is more refined. Discovery Institute fellow and law professor David DeWolf has invoked the legal concept of "viewpoint discrimination" (DeWolf et al. 1999), arguing that schools should "teach the controversy."

Neither approach is appropriate for the classroom. "Teach the controversy" has become a popular buzz phrase in ID literature; its proponents mean, "Teach that scientists dispute evolution," which is clearly not the case. Students should not be misled into thinking that there is some great turning-away from evolutionary science among scientists; it would, in fact, be quite unfair to miseducate them on this point (see Petto and Godfrey, in this vol-

9. www.sciohio.org/start.htm (accessed September 1, 2002); emphasis in original.

ume). "Viewpoint discrimination" is also irrelevant to the science classroom, for this body of case law refers to the expression of opinions in a public forum. If a community allows groups to obtain parade permits for the use of city property, for instance, it cannot allow the Knights of Columbus to parade and then deny a permit to the Knights of the Ku Klux Klan; if a school allows after-school programs unrelated to the curriculum (that is, provides a limited public forum), then it cannot discriminate between a Christian group and a Wiccan group.

But scientific discoveries are not "viewpoints." It is not a matter of opinion whether the sun goes around the earth or the earth goes around the sun, nor is it a matter of opinion whether living things descended with modification from common ancestors. What gets taught in the pre-college science class is the consensus of scientific opinion on an issue. In fact, some "viewpoints" are actively and enthusiastically discriminated against in science! Physicists discriminate against supposed perpetual-motion machines because they violate the first and second laws of thermodynamics, just as geologists discriminate against the idea that fossils are merely odd-shaped rocks rather than the remains of once-living things.

Yet equity arguments have great power to sway the public—and teachers. Although there are no reliable recent national polls of opinions of science teachers, regional and state polls suggest that the "equal time" argument appeals to approximately one-third of practicing teachers (Elgin 1983; Nickels and Drummond 1985; Zimmerman 1987; Tatina 1989; Aguillard 1999; Osif 1997). In an admittedly unrepresentative and not statistically significant national survey of science teachers, Eve and Dunn (1989) suggest that even some teachers who recognize that creation science is unscientific will opt for "equal time." To their question, "Regardless of the validity of the concepts of special creationism and evolution, proponents of each should be allowed equal time to express their views in science classes in school," 43 percent of respondents answered, "True."

To summarize, although ID claims to be qualitatively different from creation science, there are more similarities between the two forms of anti-evolutionism than differences. Under "similarities," both ID and creation science

- Are religious movements
- View evolution as entailing existential "purposelessness" and "meaninglessness"
- Reject methodological materialism and promote "theistic science"
- Equate evolution with atheism
- Uphold the "pillars of creationism," including the idea that evolution is a "theory in crisis"; that it is incompatible with Christianity; and that evolution should be balanced by some religious view out of a sense of fairness for "equal time"
- Aim at the same scientific targets: the origin of life, the Cambrian explosion, human evolution, and the insufficiency of "unguided, purposeless" natural cause to produce complex structures;

There are far fewer contrasts between creation science and ID:

- "Intelligent design" creationism is smaller, younger, and less institutionalized (that is, there are fewer formal institutions promoting ID than young-earth creationism)
- "Intelligent design" creationism has much greater access and appeal to the intelligentsia
- Intelligent design creationism lacks a consistent scientific model of history ("what happened?")

Great claims have been made for the power of the design approach. Dembski was described as the "Isaac Newton of information theory" (by Robert Koons, in a dust-jacket blurb for Dembski's *Intelligent Design: The Bridge Between Science and Theology*, 1999), and Behe claimed, "The discovery [of the design principle] rivals those of Newton and Einstein, Lavoisier and Schrödinger, Pasteur and Darwin" (1996, 232–33). Proponents of ID claim that they can detect design in nature using the concepts of irreducible complexity and the design inference, but no scientific research using this approach has been forthcoming (Gilchrist 1997; Forrest and Gross 2003; Petto and Godfrey, in this volume), and in fact, ID proponents have been unable even to describe what scientific questions *might* be illuminated under an ID "paradigm."

Although presented as a new scientific paradigm, "intelligent design" thus turns out to be a politically more sophisticated version of creation science. Since its inception in the mid-1980s, ID has claimed that great scientific insights are just around the corner, as soon as the new paradigm is accepted. Yet even after twenty years, the promissory notes are still out, with no prospect for redemption.

"Intelligent design" proponents sometimes complain that a scientific establishment committed to philosophical materialism is the source of the opposition the field has received. On the contrary, given the thinness of the science of ID, if the movement did not have grave consequences for public-school education and church-and-state separation, ID would languish in academic obscurity. Ironically, perhaps it is the cultural-renewal component of ID that keeps it from fading from the view of the scholarly public.

Over in Dover: Creation Science, ID, and the Law

Creation science, ID, and the pillars of creationism were very much on display in 2004 and 2005 in the Pennsylvania community of Dover. With the election of a religiously conservative school board, the usually uneventful activity of adopting a high-school biology textbook took on great symbolic importance during the summer of 2004.

At the June 7 school-board meeting, board member William Buckingham contended that the textbook chosen by the teachers was unacceptable because it was "laced with Darwinism"; he proposed "balancing" the book with the teaching of creationism. This generated a vigorous community response—about equally divided between supporters and opponents of the idea. Civil liberties organizations warned the district that such instruction would be unconstitutional, and the conservative Thomas More Law Center (TMLC) offered *pro bono* legal services if the district was sued. At the encouragement of Richard Thompson, director of the TMLC,

Buckingham lobbied to "balance" the offending biology textbook with the adoption of the ID textbook *Of Pandas and People* (Anonymous 2005). When the board narrowly voted not to purchase the books, sixty copies were donated anonymously to the district, for what was described as classroom reference. When the district superintendent indicated that the use of the book would be optional, the board was inspired to pass a policy on October 18:

> Students will be made aware of gaps/problems in Darwin's Theory and of other theories of evolution including, but not limited to, intelligent design. Note: Origins of life will not be taught.

There was considerable community and teacher opposition to this policy. Two school-board members resigned over the policy, and two others resigned for personal reasons. In November, the remaining board members appointed four pro-ID replacements. Amid whispers of a possible lawsuit, the board pulled back somewhat on its demand that ID be a classroom subject. They put the copies of *Pandas* in the library and on November 19 issued a "procedural statement" to be read to the students. It was later slightly modified:

> The Pennsylvania Academic Standards require students to learn about Darwin's Theory of Evolution and eventually to take a standardized test of which evolution is a part.
>
> Because Darwin's Theory is a theory, it continues to be tested as new evidence is discovered. The Theory is not a fact. Gaps in the Theory exist for which there is no evidence. A theory is defined as a well-tested explanation that unifies a broad range of observations.
>
> Intelligent Design is an explanation of the origin of life that differs from Darwin's view. The reference book, *Of Pandas and People*, is available for students who might be interested in gaining an understanding of what Intelligent Design actually involves.
>
> With respect to any theory, students are encouraged to keep an open mind. The school leaves the discussion of the Origins of Life to individual students and their families. As a Standards-driven district, class instruction focuses upon preparing students to achieve proficiency in Standards-based assessments.

The statement would be read before the evolution lesson would be taught. The Dover biology teachers uniformly refused to read the statement; instead, administrators read it. On December 14, a group of eleven parents, represented *pro bono* by the Pennsylvania ACLU, Americans United for Separation of Church and State, and the Pennsylvania law firm Pepper Hamilton, filed suit in federal district court, contending that the board's requirement to teach ID and "gaps/problems in evolution" was unconstitutional. Tammy Kitzmiller, mother of a ninth-grader, was selected as the lead plaintiff for *Kitzmiller v. Dover*, the case resulting in a six-week trial on the constitutionality of teaching ID and "evidence against evolution" in the public schools. The Thomas More Law Center announced it would defend the district. Preparations began.

In the complaint, plaintiffs' attorneys argued that ID was a religious view—a form of creationism, which the courts time and again have held unconstitutional to advocate in public schools. To defend itself, the school district would have to argue that the purpose for passing the ID policy was secular, not religious. It would have to prove that ID was scientific, and that there was a valid pedagogical reason for teaching it. It would also have to prove that the effect of making students "aware" of "gaps/problems in Darwin's theory and . . . intelligent design" was done for valid pedagogical reasons, rather than to promote religion-based anti-evolutionism.

Both sides lined up a star-studded cast of ID experts and well-known anti-ID scholars and scientists. The expert witnesses for the plaintiffs were cell biologist Kenneth R. Miller, paleontologist Kevin Padian, philosophers of science Robert Pennock and Barbara Forrest, theologian John Haught, and professor of education Brian Alters. Mathematician Jeffrey Shallit was listed and deposed as a rebuttal witness. All of the plaintiff witnesses served without pay.

The expert witnesses for the defense were biochemist Michael Behe, microbiologist Scott Minnich, communications professor John Angus Campbell, professor of education Richard M. Carpenter II, philosopher/mathematician William A. Dembski, and philosopher Warren A. Nord. Sociologist Steve Fuller and philosopher Stephen Meyer (director of the Discovery Institute's Center for Science and Culture) were listed as rebuttal witnesses. All of the witnesses for the defense were paid an hourly rate.[10]

Very few of the defense expert witnesses actually testified, however. Although all the defense witnesses wrote witness statements, and some wrote statements rebutting plaintiff experts, the defense seemed to fall into disarray once depositions began. After the deposition of Behe and Minnich, the defense began withdrawing witnesses—some precipitously. John Angus Campbell was withdrawn literally as the plaintiff lawyers and stenographer were sitting down to begin the deposition. Dembski withdrew two days before his deposition, as did Stephen Meyer. Nord and Carpenter were deposed but never put on the stand. Outsiders are not privy to details, but clearly there was friction between the TMLC and members of the Discovery Institute over how to handle the case. It's possible that the Discovery Institute may be "saving" expert witnesses for a potential future trial.

The trial, held in Harrisburg, Pennsylvania, began in late September before federal district court judge John E. Jones III, with testimony from witnesses for the plaintiffs. The plaintiffs' case began and ended with testimony from scientists to symbolize to the judge that science was critical to their argument, and that the scientific invalidity of ID—and not just the actions of the school-board members—should be considered in his opinion. In addition to the expert witnesses, there were many fact witnesses called by both sides: plaintiffs, board members, reporters, and others.

The first expert witness, Kenneth Miller, was the first to make a point that would be made repeatedly by virtually all of the plaintiffs' expert witnesses: By abandoning methodological naturalism in favor of supernatural causation, ID was abandoning the ground rules of science. Miller also dissected ID's key scientific concept, Behe's "irreducible complexity." Walking the judge through a careful analysis of the bacterial flagellum, the

10. Witness statements and depositions, as well as other documents from *Kitzmiller v. Dover* are available on the NCSE Web site, www.ncseweb.org.

immune system, the blood-clotting cascade, and other examples from the ID literature, Miller showed how the scientific foundation of ID was factually wrong. Because ID proponents claim that evolution fails to explain certain phenomena, Miller also illustrated the superiority of evolution over ID in explaining the functioning of living things. The final expert witness, paleontologist Kevin Padian, made the same points as he dissected the claims about fossils in *Of Pandas and People*, showing its factual and conceptual errors and its failure to live up to the standards of science. Part of the message of the scientists was that "real" evolutionary biology and paleontology—not the distorted version in *Pandas*—was not only valid but fascinating science. As one reporter commented, "For six weeks, the courtroom of Judge John E. Jones III was like the biology class you wish you had taken" (Talbot 2005).

Theologian John Haught provided a dignified discourse on the history of the idea of design, and he clearly placed ID in the company of William Paley. But perhaps the most devastating witness for the plaintiffs was philosopher of science Barbara Forrest, coauthor of *Creationism's Trojan Horse*, a well-researched 2003 book on the history and motivations of the ID movement. In her testimony, she traced the roots of ID to creation science, presenting devastating evidence (obtained from early subpoenaed manuscript drafts of *Pandas*) that the definition of creationism became the definition of "intelligent design" as the book progressed through rewrites. In *Biology and Creation* (1986), creation was defined as follows:

> Creation means that the various forms of life began abruptly through the agency of an *intelligent creator* with their distinctive features already intact—fish with fins and scales, birds with feathers, beaks, and wings, etc. (pp. 2–13, 2–14) [italics mine]

In the published version of *Pandas*, almost the same wording appears as the definition of "intelligent design":

> *Intelligent design* means that the various forms of life began abruptly through an *intelligent agency*, with their distinctive fea-

tures already intact—fish with fins and scales, birds with feathers, beaks, and wings, etc. (pp. 99–100) [italics mine]

Forrest also presented a graph showing word counts from early manuscripts and the published editions of *Pandas.* In 1987, there was a radical change in the number of times the terms *creationism* and *creationist* appeared, compared with the number of times the phrase *intelligent design* occurred—the same year that the Supreme Court case *Edwards v. Aguillard* declared creation science unconstitutional. This discovery was "astonishing" to the judge:

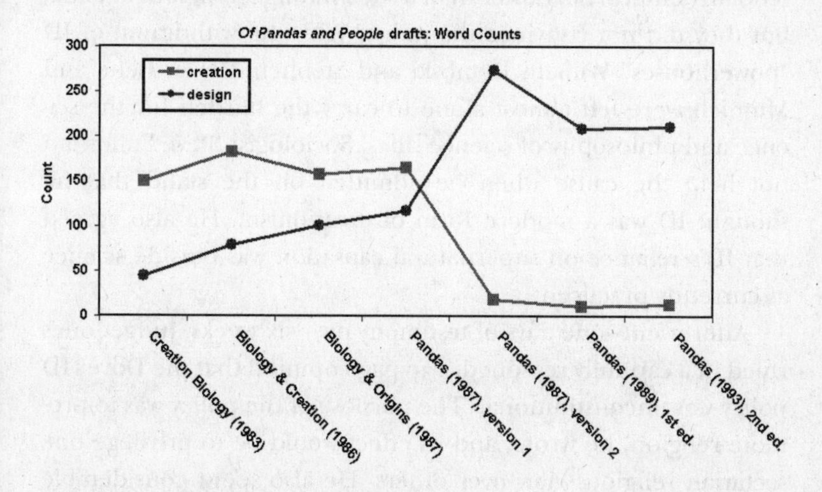

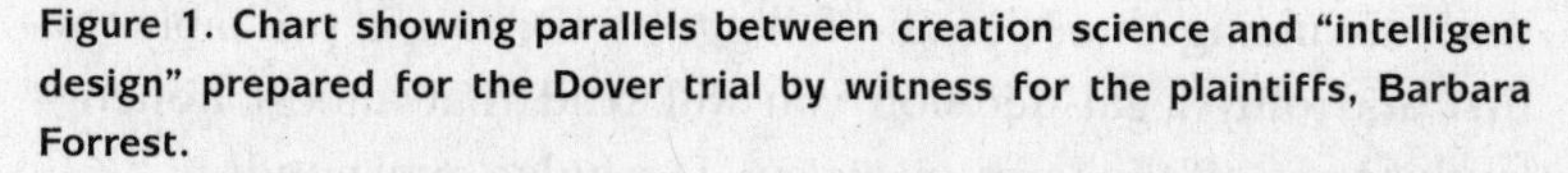

Figure 1. Chart showing parallels between creation science and "intelligent design" prepared for the Dover trial by witness for the plaintiffs, Barbara Forrest.

By comparing the pre- and post-*Edwards* drafts of *Pandas,* three astonishing points emerge: (1) The definition for creation science in early drafts is identical to the definition of ID; (2) cognates of the word *creation* (*creationism* and *creationist*), which appeared approximately 150 times. were deliberately and systematically replaced with the phrase *intelligent design*; and (3) the changes occurred shortly after the Supreme Court held in *Edwards* that creation science is religious and cannot be taught in public-school science classes. This word substitution is telling and significant and

> reveals that a purposeful change of *words* was effected without any corresponding change in *content*, which directly refutes the argument of the Foundation for Thought and Ethics that by merely disregarding the words *creation* and *creationism*, FTE expressly rejected creationism in *Pandas*. (*Kitzmiller v. Dover*, 32)

In addition, Forrest presented data showing that ID and creation science used parallel faulty arguments to discredit evolution: gaps in the fossil record, the Cambrian explosion, the rejection of naturalism, and so on. The defense expert witnesses valiantly tried to argue that ID was a legitimate scientific and scholarly enterprise, rather than a shell hiding a religious agenda, but they did not convince the judge. With the withdrawal of ID "powerhouses" William Dembski and Stephen Meyer, Behe and Minnich were left almost alone to carry the burden for the science and philosophy of science sides. Sociologist Steve Fuller did not help the cause when he admitted on the stand that he thought ID was a modern form of creationism. He also agreed that ID's reliance on supernatural causation was outside science as currently practiced.

After twenty-one days of testimony over six weeks, Judge Jones ruled in a carefully reasoned, 139-page opinion that the Dover ID policy was unconstitutional. The purpose of the policy was to promote religion, he wrote, and its effect would be to privilege one sectarian religious view over others. He also spent considerable time discussing the scientific failings of ID; this is probably the first and only legal decision you will read that almost nonchalantly tosses off the term *exaptation*. The judge explained:

> Finally, we will offer our conclusion on whether ID is science not just because it is essential to our holding that an Establishment Clause violation has occurred in this case, but also in the hope that it may prevent the obvious waste of judicial and other resources which would be occasioned by a subsequent trial involving the precise question which is before us.
>
> . . . We find that ID fails on three different levels, any one of which is sufficient to preclude a determination that ID is science. They are (1) ID violates the centuries-old ground rules of sci-

> ence by invoking and permitting supernatural causation; (2) the argument of irreducible complexity, central to ID, employs the same flawed and illogical contrived dualism that doomed creation science in the 1980s; and (3) ID's negative attacks on evolution have been refuted by the scientific community. . . . it is additionally important to note that ID has failed to gain acceptance in the scientific community, it has not generated peer-reviewed publications, nor has it been the subject of testing and research. (*Kitzmiller v. Dover*, 63–64)

Between the end of the trial and the judge's decision, an election was held in Dover in which the incumbent school board was replaced by one that had no intention of sustaining the ID policy. Thus, the judge's decision was not appealed. *Kitzmiller* is therefore precedent only in the Middle District of Pennsylvania, although it will surely be highly influential anywhere the issue of teaching ID arises in the future. The completeness of the analysis makes it likely to have the effect of blunting future efforts to pass Dover-like ID policies.

This does not mean that proponents of ID have given up. Although Discovery Institute director Bruce Chapman admitted that the trial was a disaster "as a public relations matter" (Postman 2006), William Dembski defiantly commented, "This galvanizes the Christian community. People I'm talking to say we're going to be raising a whole lot more funds now" (Zoll 2005). On the positive side, within six months after the decision, the state of Ohio rescinded not only a model curriculum lesson conceded to reflect ID but also the science education standard on which it was based (Anonymous 2006a). The New Mexico community of Rio Rancho similarly rescinded a policy that locals had promoted to permit the teaching of ID. In early January 2006, the school district in Lebec, California, attempted to teach a "philosophy of intelligent design" course that was a mishmash of creation science and ID. Parents sued, and the case was promptly settled, with the promise not to teach the subject in the future (Anonymous 2006b). In addition, none of the 2006 legislative measures promoting ID had passed as of May 2006. But although ID as a legal strategy can safely be said to be dead, it is alive and well as a movement. Similarly, although creation science failed as a legal strategy after *Edwards,* it remains a robust and active movement.

What direction will ID take next? When creation science was declared unconstitutional to teach, it was minimalized and repackaged as ID. The ID supporters in turn may try to relabel ID, perhaps seeking an agent-less form of anti-evolutionism, since employing an "intelligent agent" seemed to the judge to be too much like employing an "Intelligent Agent." In the forthcoming third edition of *Pandas,* now edited by William Dembski and retitled *The Design of Life,* appears this eerily familiar sentence:

> Sudden emergence holds that various forms of life began with their distinctive feature already intact: fish with fins and scales, birds with feathers and wings, mammals with fur and mammary glands. (*The Design of Life,* MS, William Dembski n.d., 28 [*Kitzmiller v. Dover,* plaintiffs' exhibit #775])

But the most likely future strategy is the one the Discovery Institute currently encourages: Teachers should teach evolution but "balance" it with the teaching of "evidence against evolution." This is equivalent to the component of the Dover policy calling for teaching "gaps/problems" in evolution—and, being less obviously religious, it might fare better in courts. Policies promoting "evidence against evolution" have many variants. Teachers are exhorted to "teach the controversy," or to teach "strengths and weaknesses of evolution," "both evidence for and evidence against evolution," "evolution as theory not fact," "the full range of views about evolution," or students are exhorted to "critically analyze" evolution. Common to this approach is presentation of a lot of bad science emanating from the creationist literature along with whatever amount of evolution they might be getting in the standard curriculum—and the former is probably more extensive than the latter, given the small amount of evolution (if any) actually taught in most high-school biology courses.

The Discovery Institute has a model "evidence against evolution" policy that it has circulated since at least 2004: it was sent to one of the Dover board of education members at that time. It is an interesting blend of "evidence against evolution" and a "permissive" approach to teaching ID, where ID is not required by statute or policy but is "mentioned," as if calling it to the attention of individual educators to consider teaching. A policy that does "not call for" the teaching of a subject also does not forbid it.

> Teachers, in their discretion, may encourage students to consider both the scientific strengths and weaknesses of evolutionary theory in order to better understand the assigned curriculum. This policy does not call for the study of creationism, nor does it call for the study of "intelligent design" theory. (Seth Cooper, e-mail to Alan Bonsell, December 10, 2004. [*Kitzmiller v. Dover* plaintiffs' exhibit #112])

If a district *requires* the teaching of ID, as Dover did, the policy can be challenged on its face with comparative ease, as occurred in *Kitzmiller*. A permissive policy, such as that suggested by the Discovery Institute, would be more difficult to challenge facially; it would require locating a teacher who viewed the policy as encouragement to teach ID. But to stop the teaching of ID by an individual teacher requires an "as applied" legal challenge: a parent with a student in the class must first find out what is being taught (not an easy task with high-school students!). Then, the parent must be willing to sue and must have legal standing to bring suit—both of which may be problematic. Altogether, facial challenges are much easier to mount than "as applied" challenges—which makes the Discovery Institute policy, if enacted, less likely to be challenged.

However, the "evidence against evolution" approach is not necessarily bound to succeed. The judge in *Kitzmiller* astutely recognized the connection between ID and "evidence against evolution" arguments, acknowledging their historical continuity with creation science. The reasoning of both ID and creation-science proponents is that if evolution can't do the job, they win by default. It is not necessary to provide positive evidence for special creation; it suffices that natural cause is inadequate to produce complex biological phenomena, hence the need for creation/design/Intelligent Agency. Recognizing the creationist history of these ideas, the judge spoke of ID and the "evidence against evolution" approach in the same words:

> An Objective Observer Would Know that ID and Teaching About "Gaps" and "Problems" in Evolutionary Theory are Creationist, Religious Strategies that Evolved from Earlier Forms of Creationism. (*Kitzmiller v. Dover,* 18)

An objective adult member of the Dover community would also be presumed to know that ID and teaching about supposed gaps and problems in evolutionary theory are creationist religious strategies that evolved from earlier forms of creationism, as previously detailed (*Kitzmiller v. Dover,* p. 56).

Nevertheless, the Discovery Institute's model policy is being circulated. It appeared in 2004 in Grantsville, Wisconsin, where it appeared only slightly modified:

> When theories of origin are taught, students will study various scientific models or theories of origin and identify the scientific data supporting each. Students are expected to analyze, review, and critique scientific explanations, including hypotheses and theories, as to their strengths and weaknesses using scientific evidence and information. Students shall be able to explain the scientific strengths and weaknesses of evolutionary theory. This policy does not call for the teaching of Creationism or Intelligent Design.

This policy survived a challenge by concerned parents and is now being touted as a "solution" to the "evolution problem" by the school board, which has presented its Discovery Institute–supplied materials to the annual meeting of the Wisconsin Association of School Boards. So, we can anticipate more in the future. Creationism—whether creationism classic (in the form of creation science) or creationism lite (in the form of "intelligent design")—will survive, to the detriment, unfortunately, of the teaching of evolution.

Acknowledgment

I thank Glenn Branch for very useful comments and suggestions.

References

Aguillard D. 1999. Evolution education in Louisiana public schools: A decade following *Edwards v. Aguillard. American Biology Teacher.* 61 (3): 182–88.

Anonymous. n.d. Days of praise [Web site]. Institute for Creation Research (cited April 27, 2006). Available from www.icr.org/ index.php?module=articles&action=type&ID=6.

———. 1981. Unbiased biology textbook planned. *Origins Research* 4 (2): 1.

———. 2002. Evolving banners at the Discovery Institute. NCSE Web site, www.ncseweb.org/resources/articles/8325_evolving_banners_at_the_discov_8_29_2002.asp (accessed April 29, 2006).

———. 2005. Timeline of the case [newspaper Web site]. *York* [PA] *Dispatch*, 12/21/2005 (cited May 3 2006). Available from www.yorkdispatch.com/local/ci_3330468.

———. 2006a. More on the Ohio victory [Web site]. National Center for Science Education, Inc., February 17, 2006 (cited May 4, 2006). Available from www.ncseweb.org/resources/news/2006/OH/15_more_on_the_ohio_victory_2_17_2006.asp.

———. 2006b. Settlement in *Hurst v. Newman* [Web site]. National Center for Science Education, Inc., January 17, 2006 (cited May 4, 2006). Available from www.ncseweb.org/resources/news/2006/CA/642_settlement_in_emhurst_v_new_1_17_2006.asp.

Behe, M. 1996. *Darwin's black box: The biochemical challenge to evolution.* New York: Free Press.

———. 2005. Cross-examination in *Kitzmiller v. Dover*, p. 97, lines 19–21; available at www2.ncseweb.org/kvd/trans/2005_1017_ day10_am.pdf.

Brande, S. 1989. Science textbook adoptions in Alabama: Part 1. *NCSE Reports* 9 (6): 5–7.

———. 1990. Science text adoptions in Alabama: Part II. *NCSE Reports* 10 (1): 8–10.

Brass, M. 2002. *The antiquity of man: Artifactual, fossil and gene records explored.* Baltimore, MD: America Press.

Buell, J., and V. Hearn. 1994. *Darwinism: Science or philosophy?* Richardson, TX: Foundation for Thought and Ethics.

Chapman, B. 1998. *Letter from the president.* Seattle: Discovery Institute, 3, 15.

Coyne, Jerry A. 1996. God in the details. *Nature* 383: 227–28.

———. 2001. Creationism by stealth. *Nature* 410: 745–46.

Davis, P. W., and D. H. Kenyon. 1989. *Of pandas and people: The central question of biological origins.* Dallas: Foundation for Thought and Ethics.

———. 1993. *Of pandas and people: The central question of biological origins*, 2nd ed. Dallas: Haughton Publishing Co.

Dembski, William. 1995. What every theologian should know about creation, evolution, and design. *Center for Interdisciplinary Studies Transactions* 3 (2): 1–8.

———. 1998. *The design inference: Eliminating chance through small probabilities.* New York: Cambridge University Press.

———. 2000. Mechanism, magic, and design. *Christian Research Journal*: 22–27, 44–46.

———. 2001. *No free lunch: Why specified complexity cannot be purchased without intelligence.* Lanham, MD: Rowman and Littlefield.

———. 2005. Intelligent design's contribution to the debate over evolution: A reply to Henry Morris [Web site] (cited October 5, 2005). Available from www.designinference.com/documents/2005.02 .Reply_to_Henry_Morris.htm.

DeWolf, D. K., S. C. Meyer, and M. E. DeForrest. 1999. *Intelligent design in public school science curricula: A legal guidebook.* Richardson, TX: Foundation for Thought and Ethics.

Edis, T. 1999. Cloning creationism in Turkey. *Reports of the National Center for Science Education* 19 (6): 30–35.

Elgin, P. G. 1983. *Creationism vs. evolution: A study of the opinions of Georgia teachers.* Atlanta: Georgia State University.

Evans, S. 2002. Ohio: The next Kansas? *Reports of the National Center for Science Education* 22 (1–2): 4–5.

Eve, R. A., and D. Dunn. 1989. High school biology teachers and pseudoscientific belief: Passing it on? *The Skeptical Inquirer* 13: 260–62.

Eve, R. A., and F. B. Harrold. 1991. *The creationist movement in modern America.* Boston: Twayne Publishers.

Forrest, B, and P. R. Gross. 2003. *Creationism's Trojan Horse: The wedge of intelligent design.* New York: Oxford University Press.

Foust, M. 2001. Phillip Johnson: Evolution battles at Baylor, Kan. could have been won. *Baptist Press News.* Available from www.sbcbaptistpress.org/bpnews .asp?ID=11354 (last accessed October 2005).

Futuyma, D. J. 1982. *Science on trial: The case for evolution.* New York: Pantheon (2nd ed., 1995, published by Sinauer Associates, Sunderland, MA.)

Geisler N. L., and J. K. Anderson. 1987. *Origin science: A proposal for the creation-evolution controversy.* Grand Rapids, MI: Baker Book House.

Gilchrist, G. W. 1997. The elusive scientific basis of intelligent design theory. *Reports of the National Center for Science Education* 17 (3): 14–15.

Gish, D. T. 1976. *Origin of life: Critique of early stage chemical evolution theories.* San Diego: Institute for Creation Research, i–viii.

Godfrey, L. R., ed. 1983. *Scientists confront creationism.* New York: W. W. Norton.

Gordon, B. 2001. Intelligent design movement struggles with identity crisis. *Research News and Opportunities in Science and Theology*: 9.

Gould, S. J. 1992. Impeaching a self-appointed judge. *Scientific American* (July 1992): 118-21.

Haught, J. F. 2000. *God after Darwin: A theology of evolution.* Boulder, CO: Westview Press.

Hoyle, F., and C. Wickramasinghe. 1979. *Diseases from space.* New York: Harper and Row.

Johnson, Phillip. 1990. *Evolution as dogma.* Dallas: Haughton Publishing Co.

———. 1991. *Darwin on trial.* Washington, DC: Regnery Gateway.

———. 1994a. Introduction. In *Darwinism: Science or philosophy?*, ed. J. Buell and V. Hearn, 1–3. Richardson, TX: Foundation for Thought and Ethics.

———. 1994b. Darwinism and theism. In *Darwinism: Science or philosophy?*, ed. J. Buell and V. Hearn, 42–49. Richardson, TX: Foundation for Thought and Ethics.

———. 1995. *Reason in the balance.* Downers Grove, IL: InterVarsity Press.

———. 1997. *Defeating Darwinism by opening minds.* Downers Grove, IL: InterVarsity Press.

———. 2001. *Weekly Wedge update, August 13* [Internet]. Access Research Network (cited August 15, 2001). Available from www.arn.org/docs/pjweekly/pj_weekly_010813.htm.

Kenyon, D. H. 1984. Foreword. In *The mystery of life's origin: Reassessing current theories*, ed. C. B. Thaxton, W. L. Bradley, and R. L. Olsen, v–viii. New York: Philosophical Library.

Kitcher, P. 1982. *Abusing science: The case against creationism.* Cambridge, MA: The MIT Press.

Lumsden, R. D. 1994. Not so blind a watchmaker. *Creation Research Society Quarterly* 31 (1): 13–22.

Matsumura, M., ed. 1995. *Voices for evolution.* 2nd ed. Berkeley: National Center for Science Education.

Miller, K. R. 1999. *Finding Darwin's God.* New York: HarperCollins.

Montagu, M. F. A. 1984. *Science and creationism.* New York: Oxford University Press.

Moreland, J. P. 1993. Is science a threat or help to faith? A look at the concept of theistic science. *Christian Research Journal* 15 (4): 46. Available from www.iclnet.org/pub/resources/text/cri/cri-jrnl/web/crj0180a.html (last accessed October 2005).

Morris, H. M. 1951. *The Bible and modern science,* 1st ed. Chicago: Moody Press.

———. 1963. *The twilight of evolution.* Grand Rapids, MI: Baker Book House.

———. ed. 1974a. *Scientific creationism* (Public School Edition). San Diego: Creation-Life Publishers.

———. 1974b. *The troubled waters of evolution.* San Diego: Creation-Life Publishers.

———. 1978. *That you might believe.* San Diego: Creation-Life Publishers.

———. 1999. Design is not enough! *Back to Genesis,* no. 127: a–c.

Morris, H. M., and J. D. Morris. 1996. *The modern creation trilogy. Book 2: Science and creation.* vol. 2. Green Forest, AR: Master Books, Inc.

Morris, J. D. 1996. Does it help to compromise with evolution? *Back to Genesis.* El Cajon, CA: Institute for Creation Research.

———. 1998. How did creation fare on PBS' "Firing Line"? *Back to Genesis,* no. 110b: d.

National Academy of Sciences (US). 1998. *Teaching about evolution and the nature of science.* Washington, DC: National Academy Press.

———. 1999. *Science and creationism: A view from the National Academy of Sciences,* 2nd ed. Washington, DC: National Academy Press.

Nelson, P. 2001. From the Editor. *Origins and Design* 39: 4.

———. 2002. Life in the big tent: Traditional creationism and the intelligent design controversy. *Christian Research Journal* 24 (4): 20–25, 41.

Nickels, M. K., and B. A. Drummond. 1985. Creation/evolution: Results of a survey conducted at the 1983 ISTA convention. *ISTA Spectrum* 11 (1): 11–15.

Numbers, R. 1992. *The creationists.* New York: Alfred A. Knopf.

Nussbaum, A. 2002. Creationism and geocentrism among Orthodox Jewish scientists. *Reports of the National Center for Science Education* 22 (1–2): 38–40.

Osif, B. A. 1997. Evolution and religious beliefs: A survey of Pennsylvania high school teachers. *American Biology Teacher* 59 (9): 552–56.

Ostling, R. N. 2002. Ohio school board debates teaching "intelligent design." *Washington Times* (March 14). Available from www.discovery.org/scripts/viewDB/index.php?command=view&program=CSCStories&id=1140 (last accessed October 2005).

Padian, K., and A. D. Gishlick. 2002. The talented Mr. Wells. *Quarterly Review of Biology* 77 (1): 33–37.

Pennock, R. T. 1996. Naturalism, creationism and the meaning of life: The case of Phillip Johnson revisited. *Creation/Evolution* 16 (2): 10–30.

Perloff, J. 1999. *Tornado in a junkyard: The relentless myth of Darwinism.* Arlington, MA: Refuge Books.

Pigliucci, M. 2001. Intelligent design theory. *Bioscience* 51 (5): 2257–58.

Plantinga, A. 1991. When faith and reason clash: Evolution and the Bible. *Christian Scholar's Review* 21 (1): 8–32.

Postman, D. 2006. Seattle's Discovery Institute scrambling to rebound after intelligent-design ruling. *Seattle Times* (April 26).

Russell, R. J., W. R. Stoeger, and F. J. Ayala. 1999. *Evolutionary and molecular biology: Scientific perspectives on divine action.* South Bend, IN: University of Notre Dame Press.

Scott, E. C. 1989. New creationist book on the way. *NCSE Reports* 9 (2): 21.

———. 1993. The social context of pseudoscience. In *The Natural History of Paradigms,* ed. J. H. Langdon and M. E. McGann, 338–54. Indianapolis: University of Indianapolis Press.

———. 1997. Anti-evolutionism and creationism in the United States. *Annual Review of Anthropology* 26: 263–89.

———. 1999. Bleeding Kansas: What happened? What's next? *Reports of the National Center for Science Education* 19 (4): 7–9.

———. 2001a. Fatally flawed iconoclasm. *Science* 292: 2257–58.

———. 2001b. The big tent and the camel's nose. *Reports of the National Center for Science Education* 21 (2): 39–41.

———. 2005. *Evolution vs. creationism: An introduction.* Berkeley: University of California Press.

Scott, E. C., and T. C. Sager. 1992. *Darwin on trial*: A review. *Creation/Evolution* 12 (2): 47–56.

Stafford, T. 1997. The making of a revolution: Law professor Phillip Johnson wants to overturn the scientific establishment's "creation myth." *Christianity Today*: 16.

Strahler, A. 1987. *Science and earth history: The evolution/creation controversy,* 1st ed. Buffalo: Prometheus Books.

Talbot, M. 2005. Darwin in the dock. *The New Yorker* (December 5): 66–77.

Tatina, R. 1989. South Dakota high school biology teachers and the teaching of evolution and creationism. *American Biology Teacher* 51 (5): 275–79.

Thaxton, C. B., W. L. Bradley, and R. L. Olsen. 1984. *The mystery of life's origin: Reassessing current theories.* New York: Philosophical Library.

Thomas, J. A. 1990. The Foundation for Thought and Ethics. *Reports of the National Center for Science Education* 10 (4): 18–19.

Wells, J. 2000. *Icons of evolution: Science or myth?* Washington, DC: Regnery.

Whitcomb, J. C., and H. M. Morris. 1961. *The Genesis Flood: The biblical record and its scientific implications.* Philadelphia: Presbyterian and Reformed Publishing Co.

Witham, L. 2002. *Where Darwin meets the Bible: Creationists and evolutionists in America,* 1st ed. New York: Oxford University Press.

Zimmerman, M. 1987. The evolution-creation controversy: Opinions of Ohio high school biology teachers. *Ohio Journal of Science* 7: 115–21.

Zoll, R. 2005. Weakened, but not defeated [Web site]. Associated Press, December 22, 2005 (cited May 3, 2006). Available from Corvallis (OR) *Gazette-Times* archives, www.gazettetimes.com/articles/2005/12/22/news/nation/thunato2.txt.

Wielding the Wedge: Keeping Anti-Evolutionism Alive

John R. Cole

HARDLY HAD THE THEORY OF evolution been posed than nineteenth-century scientists and theologians began the first phase of anti-evolutionism and resistance to Darwin's research. By the turn of the twentieth century, supporting scientific evidence mounted, opposition faded, and evolution became commonplace in popular textbooks. After World War I, Americans took the lead in the struggle against evolution. There were a number of reasons for this, not the least of which was the scholarly elite's relentless equation of evolution with the "doctrine of progress" and other concepts associated with "modernism." The notion that "progress" and "modernism" would be an improving force for mankind seemed to fly in the face of the recent war and ensuing economic distress. Even though the United States had prospered after the war, the European devastation was seared in the public mind. Furthermore, in the early twentieth century, the United States lacked centralized political, religious, and educational systems, a situation that left decisions on curriculum under the power of local citizenry and provided an incentive for activism on the part of anti-evolutionists. By the time of the notorious Scopes trial in 1925, laws banning the teaching of evolution in public schools had emerged.

The Scopes trial in Dayton, Tennessee, was the first large confrontation between evolutionists and anti-evolutionists. Defending John Scopes's right to teach evolution were lawyers Clarence Darrow and William Dudley Malone. Darrow was something of a professional agnostic and, like Malone, a skilled orator. On the side of the anti-evolutionists stood special prosecutor William Jennings Bryan—three-time Democratic presidential nominee, former secretary of state, and renowned orator in the cause of Christian fundamentalism. Bryan won the case. The issue, the judge insisted, was simply whether Scopes had taught the subject of evolution, and Scopes had conceded that much. Scopes was convicted and duly fined $100 for violating a new law forbidding the teaching of evolution. (The sentence was later overturned on a technicality by the state appellate court because the fine had been set by the judge, rather than by the jury as the law required.)

But the intervention of Bryan transformed a civil-liberties test case into an explosive forensic contest and revival meeting. Bryan set the scene for the trial in a speech before Seventh-day Adventists by proclaiming, "All the ills from which America suffers can be traced back to the teachings of evolution," and by portraying the trial as a battle for the survival of evolution or Christianity. Evolutionists remember the trial as a big circus—one in which Bryan was led into illogical, untenable corners time after time, and one in which he was humiliated and mocked in the press around the world. Evolution emerged victorious if the debate was to be judged forensically rather than legally, and Bryan emerged a rather tarnished defender of the faith. Yet, in effect, the trial did not challenge the Tennessee law prohibiting the teaching of evolution, and evolution remained effectively excluded from American public schools and textbooks until at least the 1960s (Hofstadter 1955, 1963; Ginger 1958; Cole 1983).

With the post-Sputnik upsurge of science education (including the teaching of evolution) during the second half of the twentieth century came a new wave of anti-evolutionism. In response to more ardently pro-evolution texts, the dominant strategy among anti-evolutionists was to promote the idea that alternatives should be given equal time whenever evolution was taught. Those "alternatives" included "scientific creationism," basically a renewed engagement for the "Flood geology" espoused by George McCready Price (1926). Even Price's work was little more than an echoing of the Seventh-day Adventist version of Ellen White's earlier anti-evolutionism (Nelkin 1977). Whitcomb and Morris's 1961 book *The Genesis Flood* was the basic text of the new "scientific creationism" movement, although there had been a steady trickle of such books throughout the century. It would take nearly three decades for the U.S. Supreme Court to reject the "equal time" arguments. In 1987, it heard an appeal of a Louisiana law demanding equal time for creationism in the curriculum. In its decision on *Edwards v. Aguillard,* the Court ruled against "equal time" for creationism, and a large number of evolutionists breathed sighs of relief, believing that they could finally drop the struggle.

Creationists, however, were undeterred. The court decision neither *required* evolution education nor *banned* anti-evolutionism. Even the majority opinion left room for "valid" challenges to evolution in school curricula. The minority opinion, written by Chief Justice William Rehnquist, was scathing in dissent, fully defending the Louisiana law's content and intent (*Edwards v. Aguillard* 1987). Critics found it startlingly ignorant of science and hostile to church-state separation. Many anti-evolutionists, in contrast, viewed even the majority opinion as friendly—an open invitation for someone to find an intellectual and legal strategy that might be sufficiently "valid" to qualify for equal time.

Anti-Evolutionism Evolves

With Rehnquist's opinion as the guiding principle, a much more subtle and sophisticated intellectual strategy emerged in the mid-1990s. Nicknamed "The Wedge" (Lankford 1999; Johnson

2000, 2002; Forrest 2001; Forrest and Gross 2003) by its designers, it sought to call into question the underpinnings of all biology. The acknowledged leader of this movement is recently retired University of California (Berkeley) law professor Phillip Johnson. Johnson's version of anti-evolutionism is known as "intelligent design." According to ID, the universe shows a type of complexity that is impossible for natural processes to produce and that therefore requires an "Intelligent Designer." (See articles in this volume by Pennock, Elsberry, Dorit, and Stenger for discussions of various aspects of ID theory and complexity in nature.) Johnson's "Wedge" strategy is simple, intended to appeal to a broad audience—far broader than that for the ID movement from which it sprang. The strategy is built around a metaphor: If one's road is blocked by a log too heavy to be moved, Johnson writes, one need only find the cracks in the log and divide it methodically using a hammer and a wedge. "The log in this metaphor is the ruling philosophy of modern culture, a philosophy called naturalism or simply modernism" (Johnson 2000, 13).

Wedge advocates laid out a pragmatic program to methodically destroy the culture of evolution, which can be outlined thus:

GOVERNING GOALS

- To defeat scientific materialism and its destructive moral, cultural and political legacies.
- To replace materialistic explanations with the theistic understanding that nature and hurnan beings are created by God.

FIVE YEAR GOALS

- To see intelligent design theory as an accepted alternative in the sciences and scientific research being done from the perspective of design theory.
- To see the beginning of the influence of design theory in spheres other than natural science.
- To see major new debates in education, life issues, legal and personal responsibility pushed to the front of the national agenda.

TWENTY YEAR GOALS

- To see intelligent design theory as the dominant perspective in science.
- To see design theory application in specific fields, including molecular

biology, biochemistry, paleontology, physics and cosmology in the natural sciences, psychology, ethics, politics, theology and philosophy in the humanities; to see its influence in the fine arts.

- To see design theory permeate our religious, cultural, moral and political life.[1]

This language bears a striking resemblance to that spouted by anti-evolutionists to evolution in the early twentieth century. Wedge proponents are convinced that "modernism" is inherently immoral and that a wedge must be thrust between the leading social, political, and educational institutions in the United States and the naturalistic worldview that dominates the natural and social sciences. For example, Willard Gatewood quotes a 1920s Louisiana clergyman whose anti-evolutionism would fit comfortably within the contemporary rhetoric of Wedge strategists:

> A modernist in government is an anarchist and Bolshevik; in science he is an evolutionist; in business he is a Communist; in art a futurist; in music his name is jazz and in religion he is an atheist and infidel. (Gatewood 1969, 6)

Similarly, Albert Johnson, a Presbyterian leader of the early twentieth century, claimed that evolution leads "to sensuality, carnality, Bolshevism, and the Red Flag" (quoted in Gatewood 1969, 24). Eighty years later, Phillip Johnson's (2000) book, *The Wedge of Truth: Splitting the Foundations of Naturalism*, pulls no punches in denouncing the same foe. On page 161, he writes that Darwinists are like Leninists—take away the dictatorial power from the elite and they will crumble, because evolution, like Leninism, has no substance. Like historians Gertrude Himmelfarb (1959, 1999) and Jacques Barzun (1941), Phillip Johnson repeatedly characterizes Marxism, Freudianism, and Darwinism as pseudoscientific remnants of the culture of the nineteenth century (Branch 2002a). To Phillip Johnson, the Enlightenment is a "parasite on

1. This text disappeared from the Discovery Institute Web site shortly after it was posted. This excerpt appeared at www.humanist.net/skeptical/wedge.html—only one of the Web sites that opted to save the original version before it was removed from the DI Web site.

Christianity" (p. 162), which, like rationalism, "is dissolving into its antagonistic positivist and relativist components" (p. 167) and needs to be "repealed." The agenda of the Wedge strategists also includes the eradication of "materialism" in science and society. Materialism arises, Johnson claims, from "the Sin of Pride . . . which refuses to respect the limitation inherent on our states as both created and fallen beings" (p. 155). On page 151, he quotes John 1:1–3, "In the beginning was the Word, and the Word was with God, and the Word was God," and advocates: "Building a New Foundation for Reason: What If We Start with the WORD?" (this is the title of his chapter 7). Writings such as these by the leader of the Wedge movement reveal clearly the religious basis of his anti-evolutionism. Such knee-jerk rejection of evolution is political, personal, and philosophical—not scientific.

Despite their prodigious output of books, Web sites, letters to the editor, and op-ed pieces, ID proponents and Wedge strategists sometimes operate with surprising stealth. Whereas scholars are typically fanatic about publishing their ideas in journals or databases available for all time—and whereas careers rise or fall depending on feedback from a wide range of other scholars, including critics—a significant number of new creationist publications are either ephemeral or secret. Many seemingly quotable pieces posted by the Discovery Institute (DI) on its Web site are flagged with notices that they are not to be referenced. Many are posted and promptly erased. National planning conferences, such at those held at Biola University in 1996 ("Mere Creation") and 1999 ("After Materialism"), are often private affairs; outsiders are not invited and papers are not published. This is a strange way to treat "scientific" communications, whether the format is a press release, a scholarly conference, or a popular article. Some documents are available only to members with passwords. For example, sample curricula and classroom lesson plans "published" by the Center for the Renewal of Science and Culture (later renamed the Center for Science and Culture—CSC) are available only in this format, firewalled from evaluation by outsiders. Ironically, some anti-evolutionist documents can be located only on the Web sites of evolution supporters who recognized their potential legal or historical import and saved them for future reference.[2]

The anonymous and somewhat secretive appearance of the "Wedge Document" itself is discussed at length in a book by Barbara Forrest and Paul Gross called *Creationism's Trojan Horse: The Wedge of Intelligent Design* (2003). These authors exhaustively document the origin of the "Wedge"—its true authorship and centrality to the new anti-evolution movement. Despite its quick removal from the public areas of the DI Web site where it first appeared, pieces of that document preserved elsewhere present a fair overview of the plans that were originally posted, demonstrating how they have been echoed in Johnson's books and newsletter and subsequently modified. The DI coyly refuses to confirm that the "Wedge Document" is its work, but virtually identical wording has appeared under their names and in the works of "intelligent design" proponents and fellows at the DI's Center for Science and Culture, including Phillip Johnson, whose online newsletter, the *Weekly Wedge Update*, is hosted by the creationist Access Research Network (ARN 2002), which links to the Discovery Institute as its "partner."

Phillip Johnson and other Wedge proponents advocate a "big tent" assault on evolution, admitting all allies, from ID philosophers to televangelists. The young-earth creationists of the Institute for Creation Research (ICR) and Ken Ham with his fire-and-brimstone evangelism are as welcome in the Wedge movement as are the PhDs from the Discovery Institute. The shared commitment to oppose evolution seems to be enough for now—the new breed of anti-evolutionists hopes to sort out internal debates after evolution has been defeated (Scott 2001). This is the "party line," fostering cooperation among the different factions. But sometimes there is slippage in the common front. Discovery Institute scholars take pride in their elite status, which angers traditional creationists. After the

2. One of these sites is www.antievolution.org/wedge.html.

2001 PBS broadcast of the miniseries *Evolution*, a spokesperson for the Discovery Institute objected to the portrayal of anti-evolutionists and the producers' inclusion of ID in the episode devoted to religious anti-evolutionism:

> "We wanted to talk about science, and they wanted us to do Sunday school," said Mark Edwards, a spokesman for the Discovery Institute. "The final episode paints a picture that the only critics of Darwinian theory are these guitar-strumming hillbillies in Kentucky who are creationists, and that's just not true. We're glad we're not part of that stereotype." (Carter 2001)

The Discovery Institute

Based in Seattle, Washington, the Discovery Institute (DI) has been instrumental in the development and promotion of "intelligent design" and the "Wedge." Founded in 1991 as an institutional home for President Reagan's economic adviser George Gilder, the institute quickly attracted funding and members, many of them former Reagan administration officials devoted to issues such as free trade, reduced environmental regulation, Social Security privatization, and other generally libertarian issues. With considerable funding from the timber industry, the DI initially focused largely on matters relating to Northwestern United States and Western Canadian policy. One corner of the DI not devoted to economics was the Center for the Renewal of Science and Culture (CRSC), which became the institutional base of the "intelligent design" and "Wedge" movements. In late 2002, in part because of criticism about the term *Renewal*, they renamed it "The Center for Science and Culture" and suggested that it henceforth be called "The Center" (Branch 2002b; Center for Science and Culture 2002; Center for the Renewal of Science and Culture 2002).

Senior fellows at the CSC include mathematician David Berlinski, theologian and molecular biologist Jonathan Wells, biophysicist Michael Behe, mathematician William Dembski, philosopher Paul Nelson, and others. Law professor Phillip Johnson and economist George Gilder are "advisers." At least one scholar critical of creationism, historian Edward Larson, was for a short time affiliated with the DI but not the CSC. Scholars with PhDs from established universities dominate CSC, which is in stark contrast to those who dominate the "scientific creationist" movement and whose degrees are sometimes literally from mail-order or unaccredited institutions. CSC goes to great pains to stress its high academic standards and tends to disparage old-style creationists, despite Wedge strategy "rules."

In the late 1990s, the CSC received a considerable boost thanks to a $1.5 million grant from Fieldstead and Company, the private foundation of Howard Fieldstead Ahmanson, Jr.; this grant was later augmented by an additional $2.8 million (Stephens 2002). He should not be confused with his late father, Howard Ahmanson, *Sr.*, who owned Home Savings of America and whose Ahmanson Foundation funds environmental causes, public radio, various liberal projects, and science education. Ahmanson, Jr., in contrast, has funded creationism projects and has funded and served as a director of the Chalcedon Institute, an organization devoted to "Christian Reconstructionism"—a movement aiming to make the United States a theocracy governed by biblical law (Anson and Cogan 1994; Benen 2000).

Beginning at about the time of the Fieldstead grant, the CSC grew dramatically in public profile and activity level. Indeed, this appendage to the DI seems a bit like the tail wagging the dog, judging from the DI Web site, which is now dominated by press releases relating to CSC initiatives and publications. The vast majority of these postings concern ID theory and how to revise science education to include ID. They focus on how to eliminate naturalism or materialism from science. Thus should science and culture be "renewed."

One of many traits shared by new and old creationists is a concentration on "Darwinism" rather than on "evolution." Their single-minded critique of nineteenth-century scholarship dismays modern evolutionary biologists, who are asked to defend ideas long since discarded or refined—from the age of the earth to the alleged lack of transitional fossils to Haeckelian embryology. Jonathan Wells's

Icons of Evolution (Wells 2000; Padian and Gishlick 2002) is a catalogue of such arguments against a straw man called "Darwinism."

Another anti-evolutionary "tradition" links the Discovery Institute with scientific creationists of the past: a selective use of the scientific and scholarly literature to create the perception that evolution is a weakened theory under assault from within the scientific community. The misrepresentation of the scientific literature by ID advocates, done expressly to weaken evolution *education,* can become central to debates over science-education standards (see, for example, Branch 2002b and 2002c, on a case in Ohio). Such tactics mirror those employed by scientific creationists in the latter half of the twentieth century (Cole 1981). Clearly, despite the rhetoric to the contrary, this technique does not distinguish the work of a CSC "scholar" from that of an Institute for Creation Research pamphleteer.

The "Wedge" movement also takes comfort from other academic—though not necessarily *scientific*—critiques of evolution. Overt *political* anti-evolutionism is deeply rooted in conservative (especially *neoconservative*) political thought, and it has recently emerged also as a minor element of some leftist thought in "postmodernism" (Gross and Levitt 1998). Postmodernism is a perspective on the nature of knowledge in some academic disciplines—predominantly in the social sciences and humanities, but with important implications for science or at least for the public perception and interpretation of science. In brief, the argument is that all knowledge is "constructed," and thus "true" only in the context in which it is constructed; one cannot find absolutes (Sokal 1966).

Because in this view scientific research and discoveries are interpretable as social phenomena, a scientific theory such as evolution can be seen as merely one of many possible "ways of knowing," no more absolutely true than any other. While there is much to be said for understanding the scientific endeavor as a social phenomenon, it is easy to see how this discussion can be misrepresented as an indictment of the scientific method. To the contrary, postmodernism can be properly seen as supporting evolution *within the constructed knowledge of the scientific method* while excluding the "intelligent design" movement, which seeks to acquire scientific legitimacy from outside the framework of scientific knowledge. Indeed, the deistic worldview that ID proponents would bring to science differs only in strategic use of vocabulary (e.g., "designer" rather than "creator") from that of old-style scientific creationism.

However, strategic word choice does have certain benefits for ID proponents. Some of the neoconservative academics from whom the new ID theorists draw support are "proper" scholars, including: renowned historian, critic, and essayist Jacques Barzun (1941), historian Gertrude Himmelfarb (1959, 1999), and William F. Buckley, Jr. (1997), host of PBS's *Firing Line* from 1966 to 1999. These individuals, as well as other prominent conservatives whose sympathetic views toward "intelligent design" have been published in the *National Review, American Spectator, Intercollegiate Review*, the *Washington Times*, and the *Wall Street Journal*, grant the movement an illusion of legitimacy. Their literature becomes a powerful resource for anti-evolutionists and for politicians who heed their call. The same politicians who might scorn the rhetoric of a Bible-thumping creationist will be open to similar ideas spouted by academics with the credentials and prestige of a Barzun or a Himmelfarb. Despite the fact that none of these commentators addresses the *scientific* aspects of evolution, their prominence as scholars produces a sort of "halo effect" that lends weight to their pronouncements about evolution that would never accrue to old-style scientific creationists.

Every "Victory" Counts

In addition to the arena of public opinion, Wedge activists have also focused on political action, perfecting the technique of going for small victories that can be represented as grand ones. For example, in 2001 Senator Rick Santorum (R–Pennsylvania) managed to attach a small, innocuous-looking item to S. 1, the bipartisan education bill; the Senate voted for the entire bill 91–8 (Branch 2002d). The Santorum amendment reads:

> It is the sense of the Senate that—(1) good science education should prepare students to distinguish the data or testable theories of science from philosophical or religious claims that are made in the name of science; and (2) where biological evolution is taught, the curriculum should help students to understand why this subject generates so much continuing controversy, and should prepare the students to be informed participants in public discussion regarding the subject.

Immediately, creationists around the country began a drumbeat of claims that the Senate had voted almost unanimously to support teaching that evolution was controversial, echoing a long-standing Wedge goal of "teaching the controversy" (Discovery Institute 2001; see Petto and Godfrey, in this volume). In essence, the argument runs: "Just teach about how controversial evolution is and let students research both sides, let them argue, and education is well served. Oh, my—we would *never* consider banning evolution; in fact, we just want more to be taught about it, such as the [alleged] fact that it has all sorts of weaknesses, and then students can decide for themselves whether or not to believe such a theory." This is a clever variation on the old equal-time argument. The onus shifts from the school board or teacher to the student. "Who wants to stand in the way of students' doing their own research?" they argue.

In the end, the Santorum amendment was dropped from the bill in conference committee (Branch 2002d). After months of work by most of the U.S. scientific organizations, led by the American Geological Institute, this should have been the end of the matter, but Senator Santorum convinced the conference committee to include some discussion material about his discarded amendment in the committee report, though not in the bill. Even though this inclusion had no *legal* significance, creationists have been hailing this "victory" ever since, quoting the committee report as if it were law to school boards and anyone who will listen (Branch 2002e). The original Santorum wording was written by Phillip Johnson (2002), and the strategy of pulling victory from defeat via rhetoric and political spin is vintage "Wedge" and "new creationist."

Anti-evolutionists also appeal to the First Amendment on behalf either of teachers forced to teach something "against their religious beliefs" or of students required to study something "violating their religious freedom to disbelieve" in evolution. However, the response of the courts has been that First Amendment rights or academic freedom does not, in effect, give teachers the right to change the content of the approved curriculum or give students the right not to take required subjects. Courts have ruled against such claims in Washington state and Minnesota as recently as 2001 (*Rodney LeVake v. Independent School District* 656 et al. 2000; Scott 2000). In 2001, high-school students in Lafayette, Indiana, may have acted as a harbinger of a similar tactic: student-led demands for teaching creationism without overt adult input. However, these students' demands and slogans repeated arguments verbatim from the nationwide adult neocreationist movement, as could be seen in the PBS television series *Evolution*. The school board resisted their demand and supported the evolution curriculum (*Evolution* 2001; Randak 2001). Because they are neither outside agitators nor rebel teachers, students demanding their alleged religious freedoms are a more complex opponent for evolutionists, and this tactic is used to advantage by ID proponents. ID clubs are popping up as student organizations in universities and secondary schools around the nation.

Another ongoing tactic is the use of warning labels or disclaimers in textbooks. In various ways, these warn students that what they are about to study may be nonsense. Most intellectuals would agree, in principle, with the labels' admonition that textbooks should be read critically, but anti-evolutionists do not support models of critical thinking (see Petto and Godfrey, in this volume). In fact, many of them become very upset with efforts to teach critical thinking, on the grounds that it teaches general questioning of authority—first textbooks, next parents and teachers, and then the Bible. Norma Gabler and her late husband, Mel, perpetual Texas textbook critics, exemplify this sentiment. The subtler approach they and many others have advocated in

Texas, Louisiana, and other states is a textbook warning that *evolution alone* requires critical assessment. In December 2001, Alabama renewed its statewide requirements for such a textbook disclaimer, singling out evolution as "controversial" and advising students to think for themselves on *this* one. The new Alabama disclaimer reads, in part:

> The Alabama Course of Study: Science includes many theories and studies of scientists' work. The work of Copernicus, Newton and Einstein, to name a few, has provided a basis of our knowledge of the world today. The theory of evolution by natural selection is a controversial theory that is included in this document. It is controversial because it states that natural selection provides the basis for the modern scientific explanation for the diversity of living things. Since natural selection has been observed to play a role in influencing small changes in a population, it is assumed, based on the study of artifacts, that it produces large changes, even though this has not been directly observed. (Anonymous 2001)

Recently, lawsuits have successfully challenged these disclaimers. Efforts by Tangipahoa Parish, Louisiana (*Freiler v. Tangipahoa* 1997), to require an oral disclaimer read by teachers in biology classes resulted in the school board's and its insurer's loss of serious cash as each appeal piled up the bills and then failed. That district seems to have dropped its efforts, for now. In Selby, Georgia, a federal judge ordered the disclaimers removed from biology textbooks. In Dover, Pennsylvania, a lawsuit (*Kitzmiller v. Dover*)[3] that dealt in part with the school board's demand that science teachers make students aware of "alternative" scientific theories such as "intelligent design" resulted in a legal repudiation of the ID argument and electoral defeat for the school-board members who had promoted ID. It also incurred severe legal expenses for the district, which was billed for court costs, although it had pro bono legal representation.

The expense of lawsuits is certainly one issue that can operate

3. The decision in this case is available at www.pamd.uscourts.gov/kitzmiller/kitzmiller_342.pdf.

in favor of anti-evolutionists. Free legal services are scarce and rationed by defenders of evolution education, often because they are drawn from organizations that take on anti-evolutionism as only one of many issues. However, a number of anti-evolution legal resources are available through organizations that focus primarily on opposing secular materialism (see Cole 2000).

For example, the Rutherford Institute is devoted to providing legal advice and litigation on behalf of conservative—some would say theocratic—Christian causes. The American Council for Legal Justice (ACLJ) was founded, directly or indirectly, by Pat Robertson and the Christian Coalition to act as a sort of mirror image and acronym look-alike to the American Civil Liberties Union (ACLU) (Newfield 2002). Both organizations concentrate on a very conservative version of "Christian" and "family" issues and have entered several of the new anti-evolution legal frays; it seems likely they will continue and expand this interest. More recently, the Thomas More Law Center (TMLC) has emerged as a source of legal support for the teaching of ID in public schools. The TMLC describes itself as "a not-for-profit public interest law firm dedicated to the defense and promotion of the religious freedom of Christians, time-honored family values, and the sanctity of human life."[4] The "science" these groups promote is not one committed to understanding the world around us in terms of the operation of natural laws. They want to eliminate from science any reliance—indeed, insistence—on a naturalistic methodology.

Concluding Remarks

The goal of anti-evolutionists has remained the same for the last eighty years, but the tactics of twenty-first-century opponents are more varied and sometimes more sophisticated. The "Wedge Document" is a strategic plan to separate public understanding of science from the naturalistic method and practice on which science has been based for more than two centuries. The Wedge is much more flexible and more sophisticated and "modern-

4. Information about TMLC comes from its Web site: www.thomasmore.org/about.html.

looking" than earlier versions of creationism. It is covert in its use of biblical language and references (but see the earlier discussion of Johnson's work and the discussion in Scott—in this volume—regarding the "ancestry" of ID in old-style creationism), even though at its heart it attacks "modernism" just as much as some evangelists did at the time of the Scopes trial—and with the same warnings about its perilous effects on society and personal salvation. If there is an exploitable weakness in this manifestation of anti-evolutionism, it is that the handful of well-trained scholars associated with the Discovery Institute's Center for Science and Culture are ill at ease with their allies in the "traditional" creationist organizations. The reverse is also true: Institute for Creation Research leaders welcome any criticism of evolution, but they fault the ID camp for not embracing biblical literalism. Indeed, young-earth creationist John Whitcomb (2006) characterizes ID as "vastly insufficient." Such tensions have led to occasional breaches in the united front that modern anti-evolutionists have pursued. A rift may be growing within the "big tent" between traditional biblical literalists and those promoting the use of less overtly religious language.

Ironically, evolutionists may be able to exploit such cracks in much the same way Wedge strategists exploit splits within the modern academic world. Showing traditional creationists why ID will never be able to move respectably toward overt support for sectarian religious positions might weaken support for ID among those who wish to use it as a first step toward reestablishing a biblical basis for public life. Furthermore, within the "big tent" are some who are quite comfortable with aspects of evolutionary theory (see Scott's discussion of "theistic evolution" and the relationship of its supporters to others in the "big tent" of ID). It may be possible to develop a "web" strategy that can draw these people closer to contemporary science through shared concerns such as support for biomedical and agricultural research—concerns that can work to "stitch coalitions together" (Carville and Begala 2001). As we have seen, the anti-evolutionist movement is itself philosophically diverse. We need to emphasize those goals and values that are shared by many evolutionary biologists and people who may have been drawn to the "big tent" but who are uncomfortable with both biblical literalism and with metaphysical naturalism (Scott 2001; and others in this volume).

What is at stake is nothing less than the public understanding of the nature of science—and an organized effort by a small group of individuals to reshape that understanding. In part, this may be accomplished through the renewed emphasis among professional educators in teacher preparation and professional development and outreach to school boards and legislatures. The "web" will show these parties a view of scientific issues that reflects the contemporary practice of science as it is understood by its professional practitioners.

This is a long-term and time-consuming strategy with many dimensions (as recognized by the original "Wedge Document"). It is clear that the issues go far beyond the legal and constitutional battles. Although constitutional barriers to overtly sectarian ideas in the sciences have served well in preserving evolution education, changes in the positioning and judicial philosophy of sitting judges could quite easily erase eighty years of support for evolution in the public schools (Newfield 2002). There is, after all, no constitutional protection against pseudoscience.

References

"After Materialism." 1999. "Intelligent design" conference held at Biola University, La Mirada, CA, cf. p. 85, December 2–5. www.biola.edu/academics/torrey/calendar/design.cfm (last accessed September 30, 2002).

Aguillard v. Edwards. 1987. 482 U.S. 578.

Anonymous. 2001. Alabama upgrades disclaimer. *Reports of the National Center for Science Education* 21 (3–4): 4–5.

Anson S.G., and D. M. Cogan. 1994. God's banker. *LA Magazine* (November) 39 (11).

ARN. 2002. Access Research Network: www.arn.org/id_faq.htm (last accessed September 27, 2002; homepage unavailable).

Barzun, J. 1941. *Darwin, Marx and Wagner: Critique of a heritage; the fatal legacy of "progress."* Boston: Little, Brown.

Benen, S. 2000. From Genesis to dominion: Fat-cat theocrat funds creationism crusade. *Church and State* (July/August).

Branch, G. 2002a. Saving us from Darwin: An interview with Fred Crews. *Reports of the National Center for Science Education* 22 (6): 27–30.

———. 2002b. Analysis of the Discovery Institute's "Bibliography of Supplementary Resources for Ohio Science Instruction." *Reports of the National Center for Science Education* 22 (4): 12–18, 23–24

———. 2002c. Quote-mining comes to Ohio. *Reports of the National Center for Science Education* 22 (4): 11–13.

———. 2002d. Farewell to the Santorum amendment. *Reports of the National Center for Science Education* 22 (1-2): 12–13.

———. 2002e. Santorum redux. Reports of the National Center for Science Education 22 (3): 4–5.

Buckley, W. F. 1997. Resolved: Evolution should acknowledge creation. *PBS Firing Line* (December 19).

Carter, M. 2001. Lab scientists challenging Darwin; "Intelligent design" theory supports a thoughtful creator. *Tri-Valley Herald*, Dublin, CA (September 26).

Carville, J., and P. Begala. 2001. 2004: How to choose the chosen one. *Esquire* (October): 176–77.

Center for the Renewal of Science and Culture. About CRSC. www.discovery.org/crsc/about.html (last accessed July 13, 2002).

Center for Science and Culture. 2002. www.discovery.org/crsc/nameChange.html (last accessed October 5, 2002).

Cole, J. R. 1981. Misquoted scientists respond. *Creation/Evolution* 6: 34–44.

———. 1983. The Scopes trial and beyond. In *Scientists confront creationism*, ed. L. R. Godfrey, 13–32. New York: W. W. Norton.

———. 2000. Money flows into anti-evolutionists' coffers. *Reports of the National Center for Science Education* 20 (1–2): 64–65.

Discovery Institute. 2001. Congress urges teaching diverse views on evolution, but Darwinists try to deny it. www.discovery.org/news/congressUrges.html (last accessed December 29, 2001).

———. 2002. www.discovery.org (last accessed September 27, 2002).

Evolution. 2001. First broadcast September 24–27, 2001, by PBS. Produced by Richard Hutton; WGBH Educational Foundation (Boston) and Clear Blue Sky Productions, Inc.

Forrest, B. 2001. The wedge at work: How intelligent design creationism is wedging its way into the cultural and academic mainstream. In *Intelligent design and its critics*, ed. R. T. Pennock, 5–54. Cambridge, MA: MIT Press.

Forrest, B, and P. Gross. 2003. *Creationism's Trojan horse: The wedge of intelligent design.* New York: Oxford University Press.

Freiler v. Tangipahoa Parish Board of Education. 1997. No. 94-3577 (E.D. La., August 8, 1997).

Gatewood, W. B., Jr., ed. 1969. *Controversy in the twenties: Fundamentalism, modernism, and evolution.* Nashville: Vanderbilt University Press.

Ginger, R. 1958. *Six days or forever? Tennessee vs. John Thomas Scopes.* Boston: Beacon Press.

Grabiner, J., and P. Miller. 1974. Effects of the Scopes trial. *Science* 85: 832–33.

Gross, P. R., and N. Levitt. 1998. *Higher superstition: The academic left and its quarrels with science,* 2nd ed. Baltimore: Johns Hopkins University Press.

Himmelfarb, G. 1959. *Darwin and the Darwinian revolution.* London: Chatto and Windus.

———. 1999. *One nation, two cultures.* New York: Alfred A. Knopf.

Hofstadter, R. 1955. *Social Darwinism and American thought.* Boston: Beacon Press.

———. 1963. *Anti-intellectualism in American life.* New York: Alfred A. Knopf.

Johnson, P. E. 2000. *The wedge of truth: Splitting the foundations of naturalism.* Downers Grove, IL: InterVarsity Press.

———. 2002. *Weekly Wedge Update.* www.arn.org/johnson/wedge.htm (last accessed September 28, 2002).

Lankford, K. 1999. The wedge: A Christian plan to overthrow modern science? *Doubting Thomas* feature story 6 (April/May). Available from www.freethought.org/ctrl/archive/thomas/wedge.htm (last accessed July 13, 2002).

"Mere Creation." 1996. Conference at Biola University, La Mirada, CA, November 14–17. www.origins.org/mc/ (last accessed September 29, 2002).

Nelkin, D. 1977. *Science textbook controversies: The politics of equal time.* Cambridge, MA: MIT Press.

Newfield, J. 2002. The Right's judicial juggernaut. *The Nation* (October 7): 11–16.

Padian, K., and A. D. Gishlick. 2002. The talented Mr. Wells. *Quarterly Review of Biology* 77 (1): 33–37.

Price, G. M. 1926. *Evolutionary geology and the new catastrophism.* Mountain View, CA: Pacific Press Publishing.

Randak, S. 2001. The children's crusade for creationism. *Reports of the National Center for Science Education* 21 (1–2): 27–28.

Rodney LeVake v. Independent School District 656 et al. 2000. Court File Nr. CX-99-793, District Court for the Third Judicial District of the State of Minnesota.

Scott, E. C. 2000. Rodney LeVake loses appeal. *Reports of the National Center for Science Education* 20 (5): 8–9.

———. 2001. The big tent and the camel's nose. *Reports of the National Center for Science Education* 21 (1-2): 39–41.

Sokal, A. D. 1966. A physicist experiments with cultural studies. *Lingua Franca* 6 (4): 62–64.

Stephens, S. 2002. Heir spends family fortune to discredit evolution theory. *Cleveland Plain Dealer* (December 23), p. A1.

Wells, J. 2000. *Icons of evolution: Science or myth?* Washington, DC: Regnery.

Whitcomb, J. C. 2006. The history and import of the book *The Genesis Flood.* Impact no. 395 (May). El Cajon, CA: Institute for Creation Research.

Whitcomb, J. C., and H. M. Morris. 1961. *The Genesis Flood.* Philadelphia: Presbyterian and Reformed Publishing Co.

PART TWO

Scientific Perspectives

Physics, Cosmology, and the New Creationism

Victor J. Stenger

The New Design Arguments

While the argument from design is an ancient one, new variations have appeared in recent years. There are three main forms of the design argument today. In one form, referred to as "intelligent design," the modern science of information theory is brought to bear supposedly to "prove" that certain biological systems cannot have arisen naturally (see Elsberry, Pennock, in this volume). In a second form, the cosmological evidence that our universe began with the Big Bang 13 to 15 billion years ago is claimed to demonstrate that the universe was divinely created. In a third form, the laws and constants of physics are said to be so finely tuned for life that they could only have arisen by the act of a Creator with the purpose of producing humanity. Each of these contributes to the rather vague concept known as "intelligent design" theory. But each of them has serious flaws and neither stands up on its own nor presents any significant challenge to modern naturalistic science.

Information Theory and "Intelligent Design"

In his extensive writings, William Dembski claims to use information theory to demonstrate that biological systems are too complex to have been formed by purely natural processes and so must have been "intelligently designed" (Dembski 1998, 1999, 2001). However, he has not applied information theory as it is conventionally practiced in that field (see Elsberry and Pennock articles in this volume for further discussion). Despite pages of formulas and complex descriptions, Dembski makes at least one elementary mistake that casts doubt on the whole enterprise: He incorrectly derives *his* definition of "the measure of information in an event" from natural sources.

Dembski defines "the measure of information in an event of probability p as $-log_2 p$." He cites as a reference *The Mathematical Theory of Communication* by Claude Shannon and Warren Weaver (1949). Shannon is regarded as the father of information theory, and his work is the foundation for the research conducted in this field for nearly sixty years. In his work at Bell Laboratories, Shannon was concerned with the efficient communication of electronic signals: how certain we could be that specific symbols were received. Let us give a summary of his theory, which is now widely applied in communications engineering.

Suppose we want to transmit a message containing a single symbol, such as a letter or number, from a set of n symbols. Shannon defined a quantity:

$$H = -\sum_i p_i \, log_2 \, p_i = -\langle log_2 \, p_i \rangle \qquad \text{[EQUATION 1]}$$

which he called "the entropy of the set of probabilities $p_1 \ldots p_n$" for the symbols in the message. That is, p_i is the probability of the presence of *ith* symbol in the list. Because of the base-2 logarithm, the units of H are *bits*—or in binary format. The angle brackets in equation 1 refer to the average of the enclosed quantity, and the fact that H is an average over an ensemble of symbols is important to keep in mind in the ensuing discussion. In today's literature on information theory, H is called the *Shannon uncertainty*.

The information R carried by a message is defined as the decrease in Shannon uncertainty when the message is transmitted. That is,

$$R = H\,(\text{before}) - H\,(\text{after}) \qquad \text{(EQUATION 2)}$$

If we consider the special case when all the probabilities p_i are equal to p, we get the simpler form:

$$H = -\log_2 p \qquad \text{[EQUATION 3]}$$

Let us illustrate the idea of information with a simple example of a single-character message that can be one of the eight letters S, T, U, V, W, X, Y, and Z with equal probability. Before the message is transmitted, the number of symbols is 8, the probability that any given symbol is picked randomly then is $p = 1/8$, and the Shannon uncertainty is H (before) $= -\log_2(1/8) = \log_2(8) = 3$. After the message is successfully transmitted, we know what the character is, so $p = 1$ and H (after) $= -\log_2(1) = 0$. Thus, $R = 3$ bits of information are received, as the uncertainty is reduced by 3 bits.

Now suppose that the message is a little garbled so that we know the symbol transmitted is either a U or a V, but we cannot tell which and they have equal probability. Then, after the message is received, the probability reduces to $p = 1/2$ and H (after) $= -\log_2(1/2) = 1$. In that case, $R = 3-1 = 2$ bits of information are received. Here, uncertainty is reduced as well, but because the *reduction* in uncertainty is less, the second transmission has less information.

Dembski's definition of information, $I_D = -\log_2 p$, is identical to the Shannon uncertainty in the special case of equal probabilities given in equation 3. In our example, the probability of each character is $p = 1/8$, so $I_D = -\log_2(1/8) = 3$ bits, as above. However, because Dembski's rendering of this definition is not conventional, it will equal R, as given in equation 2—*only* for equal probabilities among all symbols and *only* when the transmission is *perfect* so that H (after) $= 0$. While Dembski refers to Shannon, he does not mathematically derive the expression for information he uses from Shannon's expression—nor does he justify it by any other method. His examples, however, indicate that he does *not* limit himself to cases having equal probabilities within an ensemble of symbols or "events." Nor does he average the probabilities over the ensemble. In fact, his so-called information is really just another way of writing the probability *p* of an event in logarithmic form. This quantity is called *surprisal* in the literature.

Before continuing with Dembski's peculiar version of "information," let us take a closer look at the interpretation of the Shannon uncertainty *H*. Shannon notes: "The form of *H* will be recognized as that of entropy as defined in certain formulations of statistical mechanics," referring to the classic monograph *The Principles of Statistical Mechanics* by Richard Tolman (1938). Shannon explicitly states: "*H* is then, for example, the *H* in Ludwig Boltzmann's *H*-theorem."

Actually, in statistical mechanics the quantity we will call H_{SM} is defined without the minus sign and using the natural logarithm:

$$\mathrm{H}_{SM} = \sum_{i} p_i \log_e p_i \qquad \text{[EQUATION 4]}$$

However, Shannon notes that any constant multiplying factor, positive or negative, could have been used because constants do not affect the underlying relationships among variables. The main reason for the choice of multipliers is to set the units for the outcome of the calculation. The choice Shannon made in equation 1 produces a result measured in bits.

Boltzmann and Josiah Willard Gibbs found that the laws of classical continuum thermodynamics could be derived from statistical mechanics on the assumption that matter was composed of atoms (whose existence had not yet been fully confirmed by experiment at that time). In particular, the quantity H_{SM} was seen to be simply related to the thermodynamic entropy *S* by the relationship in this equation: $S = -kH_{SM}$, where *k* is Boltzmann's constant. The *H*-theorem implied that the entropy approaches maximum at equilibrium, and this gave a statistical explanation for the Second Law of Thermodynamics, which says that the entropy of an isolated system will increase with time or stay constant. The relationship between the entropy S of statistical mechanics and Shannon's uncertainty *H* is

$$S = k\, log_e(2)H \quad \text{[EQUATION 5]}$$

This shows that S and H are equal within a constant and have the same sign. So Shannon was justified in calling H the "entropy."

Summarizing the conclusions of this section: (1) Dembski's definition of information is not that used in the discipline of information theory; (2) information is conventionally defined as the change of a quantity called the Shannon uncertainty; and (3) entropy and Shannon uncertainty are comparable constructs and equal within a constant.

Conservation of Information

Dembski claims to prove a principle he calls—borrowing the phrase from the 1960 Nobel laureate in medicine Peter Medawar—the *Law of Conservation of Information.* According to Dembski's—but not Medawar's—version of this principle, the number of bits of information cannot change in any natural process such as chance or the operation of some physical law. As Dembski states it, "Chance and law working in tandem cannot generate information" (1999, 168). I show that this is incorrect when interpreted as some universal principle applying under all circumstances.

In most of his writings, Dembski focuses on a quantity he calls *complex specified information (CSI),* which I will discuss below. It would be reasonable to infer that Dembski intends his law of conservation of information to apply only to this type of information. However, he is quite inconsistent on this, and his "derivation" of conservation of information recognizes no such restriction. His "law" is meant to imply universal constraints on changes in information applicable to biological complexity.

The basic idea of conservation of information, as used by Dembski, is simple and illustrated in figure 1. Suppose we start out with a certain number of bits of information (that is, Shannon uncertainty) about a system. For example, the system might be composed of five coins. Any configuration of heads or tails is information that can be represented by five bits. For example, HTTHT = 10010. According to Dembski, two possible natural processes

"PROOF" OF CONSERVATION OF INFORMATION

Law just rearranges bits by some rule

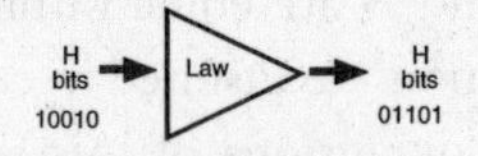

Chance just scrambles bits randomly

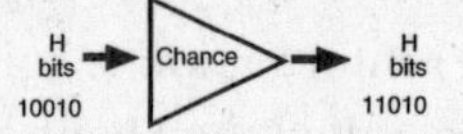

Conclusion: The generation of information cannot be natural

Figure 1. Representing the results of a sequence of coin tosses as "bits" of information illustrates Shannon's concept of uncertainty and the problems with Dembsky's concept of "conservation of information."

can act on that information. One is some well-defined operation that can be likened to the action of a physical law or computer algorithm. For example, the operation might be: Every time a flipped coin lands on the table, turn it over. Thus we have HTTHT → THHTH, or 10010 → 01101. Clearly, the number of bits has not changed and so, while the message may be different in *content*, it contains no more or less *information* than previously. In this process at least, information is conserved. In his 1984 book, *The Limits of Science*, Medawar described the impossibility of creating new information from closed logical systems: "No process of logical reasoning—no mere act of mind or computer-programmable operation—can enlarge the information content of the axioms and premises or observation statements from which it proceeds." However, unlike Dembski, Medawar did not claim this was a *universal* principle but only that it applied to *closed* systems—a limitation that Dembski does not admit. Medawar also made no claim that the same rule applied to chance processes, which Dembski includes in his version of the principle.

Let us look at what happens to our original information representing five coin tosses under the operation of random chance. Regardless of the original bit sequence, the process will produce a new one in which H and T are equally likely at any location in the sequence. Thus, HTTHT → TTHTT or HHTTH (10010 → 00100 or 11001) or any other possible permutation. Again, the number of bits does not change and so no *information* is generated or lost. Thus, Dembski claims, the very existence of information in the universe is irrefutable evidence for the existence of design. This conclusion has left many people impressed. For example, Robert Koons, an associate professor of philosophy at the University of Texas, calls Dembski "the Isaac Newton of information theory" (Kern 2000). Hardly.

First, as we have already noted, Dembski's definition of information does not correspond to that used in the field, except as a special case to which he does not limit himself. We have seen that information is conventionally defined *in this field* as the decrease in Shannon uncertainty during the transmission of a message. Furthermore, we have seen that Shannon uncertainty is equal, within a constant, to the entropy used in statistical mechanics. It has been well known in physics for more than a century that *entropy* is *not* conserved. In fact, the Second Law of Thermodynamics says that the total entropy of an isolated system of many bodies must remain constant or *increase*, as implied by Boltzmann's *H*-theorem, discussed above.

On the other hand, entropy can *decrease* for nonisolated systems, which happens when they are organized from the outside, or in any system with small numbers of particles. It is important to realize, in fact, in the example I gave above of the transmission of a message, the entropy/uncertainty *does* decrease (and information *increases*). This is an illustration of a nonisolated system, the transmitter, sending information to another nonisolated system, the receiver. The result is not the *conservation* of information but rather an increase in information as the Shannon uncertainty decreases.

Indeed, every time we rub our hands together, we are making entropy. From an information standpoint, the Shannon uncertainty is increasing (the molecular motions in our hands are becoming more irregular), so information is being lost. It is thus

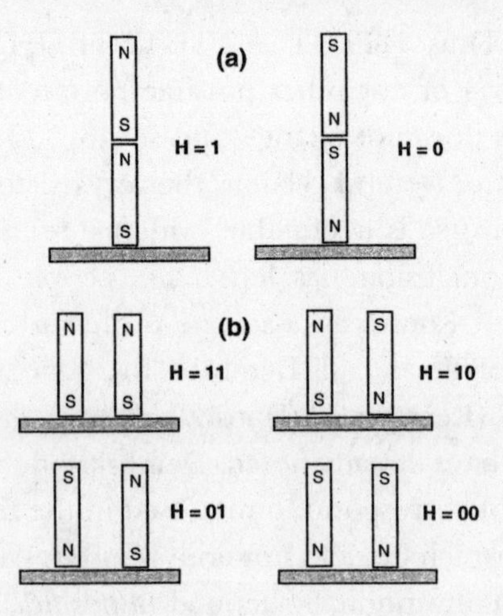

Figure 2. When one magnet sits atop the other, as in (a), information about the two possible configurations can be contained in one bit of information. But if a random event, such as a strong breeze, comes along and separates the two magnets, then the minimum number of bits necessary to describe the situation is increased.

possible to come up with many examples in which information is not conserved. So, Dembski's "proof" fails because it violates the Second Law of Thermodynamics. Actually, Dembski admits that information can degrade (Dembski 1999, 70), but this only demonstrates his inconsistency. His "law" of conservation of information does not permit this.

Now, perhaps Dembski might argue that it was not his intent to define information in terms of the Shannon uncertainty, although he uses Shannon as a reference and mentions no other source. In any case, we can still imagine natural processes adding bits of Dembski information to a system. A computer simulation can be used to illustrate this, but ID proponents can (and often do) object that such a simulation is still "designed." Let us consider instead an example that involves only chance, with no designer intervention.

Suppose we have two bar magnets, one sitting on top of the other, as shown in figure 2a. Because of their mutual attraction, only the two configurations shown, with either both north poles up or both south poles up, will be stable. This can be specified by one bit of Dembski information—say, $I_D = 1$ for both north poles up and $I_D = 0$ for both north poles down. We open the window and a random breeze comes through and knocks apart the magnets. Assume they are constrained so they cannot fall on their sides but must always land vertically.

Now, because the poles are no longer in contact, the four configurations shown in figure 2b are possible. We then need two bits to describe the situation: $I_D = 11$ for both north poles up, $I_D = 10$ for the first north up and the second down, $I_D = 01$ for the second north up and the first down, and $I_D = 00$ for both north poles down. The calculation of *H* according to equation 3 reflects the change from 1 bit to 2 bits of information in this system. Thus, the information in the system has increased by 1 bit as the result of a chance process. (We would need even more bits to describe the possible orientations for the magnets on their sides.) In this example, then, Dembski information is generated by chance, in violation of Dembski's "law" of conservation of information. This simple example shows why Dembski's concept of this "law" is not found in the standard usages and practices within the field of information theory: because it is patently incorrect.

Panning for Design

The "law" of conservation of information is not the only unloaded weapon in the arsenal of "intelligent design." Dembski attempts to show that design is evident by virtue of what, in his personal estimation, are probabilities that are too low for the natural production of order. Indeed, when one looks at all the variations in the argument from design that have appeared over the years, including the most recent, they amount to nothing more than the claim, "I cannot see how the universe and life could have happened naturally, therefore they must have been created supernaturally" (see Elsberry, Pennock, in this volume). This conclusion appears to be due simply to a failure of imagination.

Dembski introduces a series of "filters" that he applies to observed phenomena in order to determine whether or not they are designed (see Elsberry, in this volume). He tests these filters by applying them to examples of *human* design, on the assumption that any "intelligent agent" must follow the same rules. Starting with information, his filters accept only information that is both *complex* and *specified.* The resulting complex–specified information (CSI) is then interpreted as the consequence of "intelligent design." Let us consider these criteria in terms of the example of five consecutive coin tosses discussed above. We saw that this system has five bits of Dembski information. Suppose that, before the first toss, we *specify* a particular sequence—say, all heads: HHHHH. Or it could be all tails or any other sequence, such as HTTHT, as long as it is specified in advance.

Now, five heads in a row, or any other sequence of five coins, will happen frequently by chance. On average, about one in every $2^5 = 32$ tosses of five coins will land with all five heads up. However, suppose we do the experiment with 500 coins instead of five and specify in advance that all fall heads up. It would require $2^{500} = 10^{150}$ tosses of 500 coins each, again on average, to obtain 500 heads specified in advance in an event of 500 consecutive coin tosses by chance. That is, the probability for this outcome is 10^{-150}, and the Dembski information contained in the event is 500 bits. Dembski says it is impossible, for all practical purposes, to produce this *particular* prespecified array, and he uses any event containing at least 500 bits of information as his working definition of "complex" (Dembski 1999, 70). This, he notes, is a far more stringent restriction than the 166 bits implied by the "universal probability bound" of $10^{-50} = 2^{-166}$ proposed by mathematician Emile Borel (1962, 28).

However, while some *pre-specified* sequence of 500 bits—such as all heads or any other specific pattern of heads and tails selected before the fact—has this very low probability of being produced by chance, the probability for *some* pattern of heads and tails in 500 tossed coins (or any number of tosses) is 100 percent! That is, one of the 10^{-150} possible combinations *must* occur each time, even though the prior probability of producing each one of them is very, very small. So, if we look at the sequence that is produced *after* the coins are tossed, we cannot very well say that particular sequence is impossible when there it is, staring us in the face.

Physicist (and theist) Howard Van Till points out that Dembski's definition of complexity is highly unorthodox in light of how it is applied. For example, Dembski argues that any biotic system is complex if the probability for its being assembled by natural processes is less than 10^{-150}. This subtly changes the meaning of "complexity" from a property of the system (which Dembski claims to be calculating) to an inference about the means by which that system is actualized (which is closer to what he is really calculating) (Van Till 2002). Furthermore, when Dembski actually calculates the probability for a specific system—such as the flagellum in *E. coli* bacteria—he does so by assuming that the system was assembled by chance processes alone (Dembski 1999, 178). Van Till comments: "We reject that argument as a totally unrealistic caricature of how the flagellum is actualized and an approach that totally ignores the role of the bacterial genome in coding for all of the structures and functions that contribute to the nature of *E. coli*" (Van Till 2002, 23). While the chance probability might be less than the probability bound, the probability for chance plus other natural processes, such as natural selection, will always be greater.

Unfortunately, Dembski does not define *specificity* as precisely as he does *complexity* (even though his definition of *that* concept is idiosyncratic). In the coin example I have used, Dembski's approach requires that the sequence be specified in advance. This is fine, except that in nature we do not possess knowledge of the prespecified sequence. If Dembski left his models at this point, it would defeat his whole program to detect design *after the fact.* So, as a dubious and dangerous tactic, he allows specificity to be *post*-determined. This approach is presumed to capture "design information" that otherwise would be written off to chance—perhaps rightly so (see Elsberry, in this volume). However, in the coin example above, this would be equivalent to waiting until the toss was over and then saying, "Yes, that was the complex pattern that I specified." Although specificity is difficult to define, like pornography, we are supposed to know it when we see it. Once again, despite claims of mathematical rigor, it all comes down to a subjective judgment that the structure of a given biological system contains complex specified information—after the fact—that can be demonstrated *not* to result from chance or natural processes.

At the Beginning

The new creationism couched in the rhetoric of "intelligent design" theory implies more than just a continuation of the same old attacks on biological evolution. For example, biochemist Michael Behe's concept of irreducible complexity can be extended well beyond the realm of biology to the behavior of any nonlinear system (1996). "Intelligent design" can be proposed for the formation of galaxies as well as of bacterial flagella. The "intelligent design" creationists claim to see positive evidence for a Creator in the data from physics and astronomy, as well as biology (Ross 1995).

ID theorists capitalize on the common belief that one or more miracles were required to produce the universe. One such presumed miracle is the violation of *energy conservation* or what is equivalent, the First Law of Thermodynamics. Anti-evolutionists ask, "Where did the energy and matter of the universe come from?" In fact, astronomical observations strongly indicate that the mean energy density of the universe is exactly what would be expected if there were zero initial energy at the start of the Big Bang and would require no violation of the laws of energy conservation (Ostriker and Steinhardt 2001).

Another miracle widely believed to have occurred at the beginning of the universe is the violation of the Second Law of Thermodynamics. On the assumption that no forces act on it from the outside, the universe is an *isolated* system. Thus, it would appear that no gain of information (loss of entropy) with time is possible in the universe as a whole, since the Second Law is interpreted to mean that isolated systems tend to become more *dis*organized over time. However, recent research has shown that even if the universe began in complete disorder, local pockets of order can still form as time progresses without violating the Second Law. This is made possible by the expansion of the universe, which continually opens up more room for order to form (Stenger 1988, 1990).

Several cosmological scenarios have been published by established scholars in reputable scientific journals that allow for a universe to appear as an uncaused quantum event from an initial state of zero energy (Atkatz and Pagels 1982; Hawking and Moss 1982; Vilenkin 1982; Linde 1984a). Edward Tryon (1973) may have been the first to publish this idea. The published models illustrate that serious attention is being given to the possibility of an uncaused origin of the universe. Although they may turn out to be incorrect in many specifics, the proposed scenarios for a purely natural origin of the universe are consistent with existing knowledge in physics and cosmology. This serves to refute any claim that the origin of the universe currently *requires* a miraculous creation because science has no mechanism or process that could account for this event.

It is conventional to label the time of the Big Bang as $t = 0$. However, nothing we know demands that this was the beginning of time, as is often assumed by theologians (Craig 1979), or that no universe existed at earlier times. Nonspecialists often read that we cannot understand what happens very near $t = 0$ because to do so requires a theory of quantum gravity—a marriage between quantum mechanics and general relativity that does not yet exist. This is true for the origin of the universe, but it is equally true for any other point in time—such as the instant the reader's eye reaches the period at the end of this sentence.

Every point on the time axis, including "now," is surrounded by a small time interval within which our current theory of gravitation—Einstein's general theory of relativity—does not apply. General relativity is not a quantum theory, and quantum gravitational effects are important when time intervals are as small as 10^{-43} second (what is called the *Planck time*). Similarly, every point in space is surrounded by a tiny sphere of radius 10^{-35} meter—a distance called the *Planck length*—where general relativity also does not apply.

The Heisenberg uncertainty principle of quantum mechanics makes it impossible to measure any time interval smaller than the Planck time or any distance smaller than the Planck length. According to physics convention, time and distance are *operationally defined* by their measurements. Thus, unless one changes that convention, it is physically meaningless to talk about distance and time as having any quantitative values within a sphere of Planck dimensions—even with a theory of quantum gravity. Furthermore, the inability to define distance and time within these dimensions implies that no other physical quantity—at least in current physics—can be measured, since they are all defined in terms of space and time. And if no measurements are possible, neither is information. Thus we have a condition of complete uncertainty or maximum entropy.

And so, this was the state of the universe in a 10^{-43} second time interval around $t = 0$, if it was confined within a Planck sphere as Big Bang cosmology implies. The universe was then in a condition of maximum entropy and zero information — total chaos. If a supernatural creation occurred at this point, it was a creation *without design*, since the universe was without order. As already indicated, no violation of the Second Law of Thermodynamics is implied by a universe that begins with maximum entropy, since the expanding universe that follows allows increasing room for order to form.

Is the Universe Fine-Tuned for Humanity?

The latest claim of evidence for a divine cosmic plan with humans in mind is based on the fact that earthly life is so sensitive to the values of the fundamental physical constants and properties of its environment that even the tiniest changes in any of these would mean that life, as we see it around us, would not exist. This is said to reveal a universe in which physics is exquisitely fine-tuned—delicately balanced for the production of life.

The delicate connections among certain physical constants, and between those constants and life, are collectively called the *anthropic coincidences.* For a detailed history and a wide-ranging discussion of all the issues, see *The Anthropic Cosmological Principle* (Barrow and Tipler 1986).

Many theists see the anthropic coincidences as evidence for purposeful design of the universe. They ask, "How can the universe possibly have obtained the unique set of physical constants it has, so exquisitely fine tuned for life as they are, except by purposeful design—design with life and perhaps humanity in mind?" (Swinburne 1990; Ross 1995, 118). Let us examine the implicit assumptions here. Foremost—and fatal to the design argument all by itself—is the wholly unwarranted assumption that *only one type of life is possible*—the particular form of carbon-based life we have here on earth. Carbon *would* seem to be the chemical element best suited to act as the building block for the type of complex molecular systems that develop lifelike qualities. Even today, new materials assembled from carbon atoms exhibit remarkable, unexpected properties, from superconductivity to ferromagnetism. However, to assume that *only* carbon life is possible is simply "carbocentrism" that results from the fact that you and I are constructed mainly of carbon.

Given the known laws of physics and chemistry, we can imagine life based on silicon (computers, the Internet?) or other elements chemically similar to carbon. After all, it is the collection of chemical properties and interactions of carbon that make it such a useful building block for life, so why might not other elements with similar properties also serve this purpose under the proper conditions? Furthermore, nothing in anthropic reasoning indicates any special preference for *human* life, or indeed intelligent or sentient life of any sort—just carbon.

Going further, just because we can generalize about the universe that we live in and how life fits into it, we cannot rule out that forms of matter in the universe other than molecules could serve as building blocks for complex systems. The way that atoms assemble into molecular structures perhaps might be very different in a universe with different properties and laws. Those who argue that life is highly improbable except as a result of a universe uniquely and intelligently designed to produce it need to confront the possibility that life of some type might be likely with many different configurations of laws and constants of physics—without any violation of the expectation that natural processes following these laws are capable of producing living things that conform to the laws on which those universes might be based.

The Multiverse

If our universe appeared as a quantum fluctuation in a preexisting space–time void, this could have happened more than once—and probably did. The multiple-universe scenario is implied by the original suggestion of Tryon (1973) and imbedded in the cosmological model of André Linde (1984a, b, 1990, 1994). If the universe as a whole is infinite in extent in both space and time—and we have no scientific reason to think it is not—then subuniverses can be expected to pop up randomly at different positions and times. They appear as expanding bubbles that move away from one another, never colliding or coalescing.

While the multiple-universe, or *multiverse,* concept is not required to deflate the fine-tuning argument—which, as we saw above, fails on its own accord—this scenario can be used to provide a natural explanation for the so-called anthropic coincidences. The model suggests a simple mechanism by which universes of all types of structures can arise. The kinds of particles present, the forces between them, and various physical constants can be expected to be different from subuniverse to subuniverse. Some of the subuniverses will likely contain little of interest, and some may not contain structures like stars that live very long and manufacture complex elements that can serve as platforms for life. But many others can be expected to contain complex systems capable of evolving into something resembling life (or, indeed, perhaps something resembling nothing with which we are familiar and even far exceeding human life and mind in wondrous capabilities). Thus, our particular subuniverse only appears to be fine-tuned for us because it is that subuniverse which happens to contain the properties needed for our *kind* of life to evolve.

Some have argued that the multiverse scenario is less economical than one in which only a single universe exists. However, since multiple universes are suggested by existing knowledge, and since no known principle rules them out, it becomes, in fact, less economical to assume a single universe. For example, the atomic theory of matter introduced many more elements than were involved in prior physical theories, but it explains matter more parsimoniously and completely than those earlier theories. In a similar way, the multiverse scenario provides a more robust framework for studying cosmology than the models based on a single universe. What is more important for the anthropic coincidences, however, is that current research in physics and cosmology suggests that the "designed for life as we know it" argument may be exactly backward, because any life that appeared in any of the universes that are possible would have to be subject to the laws and constants that govern those universes—not the other way around.

Summary and Conclusions

Modern variations of the ancient argument from design form the basis of the new creationism—so-called "intelligent design" theory. These arguments amount to nothing really new and are just restatements—in ostensibly more sophisticated language of the common-sense view—that the universe and life appear to be too complex to have happened without supernatural intervention. However, the new creationism *poses* as science. Despite their pretense of scientific legitimacy, several of the claims of this new "science" are provably wrong. Dembski's "law of conservation of information" violates the Second Law of Thermodynamics. Furthermore, his calculation of the probability that the flagellum of *E. coli* bacteria could be assembled by natural processes alone wrongly assumes that *only* chance processes operate.

On the cosmological scale, the new creationism claims that the Big Bang provides evidence that a miraculous origin of the universe is required by current science. This is refuted by the fact that modern cosmology provides several naturalistic scenarios by which the universe began as an uncaused quantum event that violated no laws of physics. No scientific basis exists for assuming that a universe *did not exist* before the Big Bang. The new creationism also asserts that life would not exist but for an exquisite fine-tuning of the constants of physics, the so-called anthropic coincidences. This is true, but only for life as we know it; there is no scientific basis for assuming that other forms of life cannot emerge in a universe with different constants and laws.

Furthermore, the multiple-universe scenarios suggested by modern cosmology provide a means by which the so-called anthropic coincidences may have arisen naturally. Earthly species simply emerged in that universe with suitable properties. Our universe is then not fine-tuned for humanity; *humanity* is fine-tuned for our universe.

References

Atkatz, D., and H. Pagels. 1982. Origin of the universe as quantum tunneling event. *Physical Review* D25: 2065–73.

Barrow, J. D., and F. J. Tipler. 1986. *The anthropic cosmological principle.* Oxford: Oxford University Press.

Behe, M. 1996. *Darwin's black box.* New York: Free Press, 1996.

Borel, B. 1962. *Probabilities and life,* trans. M. Baudin. New York: Dover (p. 28).

Craig, W. L. 1979. *The Kalām cosmological argument.* Library of Philosophy and Religion. London: Macmillan.

Dembski, W. A. 1998. *The design inference.* Cambridge: Cambridge University Press.

———. 1999. *Intelligent design: The bridge between science and theology.* Downers Grove, IL: InterVarsity Press.

———. 2001. *No free lunch: Why specified complexity cannot be purchased without intelligence.* Lanham, MD: Rowman and Littlefield.

Hawking, S. W., and I. G. Moss. 1982. Supercooled phase transitions in the very early universe. *Physics Letters* B110: 35–38.

Kern, L. 2000. In God's country: William Dembski thought Baylor University would be the perfect place to investigate a scientific alternative to Darwinism. Little did he realize he would be crucified for his cause. *Houston Press* (December 14).

Linde, A. 1984a. Quantum creation of the inflationary universe. *Lettere al Nuovo Cimento* 39: 401–5.

———. 1984b. Chaotic inflation. *Physics Letters* 129B: 177–81.

———. 1990. *Particle physics and inflationary cosmology.* New York: Academic Press.

———. 1994. The self-reproducing inflationary universe." *Scientific American* 271 (5): 48–55.

Medawar, P. B. 1984. *The limits of science.* New York: Harper and Row.

Ostriker, J. P., and P. J. Steinhardt. 2001. The quintessential universe. *Scientific American* (January): 46–53.

Ross, H. 1995. *The Creator and the cosmos: How the greatest scientific discoveries of the century reveal God.* Colorado Springs: NavPress.

Shannon, C., and W. Weaver. 1949. *The mathematical theory of communication.* Urbana, IL: University of Illinois Press.

Stenger, V. J. 1988. *Not by design: The origin of the universe.* Amherst, NY: Prometheus Books.

———. 1990. The universe: The ultimate free lunch. *European Journal of Physics* 11 (1990): 236–43.

Swinburne, S. 1990. Argument from the fine-tuning of the universe. In *Physical cosmology and philosophy,* ed. J. Leslie, 154–73. New York: Macmillan.

Tolman, R. C. 1938. *The principles of statistical mechanics.* London: Lowe & Brydone. (Later printings available from Oxford University Press.)

Tryon, E. P. 1973. Is the universe a vacuum fluctuation? *Nature* 246: 396–97.

Van Till, H. J. 2002. *E. coli* at the no free lunchroom" [online]: www.aaas.org/spp/dser/evolution/perspectives/vantillecoli.pdf.

Vilenkin, A. 1982. Creation of universes from nothing. *Physics Letters* 117B: 25–28.

The Ages of the Earth, Solar System, Galaxy, and Universe

G. Brent Dalrymple

TODAY THERE IS UNIVERSAL AGREEMENT amongst knowledgeable scientists that the earth, the solar system, the galaxies, and the universe are billions of years old. This was not always so. During the seventeenth century, earth's age was thought to be only a few thousand years. As eighteenth- and nineteenth-century scientists began to study earth more carefully, they eventually came to the inescapable conclusion that our planet was millions or billions—not thousands—of years old (Albritton 1980; Lewis and Knell 2001). Using a family of techniques known as isotopic (also called radiometric) dating, which utilizes the decay of long-lived radioactive isotopes in rocks and minerals, twentieth-century geologists and physicists were finally able to measure accurately the ages of earth's rocks, samples of the moon, and primitive meteorites. They concluded that the solid bodies of the solar system, including the earth and presumably also the sun and gaseous planets, originated over an interval of ~50 million years about 4.5 to 4.6 Ga (Ga = billion years) ago. Other types of measurements made on stars by astronomers show that the Milky Way galaxy and the universe are 11–14 Ga and 12–15 Ga, respectively. This evidence is so conclusive and scientists are so confident about the antiquity of the earth, solar system, galaxy, and universe that it is not a subject of debate within the scientific community. Furthermore, there has been no disagreement about this subject for more than half a century, although the details of the chronologies are still being refined through active research.

Over the past several decades, a small group of evangelical Christians—who call themselves "creation scientists" but are best described as young-earth creationists (YECs)—have made a series of bizarre and erroneous assertions about the ages and histories of the earth and the universe. In contrast to other creationists who accept the antiquity of earth (for example, Young 1982; Ross 1991), YECs claim that earth is only 6,000 to 10,000 years old. They say that their beliefs about the origin and history of the natural world, which they call "scientific creationism," are every bit as scientific as those of real science, which they commonly misname "evolution science." The YECs are demonstrably wrong—earth is not young and "creation science" is not science.

The purpose of this chapter is to briefly discuss the fallacies of the YECs claims and to summarize the scientific evidence for the antiquity of earth and the universe.

Fallacious YEC Claims

Despite overwhelming scientific evidence to the contrary, YECs assert that the earth and the universe are only 6,000 to 10,000 years old. In addition, they claim that isotopic dating and other proven tools of geologists, physicists, and astronomers do not give reliable and meaningful results. Where do they get their young age for earth and what is the basis for their other claims? Let a few prominent YEC "scientists" tell us in their own words:

> Our unified premise is that observation and theory should always be subservient to a proper understanding of the Word of God. (Vardiman 2000)

> . . . the creation chapters of Genesis are marvelous and accurate accounts of the actual events of the primeval history of the universe . . . and . . . provide an intellectually satisfying framework within which to interpret the facts which science can determine. (Morris 1974)

> . . . *only* Scripture gives *specific* information about the age of the earth and the timing of its unobserved events. Rocks, fossils, isotopic arrays, and physical systems do not speak with the same clarity as Scripture. (Morris 1994)

> The Bible . . . places creation in six literal days only a few thousand years ago, with man, the "Image of God," the goal from the very start. This date derives mostly from summing up the time spans given in Scriptural genealogies. . . . Thus we can derive a "most probable" range of dates, all of which fall into the "young-earth" position. . . . (Morris 1994)

> Creation scientists must . . . work to understand isotopic systems in the light of the creationist–diluvialist paradigm. . . . (Woodmorappe 1999)

	YEARS		
	Minimum		**Maximum**
From Creation to the Flood	1656	to	2400
From the Flood to Abraham	300	to	4000
From Abraham to Christ	2000	to	4000
From Christ to Present	2000	to	2000
Total Range of Dates	6000	to	12,000

(Morris 1994)

Table 1: Creationist Geological Timeline

By their own admission, then, the YECs' age of earth and their justification for assertions on other matters of science are dictated first, foremost, and solely by their religious beliefs. Furthermore, they freely admit that their conclusions about the age of earth and the universe are derived from their reading of Scripture and that scientific data must be made to fit those conclusions. Invocation of supernatural agents and "revealed truths," and the interpretation of scientific data to fit conclusions reached beforehand, are not methods of science. Thus, "scientific creationism" is not science but is religion pure and simple—a fact also recognized by federal court rulings in both Arkansas (*McLean v. Arkansas* 1982; Gilkey 1985) and Louisiana (*Aguillard v. Treen* 1985). During the conduct of these lawsuits—whose plaintiffs included educators and mainstream religious organizations—two federal district court judges heard and read extensive legal arguments and testimony from scientists, philosophers, theologians, and educators on the one hand, and YECs on the other. In both cases, the judges ruled that "creation science" is religion, not science, and they declared unconstitutional the "equal-time for creationism" laws of Arkansas and Louisiana. Their legal reasoning was lengthy but simple: Scientific creationism is religion and requiring that it be taught in public schools amounts to the establishment of religion, which is in violation of the First Amendment to the Constitution of the United States. The U.S. National Academy of Sciences, arguably the most prestigious scientific organization in the world, recently reaffirmed its 1984 position that "creation science" is not science (Steering Committee on Science and Creationism 1999), and the U.S. Supreme Court affirmed the Louisiana decision in 1987 (*Edwards v. Aguillard*).

The truth is that YECs have no valid data or calculations to support their claim that earth is only 6000 to 10,000 years old. Instead, they call on an array of erroneous hypotheses, arguments, and calculations that they claim are evidence of a young earth but are not. Their calculations and arguments not only are fundamentally wrong; they do not produce any specific, consistent number for the age of earth at all. For example, they claim that earth's magnetic field is decaying and that the field could be no more than 10,000 years old (Barnes 1983; Humphreys 1989, 1993). They claim that the occurrence of radiation-damage halos in certain rocks shows that these rocks are "primordial" and were formed on day one of "creation week," about 6,000 years ago (Gentry 1986). They claim that the paucity of helium in earth's atmosphere shows that earth must be young (Vardiman 1985, 1990). They claim that the amount of sodium that has accumulated in the ocean shows that earth's ocean could be no more than 62 Ma (Ma = million years) (Austin and Humphreys 1991). Each of these supposed "evidences," and the many others that YECs advocate in the name of "creation science" have all been repeatedly shown to be meaningless as indicators of earth's age (for example, Brush 1982; Dalrymple 1983, 1984; Strahler 1987; Wakefield 1988; Morton 1998; Matson 1999; Stassen 1997).

YECs admit that isotopic dating is a significant challenge for them (Vardiman 2000), presumably because it has produced credible, abundant, and consistent scientific evidence for earth's antiquity. To answer this challenge, they have attempted to show that isotopic dating is invalid on theoretical grounds (for example, Arndts and Overn 1981; Slusher 1981; Gill 1996), but such attempts invariably have fatal and elementary flaws (see Dalrymple 1984; York and Dalrymple 2000).

Isotopic dating is viewed as such a serious threat to YEC claims that a group of YECs has recently embarked on a program—sponsored by the Institute for Creation Research (ICR) and the Creation Research Society (CRS)—to ". . . investigate the basis of these claims [of isotopic dating] and offer an alternative young-earth explanation" (Vardiman 2000). This group calls itself RATE (Radioisotopes and the Age of The Earth) and has formulated several hypotheses to explain why the results of isotopic dating are in error. As an example of the type of research that RATE is pursuing, one of their hypotheses is that radioactive decay occurred much faster at times in the past. And when is this accelerated decay supposed to have happened?

> I propose that since Creation, one or more episodes occurred when nuclear decay rates were billions of times greater than today's rates. Possibly there were three episodes: one in the early part of the Creation week, another between the Fall and the Flood, and the third during the year of the Genesis Flood. (Humphreys 2000)

It is difficult to see what kind of scientific experiments or observations might be conducted to test whether or not radioactive-decay constants were significantly different between "the Fall" and "the Flood" than they are now. Not only is this hypothesis unscientific, it is incredibly naïve. A significant change in radioactive-decay rates requires changes in fundamental and delicately balanced physical constants, such as Planck's constant and the speed of light. These, in turn, affect the very properties of light and many other important physical and chemical relationships, including Einstein's famous mass-energy relationship ($E = mc^2$). The result is a universe that no longer works—or at least one that works much differently than the one in which we live. Whatever Russ Humphreys and the RATE group have in mind here, it is certainly not science.

Another YEC tactic is to focus on instances in which isotopic dating seems to yield incorrect results (for example, Woodmorappe 1979, 1999; Morris 1985; Morris 1994). In most instances, these efforts are flawed because the authors have misunderstood or misrepresented the data they claim to analyze (Dalrymple 1984; Schimmrich 1998). To be sure, there are instances where isotopic dating gives incorrect results, but only rarely does a creationist actually find an incorrect isotopic result not already identified by scientists (Austin 1996; Rugg and Austin 1998), and those instances invariably are not fundamentally different from examples already discussed at length in the scientific literature.

This latter approach—in which YECs focus on instances where isotopic dating yields incorrect results—is a curious one for two reasons. First, it provides no evidence whatsoever to support their claim that earth is very young. If earth were only 6,000 to 10,000 years old, then surely there would be some scientific evidence to confirm that hypothesis, yet the YECs have produced none so far. Where are the data and age calculations that result in a consistent set of ages for all rocks on earth, as well as those from the moon and the meteorites, no greater than 10,000 years? Such evidence cannot be provided by YECs because such evidence does not exist.

Second, it is an approach doomed to failure at the outset. YECs seem to think that a few examples of incorrect isotopic ages invalidate all of the results of isotopic dating, but such a conclusion is illogical. Everyday experience makes it obvious that even things that work well do not work well all of the time or under all circumstances. Has your automobile, your computer, or your television set ever failed to work? What did you conclude from that experience? That automobiles, computers, and television sets do not work? Of course not, because there is abundant evidence that these devices do work and work well most of the time as long as they are maintained and not misused. Just as devices are not infallible, neither are generally reliable scientific techniques.

In the same way, a few examples of incorrect isotopic ages are insufficient to prove that isotopic dating is invalid. After all, literally tens of thousands of isotopic ages have been obtained in many scientific laboratories around the world, and it would be miraculous if some of them were not wrong. All that indicates, however, is that the methods and the people who use them are not infallible. Scientists who develop and use dating techniques to solve scientific problems are well aware that the systems are not perfect under all circumstances, and they have even published numerous examples in which the techniques fail. They often test them under controlled conditions to learn when and why particular techniques fail, in order to avoid using them incorrectly. Methods that have proven unreliable have been abandoned. For example, after extensive testing over many years, it was concluded that uranium-helium dating is unreliable because the small helium atoms diffuse easily out of minerals over geologic time. As a result, this method is only used in rare and highly specialized applications that do not include measuring the formation ages of rocks.

Other dating techniques have stood the test of time. These methods provide valuable and valid age data in most instances, although there is a small percentage of instances in which even these generally reliable methods yield incorrect results. Such failures may be due to laboratory errors (mistakes happen), unrecognized geologic factors (nature sometimes fools us), or misapplication of the techniques (no one is perfect). In order to accomplish their goal of discrediting isotopic dating, however, YECs are faced with the impossible task of showing that all isotopic ages are wrong and that the methods are untrustworthy all of the time. This is a tall order. Despite their frequent claims of success, the YECs have made no progress so far and most likely never will.

In summary, the YECs' calculations, hypotheses, and arguments for a young earth, and their objections to the scientific evidence for an ancient earth and universe have no scientific basis and are unworthy of serious consideration. It is for this very reason that their "evidence" for a young earth and their criticisms of the proven methods of science are published almost exclusively in creationist publications rather than in the legitimate scientific literature, where manuscripts are subjected to careful and independent scientific peer review. Even when speaking through their own media, some YECs who claim to be scientists use pseudonyms to hide their identities, which makes it impossible to check their credentials (John Woodmorappe, for example, is a pseudonym). Neither of these behaviors—avoiding the scientific literature and concealing one's identity—is indicative of real scientists engaged in legitimate science.

Isotopic Dating—Geology's Timekeeper

Scientists have more than half a dozen good ways of measuring the ages of ancient rocks and minerals (Dalrymple 1991, 2004). These methods (table 2) are collectively called isotopic (or radiometric) dating. They rely on the decay of long-lived radioactive nuclides (parent nuclides) that occur naturally in most rocks and minerals and the corresponding accumulation of the stable nuclides produced by the decay (daughter nuclides). (Each *isotope* of an element has a different number of neutrons, and hence a different mass number, than the other isotopes of that element. The more general term *nuclides* refer to all isotopes of all elements, so all isotopes are also nuclides.) These nuclear clocks work reliably, provided that two conditions are met. First, radioactive decay must be constant over the range of conditions experienced by rocks to which scientists have access; and second, the rocks and minerals to which a method is applied must have remained closed systems to the parent and daughter nuclides since the rock was formed.

The constancy of radioactive decay has been tested both theoretically and experimentally. It has proven to be constant for all conditions experienced on earth—and other solid planets—and for the particular isotopes used in dating. The one exception is

Method	Parent Isotope (radioactive)	Daughter Isotope (stable)	Half-life (Ma)
K–Ar (potassium-argon	^{40}K	^{40}Ar	1.25
Ar–Ar (argon-argon or $^{40}Ar/^{39}Ar$)	^{40}K	^{40}Ar	1.25
Rb–Sr (rubidium-strontium)	^{87}Rb	^{87}Sr	48.8
Sm–Nd (samarium-neodymium)	^{147}Sm	^{143}Nd	106.
Lu–Hf (lutetium-hafnium)	^{176}Lu	^{176}Hf	35.9
Re–Os (rhenium-osmium)	^{187}Re	^{187}Os	43.0
Th–Pb (thorium-lead)	^{232}Th	^{208}Pb	14.0
U–Pb (uranium-lead)	^{235}U	^{207}Pb	0.704
U–Pb (uranium-lead)	^{238}U	^{206}Pb	4.47
Pb–Pb (lead-lead)	^{235}U & ^{238}U	^{207}Pb & ^{206}Pb	

Ar–Ar is an analytical variation of K–Ar. Pb–Pb uses both U–Pb decays. All methods are explained in Dalrymple (1991). Superscripts indicate the mass number (neurons + protons) of the isotopes.

Table 2. Principal Isotopic-Dating Methods Used to Determine the Ages of Rocks and Minerals

for a type of decay called *electron capture*, which has been found to vary slightly (much less than 1 percent) within different chemical compounds and at different pressures, but only for the beryllium nuclide (^7Be), which is not used for dating (Dalrymple 1991; Dicken 1995). Only one of the parent nuclides used in isotopic dating (^{40}K=potassium) decays by electron capture, and no indications of variable decay by that nuclide have ever been found.

The requirement that a rock or mineral remain a closed system is more difficult to ensure for all rocks because sometimes things happen to rocks during their histories that can cause nuclides to enter or leave the rock. Post-crystallization heating or chemical alteration can disturb an isotopic-dating system and cause the results to reflect not the age of the rock but some other event, such as the time of heating, or perhaps no specific event at all. Furthermore, some rocks are subjected to multiple episodes of heating and alteration, making these rocks difficult or impossible to date reliably. Scientists who measure the ages of rocks, however, have ways to check for analytical errors and to assess whether or not isotopic-dating results are geologically meaningful.

It is rare for a study involving isotopic dating to contain a single age measurement. Usually, age measurements are repeated to avoid laboratory errors, are obtained on more than one rock unit or more than one mineral from a rock unit, or are made using more than one dating method. Moreover, age measurements are usually evaluated against other independent information—such as the relative order of rock units as observed in the field by geologic mapping—to test and corroborate further the validity of the isotopic ages. In addition, there are mathematical and graphical techniques (called isochron and age-spectrum methods) that are applied to the data from multiple measurements of the same rock or mineral, on multiple minerals from the same rock, or on multiple rock samples from the same rock body. These methods include internal tests for validity of the closed system and do not require any assumption about the presence or absence of an initial quantity of the daughter isotope. Scientists typically use every means at their disposal to check, recheck, and verify their results. As a rule, the more important the results, the more they are apt to be checked and rechecked by other scientists working in other laboratories. Because of this rechecking, it is nearly impossible to be fooled by a good set of isotopic age data from a well-designed experiment. A few examples of the many consistent data sets from isotopic-dating studies are given below. Other examples can be found in Dalrymple (1991, 2000, 2004), Meert (2000), and the extensive scientific literature of isotopic dating.

Ages of the Earth, Moon, and Meteorites

Earth's Oldest Rocks

Dating the origin of earth precisely is somewhat of a problem because it is a dynamic planet. The earth's crust is continually being added to, modified, and destroyed. As a result, rocks that record earth's earliest history have not been found and may no longer exist. Nevertheless, the oldest rocks on earth at least provide a minimum estimate of the planet's age. Rocks older than 3.5 Ga have been found, studied, and dated on all the continents. Some of the oldest well-studied rocks are in Greenland and in Canada.

Field studies, laboratory analyses, and evidence from plate tec-

Godthaab, W. Greenland	Isua, W. Greenland	Labrador	Approximate Age (Ga)
mafic dikes	mafic dikes	mafic dikes	?
Qôrqut granite		post-tectonic granite	2.5–2.6
Nûk gneisses		Ikarut, Kammersuit, and Kiyuktok gneisses	2.7–3.0
Layered anorthosites and gabbros		Anorthosites and gabbros	2.8
Malene supracrustals		Upernavik supracrustals	?
Ameralik mafic dikes	Tarssartoq mafic dikes	Saglek mafic dikes	3.2–3.4
Amîtsoq gneisses	Amîtsoq gneisses	Uivak gneisses	3.6–3.7
Akilia supracrustals	Isua supracrustals	Nulliak assemblage	3.6–3.8

The age ranges shown in the last column are based on hundreds of isotopic-dating measurements in several laboratories on rocks in the three areas. Data from summary in Dalrymple (1991).

Table 3. Simplified Sequence of Major Rock Units in the Godthaab and Isua Areas of West Greenland and in Coastal Labrador, as Determined by Field Mapping

tonics show that Greenland and Labrador were once connected, so the Greenland rocks have counterparts in Labrador (table 3). The oldest rocks in the Greenland-Labrador sequence have been dated many times by four independent isotopic-dating methods—uranium–lead (U–Pb), lead–lead (Pb–Pb), rubidium–strontium (Rb–Sr), and samarium–neodymium (Sm–Nd)—and give ages that fall within the range 3.65–3.85 Ga. Many of these rocks, called *supracrustals*, are metamorphic rocks that were originally sediments and lava flows. The existence of sediments shows that even older rocks, the sources for the sediments, must have existed at one time. Because many samples of these ancient rocks have been dated repeatedly, by several isotopic methods, in different laboratories, by different investigators—and because the dating results agree with the known order of formation of these rocks as determined independently by field mapping—there is little likelihood that the dating is greatly in error.

The oldest rocks found on earth so far are in Canada's Northwest Territories, near Great Slave Lake, and are just slightly over 4.0 Ga in age (Bowring and Williams 1999). These rocks, found within the Acasta Gneiss Complex, were originally formed as intrusive rocks and are similar to granite but have a slightly different composition. The three oldest rock units, dated by U–Pb analysis of multiple zircon crystals using an ion microprobe, gave ages of 4002 ± 4 Ma, 4012 ± 6 Ma, and 4031 ± 3 Ma.

The Acasta gneisses, however, are not the oldest earth materials that have ever been found. In Australia tiny grains of zircon found in younger (~3.0 Ga) sedimentary rocks have U–Pb ages, measured using an ion microprobe, in excess of 4.0 Ga. The oldest crystal found so far gives an age of 4.404 ± .008 Ga (Wilde et al. 2001). The source of these ancient crystals has not been found and may never be, but their existence shows that earth had crustal rocks as early as 4.4 Ga.

In summary, there have been hundreds of isotopic age tests made on earth's ancient rocks from around the world (see Dalrymple 1991, 2004, for more examples). Many have ages in the range 3.5–3.9 Ga and a few have ages that exceed 4.0 Ga. Minerals as old as 4.4 Ga have been found in Australia. While these data do not provide an age for earth's formation, they do make it clear that earth is at least 4.4 Ga in age.

Oldest Lunar Rocks

The moon is a small planet. Initially, it was hot owing to the energy produced by gravity during accretion and by short-lived radioactive nuclides. This internal heat, which is necessary for sustained crustal recycling, was radiated into space long ago. Thus, the moon has been a more-or-less-dead planet for some time. Because of this, the moon has many more old rocks than earth.

Nearly 400 kilograms (about 880 pounds) of rocks were returned from nine different localities on the near side of the moon by the U.S. *Apollo* and USSR *Luna* missions. Samples were collected from both the lunar highlands and the *mare* ("seas")—giant impact basins, formed by asteroid-size objects, that were filled with lava flows well after the basin-forming impacts. From telescopic mapping, it was known before the lunar landings that the *mare* are younger than the highlands. This is most evident in the crater density: the highlands appear light-colored because the many small craters reflect sunlight more efficiently than the less heavily cratered surface of the *mare*, which appear dark.

Several hundred of the returned lunar samples have been dated, using primarily the K–Ar and Rb–Sr methods, and most are more than 3 Ga. Samples of lava flows from the *mare* range from slightly more than 3.0 Ga (*Apollo 12—Oceanus Procellarum*) to 3.9 Ga (*Apollo 11—Mare Tranquillitatis*), whereas samples from the highlands sites range from more than 3.4 to 4.5 Ga in age (Dalrymple 1991, 2004). Only a few highlands rocks have ages as old as 4.5 Ga, but there are many whose ages exceed 4.0 Ga. This shows that the moon was in existence 4.5 Ga ago. It is currently thought, and supported by substantial evidence, that the moon was formed from debris caused by the collision of a large planetoid, about the size of Mars or larger, with earth very early in earth's history. If so, then earth also must be at least 4.5 Ga in age.

Ages of Meteorites

The most primitive rocks in the solar system are the meteorites—fragments of asteroids that form when asteroids collide. Initially, the fragments orbit the sun just like the asteroids, but some of the fragments get into gravitational resonance with Jupiter and may be thrown into earth-crossing orbits; a few eventually collide with earth. Thousands of meteorites have been recovered on earth, and more than 100 have been dated by different isotopic methods; a number of meteorites have been dated by two or more methods (table 4).[1] The table shows isotopic ages obtained in different laboratories using different methods on different meteorites, yet the results clearly indicate that the meteorites are ancient objects that formed more than 4.5 Ga ago. All meteorites are not expected to be exactly the same age. The chondrites, for example, are from small, undifferentiated asteroids that did not form cores or erupt lava flows. They are

1. Additional examples can be found in Dalrymple (1991, 2004) and Allègre et al. (1995).

Meteorite	Method	Age (Ga)	Lab
	CHONDRITES		
Allende	$^{40}Ar/^{39}Ar$ age spectrum	4.52 ± 0.02	1
	$^{40}Ar/^{39}Ar$ age spectrum	4.53 ± 0.02	1
	$^{40}Ar/^{39}Ar$ age spectrum	4.48 ± 0.02	1
	$^{40}Ar/^{39}Ar$ age spectrum	4.55 ± 0.03	1
	$^{40}Ar/^{39}Ar$ age spectrum	4.55 ± 0.03	1
	$^{40}Ar/^{39}Ar$ age spectrum	4.57 ± 0.03	1
	$^{40}Ar/^{39}Ar$ age spectrum	4.50 ± 0.02	1
	$^{40}Ar/^{39}Ar$ age spectrum	4.56 ± 0.05	1
	Pb–Pb isochron (27 points)	4.553 ± 0.004	7
Guarena	$^{40}Ar/^{39}Ar$ age spectrum	4.44 ± 0.06	2
	Rb–Sr isochron (13 points)	4.46 ± 0.08	4
Indarch	Rb–Sr isochron (9 points)	4.46 ± 0.08	4
	Rb–Sr isochron (12 points)	4.39 ± 0.04	3
Olivenza	Rb–Sr isochron (18 points)	4.53 ± 0.16	4
	A–Ar age spectrum	4.49 ± 0.06	2
Saint Severin	$^{40}Ar/^{39}Ar$ age spectrum	4.43 ± 0.04	5
	$^{40}Ar/^{39}Ar$ age spectrum	4.38 ± 0.04	6
	$^{40}Ar/^{39}Ar$ age spectrum	4.42 ± 0.04	6
	Rb–Sr isochron (10 points)	4.51 ± 0.15	3
	Sm–Nd isochron (4 points)	4.55 ± 0.33	4
	Pb–Pb isochron (5 points)	4.543 ± 0.019	3
	ACHONDRITES		
Juvinas	Sm–Nd isochron (5 points)	4.56 ± 0.08	4
	Rb–Sr isochron (5 points)	4.50 ± 0.07	3
	Pb–Pb isochron (8 points)	4.556 ± 0.012	7
	Pb–Pb isochron (9 points)	4.540 ± 0.001	3
Y-75011	Rb–Sr isochron (9 points)	4.50 ± 0.05	8
	Sm–Nd isochron (7 points)	4.52 ± 0.05	8
Angra dos Reis	Sm–Nd isochron (7 points)	4.55 ± 0.04	4
	Sm–Nd isochron (3 points)	4.56 ± 0.04	4
	IRONS		
Mundrabilla	Ar–Ar age spectrum	4.57 ± 0.06	4
	Ar-Ar age spectrum	4.54 ± 0.04	1
	Ar-Ar age spectrum	4.50 ± 0.04	1
Weekeroo Station	Rb-Sr (4 points)	4.39 ± 0.07	2
	Ar-Ar age spectrum	4.54 ± 0.03	4

The chondrite meteorites are fragments of undifferentiated (smaller) asteroids; the achondrites are fragments of lava flows from the surfaces of larger asteroids; and the irons are fragments of asteroid cores. From compilation in Dalrymple (1991). Data from university laboratories in Germany (1), Great Britain (2), France (3), California (4), Minnesota (5), Missouri (6), the USGS in Colorado (7), and NASA in Texas (8).

Table 4. Isotopic Ages of Some Meteorites

expected to be slightly older than the achondrites and the irons, both of which are from asteroids that were large enough to undergo significant internal heating, form iron-nickel cores, and erupt lava flows onto their surfaces. Analytical errors and the post-formation histories of the meteorites also account for some small variations in the calculated ages.

It is thought that refractory calcium-aluminum inclusions found in the Allende meteorite (which landed in a Mexican village in 1969) may be the oldest objects in the solar system. They represent material that condensed directly from the solar nebula at the time of its isolation from the other material in the Milky Way galaxy. Four of these inclusions have been dated by U–Pb and give ages that fall within the narrow range of 4565 ± 9 to 4568 ± 31 Ma (Allègre et al. 1995).

Other Relevant Evidence

The most commonly used age for earth is 4.54 ± 0.02 Ga. This value comes from modeling the evolution of lead isotopes in four ancient terrestrial lead ores (galena), beginning from a primordial composition recorded in a non–uranium-bearing phase (FeS) of the Canyon Diablo iron meteorite and carried forward to their present values (Tera 1980, 1981). Ancient galenas from three continents seem to define a common source with a common lead composition. The age calculated for these ores is similar to the age determined for meteorites, but the precise nature of the 4.54 Ga "event" is not clear (Hofmann 2001). A recent analysis of more than a dozen terrestrial lead ores suggests that the time of origin of the lead system from which they evolved represents the time of formation of earth's core at 4.49 Ga (McCulloch 1996).

A better-understood and more precise age for earth is emerging from studies of extinct radioactive isotope systems with short (<20 million years) half-lives, such as iodine–xenon ($^{129}I/^{129}Xe$), manganese–chromium ($^{53}Mn/^{53}Cr$), and hafnium–tungsten ($^{182}Hf/^{182}W$). (A half-life is the time it takes for a radioisotope to decay, so 50 percent of a nuclide with a 20 Ma half-life will remain after 20 Ma; 25 percent remains after 40 Ma; 12.5 percent remains after 60 Ma; and so on.) These methods allow the measurement of precise age differences between objects in the solar system. If the age of one object is known, the ages of the others can be determined by addition or subtraction of the small age differences. Recent measurements on meteorites as well as on lunar and terrestrial samples, combined with increasingly precise lead-isotope studies of meteorites, have begun to clarify the timing of events in the early solar system (e.g., Allègre et al. 1995; Halliday and Lee 1999; Tera and Carlson 1999). Such studies show the sequence of events beginning with the condensation of solid matter from the solar nebula at 4.566 Ga—which is the age of the calcium-aluminum inclusions in the Allende meteorite—and ending with the final accretion of earth. The segregation of earth's core, and the formation of the moon, occurred within an interval of about 50 ± 10 million years, with the latter events occurring about 4.51 Ga (Halliday and Lee 1999). This would place the "age of earth" at somewhat less than the model lead age of 4.54 Ga. Despite the present uncertainties, which are relatively small, there is little doubt that the age of earth (or at least the material from which it formed) and the solar system exceed, by some small amount, 4.5 Ga.

Another observation does not produce a specific age for earth or the solar system but does provide semiquantitative evidence that it is old: With few exceptions, only radioactive nuclides with half-lives greater than about 80 Ma occur in the solar system. (The exceptions are those nuclides that are continually produced by natural nuclear reactions, such as the production of ^{14}C [carbon] from ^{14}N [nitrogen] in the upper atmosphere.) All of the "missing" nuclides (i.e., those with shorter half-lives) can be produced in nuclear reactors, and theoretical considerations show that they are and were produced, like the other nuclides, in stars and supernovae, but none of these are found on earth. So why are they not found on earth? The simplest and most logical answer is that such nuclides *did* exist at one time, but earth is old and they have decayed away over time. Even an age of 1 Ga for

Earth is insufficient to explain the absence of the handful of missing nuclides with half-lives between 10 and 80 Ma (Dalrymple 1991), but the observation is perfectly consistent with an earth whose age is several billion years or more.

Finally, the age of the sun has been determined. As the nuclear reactions in the sun proceed, the resulting changes in elemental composition lead to predictable changes in the internal structure of the sun over time. The sun's internal structure is revealed by the nature of the seismic waves that continually penetrate its interior; the waves cause disturbances on the sun's surface that can be measured from earth. From the present structure of the sun, Guenther and Demarque (1997) have calculated that nuclear reactions in the sun began at 4.5 ± 0.1 Ga.

More Evidence of an Ancient Earth

The proof of earth's antiquity does not rest solely on the dating of the solar system's oldest rocks. For more than 300 years, scientists have been carefully studying the rocks of earth, and in that time they have learned a lot about earth's composition and history and the processes that form and transform rocks. Virtually everything that has been learned indicates that earth is very old—a conclusion about which there is no debate within the scientific community. Everywhere we turn, there is ample evidence of earth's antiquity. Following are three examples; there are hundreds, if not thousands, more described in the vast literature of science.

The Geologic Time Scale

The geologic time scale (figure 1) was constructed by many geologists who studied the relative age sequence of sedimentary rocks and the fossils they contain (Harland et al. 1989). Rocks representing all of geologic time do not occur in any one place, but it was possible to "piece together" a worldwide relative time scale by using distinctive groups of fossils to determine which sedimentary rocks in one locality were formed at the same time as those in other localities. Contrary to the assertions of some YECs

Era	Period	Epoch	Age (Ma)
Cenozoic [1841]	Quaternary [1854]	Holocene [1885]	0.01
		Pleistocene [1839]	1.6
	Tertiary [1759]	Pliocene [1833]	5.2
		Miocene [1833]	24
		Oligocene [1854]	36
		Eocene [1833]	56
		Paleocene [1874]	65
Mesozoic [1841]	Cretaceous [1823]		132
	Jurassic [1795]		208
	Triassic [1834]		245
Paleozoic [1840]	Permian [1841]		290
	Carboniferous [1822]		363
	Devonian [1839]		409
	Silurian [1835]		439
	Orovician [1879]		510
	Cambrian [1835]		570
(Precambrian)			

Figure 1. The geologic time scale is created by using sequences of sedimentary rocks and their fossils as an "index" to match strata in different parts of the world so that a complete sequence can be constructed.

(for example, Morris 1994), no assumption of evolution was involved in constructing the geologic time scale. Indeed, most of the major subdivisions of geologic time (eras, periods, epochs) were known, named, and placed in relative order before Darwin published *On the Origin of Species* in 1859. Early geologists simply observed which fossils were younger or older than others from their relative positions in the sedimentary rocks; they then used this information to determine the time-equivalencies of stratigraphic sequences in other areas. The geologic time scale is and always will be a work in progress; new details are being provided continuously as studies continue. But the major geological subdivisions after the Precambrian (that is, the Phanerozoic) and their relative order have not changed for more than a century. Precambrian time may never be subdividable to the same precision as Phanerozoic time, but we do know that fossils are rare in Precambrian rocks because early animals lacked hard parts and so were rarely preserved. There are, however, fossil Archaea and stromatolites (structures formed by a type of algae) in rocks as old as 3.5 Ga (Schopf 1999).

Because the geologic time scale is a relative scale, it contains no assumptions or information about the length of geologic time or the age of earth. The geologists who constructed it were able to determine the relative ages of groups of sedimentary rocks, but they had no way to assign numerical ages to either the subdivision boundaries or their lengths. It was fairly obvious to early (and present) geologists, however, that many types of sedimentary rocks must have been formed over long periods of time—a conclusion reached by observing how these same rock types form today. The great extent to which many of these rocks have been uplifted, eroded, tipped, folded, faulted, and chemically changed also indicates that long periods of time were required for these changes to have occurred.

Isotopic-dating methods were not perfected until the latter half of the twentieth century, and one of the first problems to which they were applied was the geologic time scale. The ages of the eras, periods, epochs, and their numerous minor subdivisions are still being refined today, but they had been pretty well dated by the mid-1960s. One of the significant results of this work was that isotopic dating found no errors in the relative order of the major subdivisions of the geologic time scale. The isotopic ages of the time-scale subdivisions fall in the same sequence as their observed relative order. If the geologic time scale had errors in the relative placement of its subdivisions, or if isotopic dating did not work, *or* if earth *were* only 6,000 to 10,000 years old, as the YECs claim, then this result (figure 1) would not have been possible.

YECs claim that, by and large, all of the fossiliferous sedimentary rocks on earth were deposited in the year of the biblical Flood (Whitcomb and Morris 1961; Morris 1974; Morris 1994). But based on three centuries of geologic study, scientific data on the formation of rocks, and the orderly and consistent sequence in which fossils occur in the geologic column, this is an absurd idea. The successful formulation and dating of the geologic time scale, showing that Phanerozoic time began more than half a billion years ago, make the absurdity of this YEC assertion even more apparent.

Geomagnetic Reversals

There was some evidence in the early 1900s that earth's magnetic field had reversed its polarity in the past (that is, exchanged north and south magnetic poles). It was not until the 1960s, however, that the tools were available to test that hypothesis (see Glen 1982, for a detailed history of this work and its consequences). Two small groups of scientists in the United States and in Australia proceeded by measuring the magnetization in young lava flows, which record the direction of earth's magnetic field as they cool; they measured the ages of the same flows by K–Ar dating and plotted the magnetic polarities as a function of age. By 1966, they had not only proved that earth's magnetic field had reversed in the past but had constructed a reversal time scale for the past four million years (figure 2a).

Quite independently, and at about the same time, other groups of scientists measured the magnetic anomalies near the midocean ridges and found that the same pattern of normal and reversed magnetization occurred on both sides of the ridges (figure 2b). It was this correlation between the sea-floor magnetic anomalies and the land-based geomagnetic-reversal time scale that led to the theory of plate tectonics and proved that new sea floor is continually being created at the midocean ridges, recording earth's magnetic field as it cools and then spreading outward

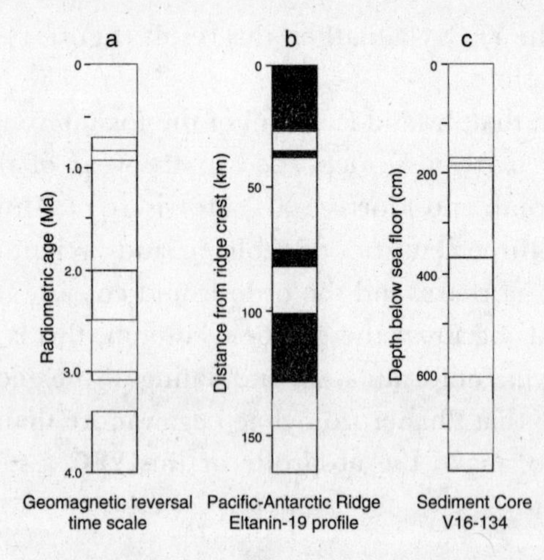

Figure 2. The earth's magnetic field has reversed several times: this information can be used to measure the rates of spreading between geologic plates.

symmetrically on either side. Using the time scale to determine spreading rates for the past four million years, along with the magnetic anomalies of even older sea floor, scientists have been able to extend the reversal time scale back to 160 million years.

As if the correspondence of the geomagnetic-reversal time scale with the sea-floor magnetic anomalies were not spectacular enough, other scientists, measuring the magnetic field directions recorded in deep-sea cores, again found precisely the same pattern of normal and reversed magnetization (figure 2c). Thus, lava flows on land, the igneous rocks that form the sea floor, and deep-sea sediments all record the same history of reversals of earth's magnetic field. These remarkable findings from three independent lines of research confirm that the sea floor has been spreading and that earth's magnetic field has reversed itself. These findings prove that some areas of the sea floors are at least 160 million years old, and they demonstrate that the dating and magnetic methods used in these studies yield valid results.

What do YECs say about geomagnetic reversals? While some YECs now admit that earth's magnetic field has reversed, they propose that the reversals occurred rapidly and all within a very short period of time in the recent past. They hypothesize that this was caused by ". . . plate movements and the breaking up of the 'fountains of the *great* deep' (Genesis 7:11) . . ." at the onset of the Flood (Morris 1994). This explanation, however, is pseudoscientific nonsense for which there is not a shred of credible scientific evidence.

Hawaiian–Emperor Volcanic Chain

The Hawaiian islands were constructed by massive volcanic eruptions that built sea-floor mountains now rising well above sea level. Even a casual study of these islands shows that they increase in age from southeast to northwest. The active volcanoes of Kilauea and Mauna Loa (and Loihi Seamount), on the Big Island of Hawaii, are at the southeast end of the Hawaiian Ridge. To the north and west, the volcanoes are extinct and progressively more eroded. But the familiar islands from Hawaii to Kauai are only the southeastern tip of the Hawaiian Ridge–Emperor Seamounts volcanic chain, which stretches nearly 3,600 miles across the North Pacific Ocean (figure 3). About 2,075 miles northwest of Kilauea, the Hawaiian Ridge takes a sharp bend to the north and continues as far as the Aleutian Trench, where the Emperor Seamounts chain disappears into the trench. Progressing to the northwest beyond Kauai, there are only a few rocky islets and then coral reefs that completely cap the hidden volcanic rocks beneath, but the chain as a whole contains 107 individual volcanoes. Beyond Midway and Kure, the westernmost islands in the chain, all of the volcanoes are beneath the sea (summarized by Clague and Dalrymple 1987).

One hypothesis for the origin of the Hawaiian islands, first proposed in 1963 before the theory of plate tectonics had entered the discourse, was that these islands were formed as the Pacific lithospheric plate moved over a fixed source of lava, or "hot spot," in earth's mantle. As the plate moved first to the north and then changed direction to the west-northwest, eruptions from the hot spot left a trail of volcanoes on the sea floor, a process that is continuing today. Even before each volcano moves past the hot-spot source, it slowly begins to sink because of its weight on the oceanic crust and because of the thermal aging (contraction) of the crust. As the volcano continues to sink, the

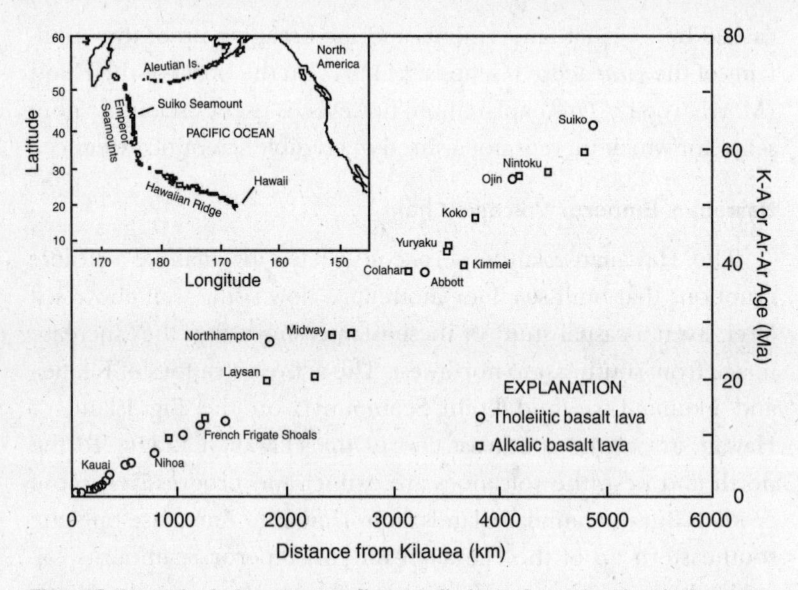

Figure 3. Isotopic-dating ages confirm predictions about the chemical composition, sequence, and movement rate of volcanoes in the Hawaiian–Emperor volcanic chain.

exposed top is eroded and a coral cap eventually forms. Coral growth keeps up with the sinking for a time, but eventually the sinking—combined with the northward movement into colder water—overwhelms coral growth, so the volcano, now a seamount, ends up several miles beneath the sea.

This hypothesis was tested in the 1970s and 1980s. Scientists analyzed the chemical compositions of lava flows from more than thirty volcanoes in the Hawaiian–Emperor chain to see if they were of the same type, and they performed isotopic dating on samples from the volcanoes. The chemical analyses showed that the volcanoes in the chain were of the same type and composition, and occurred in the same sequence, as those in the main Hawaiian islands: tholeiitic basalt formed the cores of the volcanoes and alkalic basalt formed a thin veneer on their tops. In addition, K–Ar and Ar–Ar dating showed that the volcanoes sampled are progressively older to the west and north, away from the active Kilauea volcano (figure 3). The data clearly indicate that the Hawaiian–Emperor volcanic chain began more than 64 Ma ago, and there are more volcanoes beyond Suiko Seamount that are unsampled. Furthermore, this direction and rate of volcanic propagation along the chain are consistent with what is known about the motion of the Pacific plate from magnetic anomalies and other studies of the oceanic crust (Clague and Dalrymple 1987). After decades of scientific study, nothing of what is known about the formation and history of the Hawaiian–Emperor volcanic chain, or about the history of the Pacific plate, is consistent with an earth whose age is only 6,000 to 10,000 years old.

Ages of the Galaxy and the Universe

Hubble Expansion

All of the objects that can be dated by isotopic techniques come from the rocky bodies of the inner solar system, so quite different methods have to be employed to measure the ages of the galaxy and universe. One of these takes advantage of the expanding universe.

In the late 1920s, astronomer Edwin Hubble observed that distant galaxies were moving away from our Milky Way in every direction, and the more distant the galaxies, the faster they were receding (Hubble 1929). This indicated to Hubble that the universe was expanding (or inflating). Hubble realized that the velocity-distance relationship—subsequently named "the Hubble constant" in his honor—could be extrapolated backward to the time when all matter was at a single point, and that such a calculation would provide an estimate of the time that has passed since the expansion began (the Big Bang). Since Hubble's discovery, many astronomers have verified and refined his observations so that the expansion of the universe is not in doubt.

There are several uncertainties involved in calculating an age for the universe from velocity-distance measurements. Galaxies do not move in a single direction but have "local motions" imposed on their expansion vectors. Thus, the measurements can only be made on distant galaxies, where the perturbations caused by local motions are negligible. Velocity measurements are relatively easy to make accurately using the progressive shift in the color of light from receding stars toward the red with increasing velocity (the "red shift," or Doppler effect), but distances in the universe are very hard to measure with similar accuracy. Astronomers use a variety of "distance rulers," including the brightness-period relationship of Cepheid variable stars, the brightness of Type Ia supernovas, and the double image caused when light passes through a strong gravitational field (gravitational lensing)—to name a few. None of the distance rulers in use, however, is perfect, and even the best distance measurements have errors of 10 to 20 percent or more.

The calculation of an age for the universe based on the Hubble constant involves other uncertainties as well. One is whether the expansion has been uniform over time or has been accelerating or decelerating—which depends on the gravity produced by the mass of the universe, a quantity that is not perfectly known. Naturally, gravity acts to slow the expansion. Another important variable is the magnitude of the vacuum energy (or dark energy) of the universe, a pervasive and constant repulsive force that opposes gravity and acts to accelerate expansion. Current data indicate that the expansion of the universe initially decelerated but then began to accelerate about six billion years ago as gravitational attraction lessened and vacuum energy began to dominate.

Despite these uncertainties, the Hubble expansion provides a basis for a good estimate of the age of the universe. Over the past few decades, there have been many estimates made of the age of the universe using the Hubble expansion, resulting in ages of 7 to 20 Ga. The most recent estimates, however, are in the range of 13 to 15 (± 2) Ga (Freedman 1998; Lineweaver 1999). Obviously, none of these estimates is even close to the ages proposed by YECs.

Age of the Milky Way Galaxy

The age of the Milky Way galaxy, in which our solar system resides, has been estimated in a couple of ways. One involves the ages of the oldest stars. Surrounding the center of the galaxy are several hundred star clusters—called globular clusters because of their quasi-spherical shape—each containing thousands of stars. Because older stars contain mostly hydrogen and helium, the low amount of heavy elements in globular cluster stars tell us that these are some of the oldest stars in the galaxy. Stars of different mass consume their nuclear fuel and evolve at different rates. The most massive stars last less than 10 million years, whereas ordinary stars like the sun last billions of years. As stars evolve, their brightness and color also change, so astronomers can measure the extent of evolution of stars of different mass in a globular cluster. This information can then be compared with theoretical models of star evolution to estimate the age of the globular cluster. Recent estimates of the ages of the oldest globular clusters are about 13.5 ± 2 Ga, which is a minimum age for the Milky Way galaxy (Jimenez 1998).

Another minimum age for the galaxy is based on the temperatures of white dwarfs in the globular cluster M4. White dwarfs are very small, dense, hot, and dim former stars that have exhausted their nuclear fuel and are now cooling. Based on the time required for the coolest white dwarfs in M4 to evolve to their present temperatures, the Milky Way galaxy must be at least 12.7 ± 0.7 Ga (Hansen et al. 2002)

The relative abundances in the solar system of certain long-lived radioactive isotopes, particularly those of uranium and thorium, can be used to estimate the age of the galaxy. These abundances are a function of their production rates (nucleosynthesis) in supernovas, which occur as certain stars "explode." Supernovas not only manufacture but also distribute elements widely, where later they may be incorporated into other stars. Recent estimates of the age of the elements from such calculations—assuming an approximately uniform rate of nucleosynthesis since the Milky Way galaxy formed—yield a value of 12.8 ± 3 Ga (Truran 1998).

Summary

It should be clear from this brief summary that earth, the solar system, the Milky Way galaxy, and the universe are billions of years old. The evidence is abundant, consistent, and convincing that Earth is slightly older than 4.5 Ga; that the moon is of similar age and the sun is 4.5 Ga; that the oldest meteorites are 4.56 Ga; that the elements in the galaxy are 13 ± 3 Ga and the oldest stars in our galaxy's globular clusters are 13.5 ± 2 Ga; and that the universe is 14 ± 3 Ga. In addition, we know a great deal about the geologic history of earth and moon from three centuries of careful scientific study, and *that* history is consistent with an earth billions of years old and inconsistent with the notion that earth is only a few thousand years old. Although geologists, physicists, and astronomers are still working to improve our knowledge of the universe, there is no doubt within the scientific community that earth and our cosmic surroundings are ancient.

Despite all of the scientific evidence, the YECs continue to insist that earth and the universe are no more than 6,000 to 10,000 years old. This differs from the ages measured by scientists by factors of more than 750,000 for earth and more than 2,000,000 for the universe. And what is their scientific evidence for this astounding claim? They have none. The best they can do is to attempt to discredit the legitimate findings and methods of science—an illogical task for which failure is inevitable—and to advance ad hoc pseudoscientific scenarios to explain the facts and findings of science that offend them. While it is merely ignorant to hold a belief that earth and the universe are only a few thousand years old, it is fraudulent to insist, as the YECs do, that such a belief is scientific or is supported by scientific data when it clearly is not. The public—and especially young science students—should not be victimized by teaching "creation science" as if it were legitimate science. Our children deserve better.

Acknowledgments

I am grateful to my colleagues Evelyn Sherr, Barry Sherr, and Robert Fleck, who reviewed the manuscript and suggested numerous improvements.

References

Aguillard v. Treen 1985. 634 F. Supp. 436, E.D. Louisiana, 1985.

Albritton, C. C., Jr. 1980. *The abyss of time.* San Francisco: Freeman, Cooper.

Allègre, C. J., G. Manhès, and C. Göpel. 1995. The age of the earth. *Geochimica et Cosmochimica Acta* 59: 1445–56.

Arndts, R., and W. Overn. 1981. Isochrons. *Bible-Science Newsletter* 14: 5–6.

Austin, S. A. 1996. Excess argon within mineral concentrates from the new dacite lava dome at Mount St. Helens volcano. *Creation Ex Nihilo Technical Journal* 10: 335–43.

Austin, S. A., and R. D. Humphreys. 1991. The sea's missing salt: A dilemma for evolutionists. *Proceedings of the Second International Conference on Creationism* 2: 17–33.

Barnes, T. G. 1983. *Origin and destiny of the earth's magnetic field.* ICR Technical Monograph No. 4. El Cajon, CA: Institute for Creation Research.

Bowring, S. A., and I. S. Williams. 1999. Priscoan (4.00–4.03 Ga) orthogneisses from northwestern Canada. *Contributions to Mineralogy and Petrology* 134: 3–16.

Brush, S. G. 1982. Finding the age of the earth, by physics or by faith? *Journal of Geological Education* 30: 34–58.

Clague, D. A., and G. B. Dalrymple. 1987. The Hawaiian–Emperor volcanic chain. Part I: Geological Evolution. *U.S. Geological Survey Professional Paper* 1350: 5–54.

Dalrymple, G. B. 1983. Can the earth be dated from decay of its magnetic field? *Journal of Geological Education* 31: 124–33.

———. 1984. How old is the earth? A reply to "scientific" creationism. In *Proceedings of the 63rd Annual Meeting, Pacific Division, American Association for the Advancement of Science*, vol. 1, ed. F. Awbrey and W. Thwaites, 67–131.

———. 1991. *The age of the earth.* Stanford, CA: Stanford University Press.

———. 2000. Radiometric dating does work! Some examples and a critique of a failed creationist strategy. *Reports of the National Center for Science Education* 20 (3): 14–19.

———. 2004. *Ancient earth, ancient skies.* Stanford, CA: Stanford University Press.

Dicken, A. P. 1995. *Radiogenic isotope geology.* Cambridge: Cambridge University Press.

Edwards v. Aguillard. 1987. 482 U.S. 578.

Freedman, W. L. 1998. Measuring cosmological parameters. *Proceedings of the National Academy of Sciences USA*, 95: 2–7.

Gentry, R. V. 1986. *Creation's tiny mystery.* Knoxville, TN: Earth Science Associates.

Gilkey, L. 1985. *Creationism on trial.* Minneapolis: Winston Press.

Gill, C. H. 1996. A sufficient reason for false Rb–Sr isochrons." *Creation Research Society Quarterly* 33: 105–8.

Glen, W. 1982. *The road to Jaramillo.* Stanford, CA: Stanford University Press.

Guenther, D.B., and P. Demarque. 1997. Seismic tests of the sun's interior structure, composition, and age, and implications for solar neutrinos. *The Astrophysical Journal* 484: 937–59.

Halliday, A.N., and D.-C. Lee. 1999. Tungsten isotopes and the early development of the earth and moon. *Geochimica et Cosmochimica Acta* 63: 4157–79.

Hansen, B.M.S., J. Brauer, G. G. Fahlman, B. K. Gibson, R. Ibata, M. Limongi, F. M. Rich, H. B. Richer, M. M. Shara, and P. B. Statson. 2002. The white dwarf cooling sequence of the globular cluster Messier 4. Available at http://xxx.lanl.gov/pdf/astro-ph/0205087.

Harland, W.B., R. L. Armstrong, A. V. Cox, L. E. Craig, A. G. Smith, and D. G. Smith. 1989. *A geologic time scale.* Cambridge and New York: Cambridge University Press.

Hofmann, A. W. 2001. Lead isotopes and the age of the earth—a geochemical accident. In *The age of the earth: From 4004 B.C. to A.D. 2002*, ed. C. L. E. Lewis and S. J. Knell, 223–36. London: Geological Society [London] Special Publication 190.

Hubble, E. P. 1929. A relation between distance and radial velocity among extragalactic nebulae. *Proceedings of the National Academy of Sciences* 15: 168–73.

Humphreys, D. R. 1989. The mystery of the earth's magnetic field. *Institute for Creation Research Impact* 188: i–iv.

———. 1993. The earth's magnetic field is young. *Institute for Creation Research Impact* 242: i–iv.

———. 2000. Accelerated nuclear decay: A viable hypothesis? In *Radioisotopes and the age of the earth,* ed. L. Vardiman, A. A. Snelling, and E. F. Chaffin. El Cajon, CA: Institute for Creation Research; St. Joseph, MO: Creation Research Society.

Jimenez, R. 1998. Globular cluster ages. *Proceedings of the National Academy of Sciences USA* 95: 13–17.

Lewis, C. L. E., and S. J. Knell, eds. 2001. *The age of the earth: From 4004 B.C. to A.D. 2002.* Geological Society [London] Special Publication 190.

Lineweaver, C. H. 1999. A younger age for the universe. Science 284: 1503–7.

Matson, D. E. 1999. How good are those young-earth arguments? Available from www.infidels.org/library/modern/dave_matson/ young-earth/.

McCulloch, M. T. 1996. Isotopic constraints on the age and early differentiation of the earth. *Journal of the Royal Society of Western Australia* 79: 131–39.

McLean v. Arkansas. 1982. 529 F. Supp. 1255, E.D. Arkansas.

Meert, J. 2000. Consistent radiometric dates. Available from http://gondwanaresearch.com/radiomet.htm.

Morris, H. M. 1974. *Scientific creationism* (General Edition). San Diego: Master Books.

———. 1985. *Scientific creationism,* 2nd ed. San Diego: Creation-Life Publishers.

Morris, J. D. 1994. *The young earth.* Colorado Springs: Creation-Life Publishers.

Morton, G. R. 1998. The sea's salt. Available from http://home.entouch.net/dmd/salt.htm.

Ross, H. 1991. *The fingerprint of God,* 2nd ed. Orange, CA: Promise Publishing Co.

Rugg, S., and S. A. Austin. 1998. Evidence for rapid formation and failure of pleistocene "lava dams" of the western Grand Canyon, Arizona. In *Proceedings of the Fourth International Conference on Creationism,* ed. R. E. Walsh, 475–86. Pittsburgh: Creation Science Fellowship.

Schimmrich, S. H. 1998. Geochronology *kata* John Woodmorappe. Available from www.talkorigins.org/faqs/woodmorappe-geochrono logy.html.

Schopf, J. W. 1999. *Cradle of life.* Princeton: Princeton University Press.

Slusher, H. S. 1981. *Critique of radiometric dating,* 2nd ed. San Diego: Institute for Creation Research.

Stassen, C. 1997. *The age of the earth.* Available from www.talkorigins.org/faqs /faq-age-of-earth.html.

Steering Committee on Science and Creationism. 1999. *Science and Creationism: A View from the National Academy of Sciences.* Washington, DC: National Academy Press.

Strahler, A. N. 1987. Science and earth history: The evolution/creation controversy. Buffalo: Prometheus Books.

Tera, F. 1980. Reassessment of the "age of the earth." *Carnegie Institution of Washington Year Book* 79: 524–31.

———. 1981. Aspects of isochronism in Pb isotope systematics—application to planetary evolution. *Geochimica et Cosmochimica Acta* 45: 1439–48.

Tera, F., and R. W. Carlson. 1999. Assessment of the Pb–Pb and U–Pb chronometry of the early solar system. *Geochimica et Cosmochimica Acta* 63: 1877–89.

Truran, J. W. 1998. The age of the universe from nuclear chronometers. *Proceedings of the National Academy of Sciences USA* 95: 18–21.

Vardiman, L. 1985. Up, up and away! The helium escape problem. *Institute for Creation Research Impact* 143: i–iv.

———. 1990. *The age of earth's atmosphere.* El Cajon, CA: Institute for Creation Research.

———. 2000. Introduction. In *Radioisotopes and the age of the earth,* ed. L. Vardiman, A. A. Snelling, and E. F. Chaffin. El Cajon, CA: Institute for Creation Research; St. Joseph, MO: Creation Research Society.

Wakefield, J. R. 1988. The geology of Gentry's "tiny mystery." *Journal of Geological Education* 36: 161–75.

Whitcomb, J. C., and H. M. Morris. 1961. *The Genesis Flood.* Philadelphia: Presbyterian and Reformed Publishing Co.

Wilde, S. A., J. W. Valley, H. W. Peck, and C. M. Graham. 2001. Evidence from detrital zircons for the existence of continental crust and oceans on the earth 4.4 ga ago. *Nature* 409: 175–78.

Woodmorappe, J. 1979. Radiometric geochronology reappraised. *Creation Research Society Quarterly* 16: 102–29, 147.

———. 1999. *The mythology of modern dating methods.* El Cajon, CA: Institute for Creation Research.

York, D., and G. B. Dalrymple. 2000. Comments on a creationist's irrelevant discussion of isochrons. *Reports of the National Center for Science Education* 20 (3): 18–20, 25–27.

Young, D. A. 1982. *Christianity and the age of the earth.* Grand Rapids, MI: Zondervan Corp.

Creationism and the Origin of Life: Did It All Begin in a "Warm Little Pond"?[1]

Antonio Lazcano

Introduction: Creationism and Evolutionary Theory

"For more than a thousand years," wrote Thomas H. Huxley in 1843 in the preface of his book *Science and Hebrew Tradition*:

> . . . the great majority of the most highly civilised and instructed nations in the world have confidently believed and passionately maintained that certain writings, which they entitle sacred, occupy a unique position in literature, in that they possess an authority, different in kind, and immeasurably superior in weight, to that of all other books. Age after age, they have held it to be an indisputable truth that, whoever may be the ostensible writers of the Jewish, Christian, and Mahometan scriptures, God Himself is their real author; and, since their conception of the attributes of the Deity excludes the possibility of error and—at least in relation to this particular matter—of wilful deception, they have drawn the logical conclusion that the denier of the accuracy of any statement, the questioner of the binding force of any command, to be found in these documents

1. Some of the material used in this chapter was derived from an article in the journal *Science*: Teaching Evolution in Mexico: Preaching to the Choir. *Science* (November 4, 2005) 310: 787–89.

> is not merely a fool, but a blasphemer. From the point of view of mere reason he grossly blunders; from that of religion he grievously sins.

What Huxley wrote in the nineteenth century still holds true: Literalism is found in every contemporary society. In no place, however, is this more evident than in the United States, where for the past decades modern evolutionary theory has been challenged over and over again by the sectarianism and religious fundamentalism of an outspoken and politically active creationist movement. Such attitudes are found in Australia, the United Kingdom, and in the Islamic world (Numbers 1998). Attempts to understand the emergence and evolution of life by teaching beliefs derived from the first two chapters of Genesis in the Bible (as opposed to current scientific descriptions) are also found among the more conservative American churchgoers, even when their own denomination has officially come out in support of evolution (for example, see Matsumura 1995). Of course, the idea of a supernatural origin of life is shared by many believers who are far from being strict creationists, but it is also true that in many Spanish-speaking countries of Latin America, most Catholics follow a tradition that goes back to Augustine of Hippo, and they take the Bible's work not as the literal truth but as a depiction of the ways in which divine creation took place.

Creationism and the Origin of Life

From the unsophisticated anti-Darwinian speeches of the 1920s to the general denunciation of the links of evolution as atheist, com munist, and fascist during the Cold War (Habgood 1982; Gilbert 1997), fundamentalist creationism has changed over past decades, but its essence remains unchanged. In its newer, recycled version, it has eliminated overt references to Christianity (Pennock 1999), introduced new labels such as "intelligent design" (see Pennock, in this volume), and addressed evolutionary issues formerly ignored, particularly the origin of life and biochemical evolution.

Indeed, the emergence of life is becoming a central target for the anti-evolutionists' new attacks. Although the "intelligent design" argument took form recently, it was twenty years ago that the proponents of anti-evolutionary positions began to argue that the intricacies of cellular molecular components could not be explained by purely natural processes—an extension of their earlier, failed objections to complex functional anatomical structures. Perhaps one of the first examples was *The Mystery of Life's Origin: Reassessing Current Theories* (Thaxton, Bradley, and Olsen 1984). It was followed a dozen years later by Michael J. Behe's *Darwin's Black Box: The Biochemical Challenge to Evolution* (1996), which argued that the finely-tuned metabolic pathways found in extant cells could not have been produced by an evolutionary process involving small, successive modifications of simpler systems, since any precursor route that was missing a single component could not function. The existence of such "irreducible complexity"—that is, "the purposeful arrangements of parts"—hastily concluded Behe, implied an intelligent designer.

In a remarkably candid statement, Francis H. C. Crick once wrote in an application for a student research fellowship that he was impressed by the possibility of explaining fundamental biological problems based solely on physics and chemistry and that, being an atheist, he wanted "to try to show that areas apparently too mysterious to be explained by physics and chemistry, could in fact be so explained" (Clark 1986). This goal has not yet been fulfilled, but there is no evidence indicating an ultimate supernatural origin of the basic biological phenomena. As the 1967 Nobel laureate Manfred Eigen wrote many years afterward:

> the biosynthesis of the living cell . . . is admittedly complex, but it is completely interpretable within the scope of our present-day physical and chemical knowledge. This does not mean that everything in this realm has been investigated exhaustively; instead, it means that what has been investigated has turned out not to be mysterious, but to be explicable by physics and chemistry. What we know makes it probable that there were also sim-

> pler, less efficient mechanisms that were realizable under prebiotic conditions. (Eigen 1992)

But how can a theory on the origin of life be validated? By necessity, work on the appearance of the first life forms should be regarded as inquiring and explanatory rather than definitive and conclusive. This does not imply that our theories and explanations can be dismissed as pure speculation, but rather that the issue should be addressed conjecturally, in an attempt to construct a coherent, nonteleological historical narrative (Kamminga 1991). It is unlikely that the origin of life will ever be described in full detail; at best, a sketchy outline will be constructed—one that is consistent with conditions on the prebiotic earth (such as its anoxic environment) and the physicochemical properties of the likely molecular precursors of living systems.

Can the Emergence of Life Be Understood?

"All the organic beings which have ever lived on this earth," wrote Charles Darwin in *On the Origin of Species*, "may be descended from some primordial form." Although the placement of the root of universal trees is a matter of debate, the development of molecular cladistics has shown that, despite their overwhelming diversity and tremendous differences, all organisms are ultimately related and descend from Darwin's primordial ancestor. But what was the nature of this progenitor? How and when did it come into being?

As shown by the recent debates, determination of the biological origin of what have been considered the earliest traces of life can be a rather contentious issue (van Zullen et al. 2002). The geological record of the early Archaean period is scarce, and most of the rocks that have been preserved have been metamorphosed to a considerable extent. However, there is evidence that life emerged on earth as soon as it was possible to do so. It is true that the biological origin of the microstructures interpreted as cyanobacterial remnants in the 3.5-billion-year-old Apex sediments of the Australian Warrawoona formation (Schopf 1993) have been challenged (Brasier et al. 2002), and it is possible that they may be the outcome of nonbiological processes. However, several Archaean stromatolitic horizons and other microstructures clearly exhibit the diagnostic features of a microbial community associated with a sea-floor ecosystem characteristic of those supported by hydrothermal vents (Van Kranendonk 2002).

Although traditionally it had been assumed that the origin and early evolution of life involved several billions of years (Oparin 1938; Dickerson 1978), such views are no longer tenable. While it is true that it is not possible to assign a precise chronology to the appearance of life, estimates of the available time for this to have occurred have been considerably reduced in the last few years. The planet is generally thought to have remained molten for several hundred million years after its formation 4.6 billion years ago (Wetherill 1990), and late accretion impacts may have boiled-off the oceans and destroyed all life in the planet as late as 3.8 billion years ago (Sleep et al. 1989). However, there is compelling paleontological evidence that highly diverse microbial communities were thriving during the early and middle Archaean periods, 3.3 to 3.8 billion years ago (Nisbet and Sleep 2001; Schopf et al. 2002; Van Kranendonk 2002).

Such rapid development speaks for the relatively short time scale required for the origin and early evolution of life on earth and suggests that the critical factor may have been the presence of liquid water—which became available as soon as the planet's surface finally cooled down. However, it is unlikely that data on how life actually originated will be provided by the paleontological record. There is no geological evidence of the environmental conditions on the earth at the time of the origin of life, nor any fossil register of the evolutionary processes that preceded the appearance of the first cells. Direct information is lacking not only on the composition of the terrestrial atmosphere during the period of the origin of life but also on the temperature, ocean pH values, and other general and local environmental conditions that may or may not have been important for the emergence of living systems.

Furthermore, the lack of an all-embracing, generally agreed definition of life sometimes gives the impression that the study of its origin is couched in somewhat imprecise terms, and that this lack of precision renders such research both meaningless and unscientific—especially when new research contradicts or overturns older models and hypotheses. For instance, until a few years ago, the origins of the genetic code and of protein synthesis were considered synonymous with the appearance of life itself. This is no longer a dominant point of view; the discovery of catalytic RNA (Kruger et al. 1982: Guerrier-Takada et al. 1983) and the *in vitro* development of ribozymes (Joyce 2002) have given considerable support to the idea of the "RNA world"—a hypothetical stage before the development of proteins and DNA genomes during which alternative life forms based on ribozymes existed (Gesteland and Atkins 1993). This has led many to argue that the starting point for the history of life on earth was the *de novo* emergence of the RNA world from a nucleotide-rich prebiotic soup. Others with a more skeptical view believe that it lies in the origin of cryptic and largely unknown pre-RNA worlds.

Despite the seemingly insurmountable obstacles surrounding the understanding of the origin of life (or perhaps because of them), there has been no shortage of discussion about how it

I. ABIOTIC SYNTHESIS AND HETEROTROPHIC ORIGIN	
Oparin (1938)	Primitive soup and primordial fermentation
Corliss et al. (1981)	Submarine hot-spring thermophilic heterotroph
Gilbert (1986)	The RNA world
De Duve (1991)	Thioesther world
Kauffman (1993)	Self-organization and complexity theory
II. PRIMORDIAL CO_2 FIXATION AND AUTOTROPHIC ORIGIN	
Wächtershäuser (1988)	Pyrite-based chemolithotrophic metabolic networks

Note: See references at the end of chapter for full bibliographic details.

Table 1. Some Current Hypotheses on the Origin of Life

took place (Lazcano 2001). Not surprisingly, several alternative and competing theories attempting to explain the origin of life coexist today, including proposals on RNA or thioesther worlds; on the role of submarine hydrothermal vents; and on the ultimate extraterrestrial origin of organic compounds (table 1). Günther Wächtershäuser (1988, 1992) favors the possibility that life began with the appearance of self-sustaining autocatalytic metabolic reaction networks in which no genetic material was present but that were capable of fixing CO_2, based on the formation of the highly insoluble mineral pyrite (FeS_2). With the exception of this autotrophic proposal, most researchers favor the idea of the prebiotic synthesis of organic compounds and the formation of the so-called primitive soup.

Darwin's Warm Little Pond: Have Too Many Cooks Spoiled the Soup?

On February 1, 1871, Charles Darwin wrote to his friend Joseph Dalton Hooker:

> . . . it is often said that all the conditions for the first production of a living organism are now present, which could ever have been present. But if (and oh! what a big if!) we could conceive in some warm little pond with all sorts of ammonia and phosphoric salts, light, heat, electricity &c present, that a protein compound was chemically formed ready to undergo still more complex changes, at the present day such matter would be instantly devoured or absorbed, which would not have been the case before living creatures were formed. . . . (van Wyhe 2002)

By the time Darwin wrote to Hooker, DNA had already been discovered, although its central role in genetic processes would remain unknown for many years to come. In contrast, the part that proteins play in manifold biological processes had already been firmly established, and major advances had been made in the discovery and chemical characterization of many of the building blocks of life. Equally significant, at the time that Darwin wrote his letter, a number of chemical syntheses—beginning from the work of Friedrich Wöhler on the formation of urea from ammonium cyanate—had shown that the chemical gap separating organisms from the nonliving had been bridged at least in part by the laboratory syntheses of organic molecules, which for a long time had been considered to be fundamentally different from inorganic compounds. However, neither Wöhler nor other scientists involved in these syntheses, such as Strecker, Butlerow, or Mendeleyev, recognized the evolutionary significance of their investigations.

Although the idea of life as an emergent feature of nature has been widespread since the nineteenth century, the field really began to take shape in the 1920s. The major methodological breakthrough that transformed origin-of-life studies from a purely speculative field into a workable research program was the outcome of the independent proposals by A. I. Oparin and J. B. S. Haldane. This work was based on the idea that the first life forms were the outcome of a slow, multistep process that began with the abiotic synthesis of organic compounds and the formation of a "primitive soup." There followed the formation of colloidal gel-like systems, from which anaerobic heterotrophs evolved that could take up surrounding organic compounds and use them directly for growth and reproduction.

Many of Oparin's and Haldane's original ideas have been superseded, but their hypothesis provided a conceptual framework for the development of this field. This proposal became widely accepted—not only because it is simpler to envision a heterotrophic organism originating from organic molecules of abiotic origin than an autotroph, but also because laboratory experiments have shown how easy it is to produce a number of biochemical monomers under reducing conditions. Such empirical support began to accumulate in 1953, when Stanley L. Miller, then a graduate student working with Harold C. Urey, achieved the first successful synthesis of organic compounds under plausible primordial conditions. The action of electric discharges acting for a week over a mixture of methane (CH_4), ammonia (NH_3), hydrogen (H_2), and water (H_2O) produced racemic mixtures of several proteinic amino acids, as well as hydroxy acids, urea, and other organic molecules (Miller 1953). This was followed a few years later by the work of Juan Oró, then at the University of Houston, who demonstrated the rapid adenine synthesis by the aqueous polymerization of hydrogen cyanide (HCN) (Oró 1961). The potential role of HCN as a precursor in prebiotic chemistry has been supported by the discovery that the hydrolytic products of its polymers include amino acids, purines, and orotic acid—a biosynthetic precursor of uracil, a nucleotide that makes up RNA. A potential prebiotic route for the synthesis of cytosine in high yields is provided by the reaction of cyanoacetylene with urea, specially when the concentration of the latter is increased by evaporation—the laboratory simulation, in fact, of Darwin's "warm little pond" (cf. Miller and Lazcano 2002).

The ease of formation of amino acids (to produce proteins), and of purines and pyrimidines (the chief components of RNA and DNA) under reducing atmospheres ($CH_4 + N_2$, $NH_3 + H_2O$, or $CO_2 + H_2 + N_2$) in one-pot reactions, strongly suggested that these molecules were components of the "prebiotic broth." These compounds are essential to the construction of proteins and nucleic acids. These experiments also suggested that it is highly likely that these would be joined by many other compounds found in living things, such as urea and carboxylic acids; sugars formed by the nonenzymatic condensation of formaldehyde; a wide variety of aliphatic and aromatic hydrocarbons; alcohols; and branched and straight fatty acids, including some that are membrane-forming compounds.

The above reactions are effective under reducing conditions, but not if a neutral atmosphere is employed. The possibility that the prebiotic atmosphere was nonreducing ($CO_2 + N_2 + H_2O$) does not create insurmountable problems, since the primitive "soup" could still form, albeit from other sources. For instance, geologically generated hydrogen may have been available: in the presence of ferrous iron, a sulfide ion (SH^-) would have been converted to a disulfide ion (S_{2-}), thereby releasing molecular hydrogen (Wächtershäuser 1988, 1992; Maden 1995). It is also possible that the impacts of iron-rich asteroids enhanced the reducing conditions and that cometary collisions created localized environments favoring organic synthesis. Based on what is known about prebiotic chemistry and meteorite composition, if the primitive earth had been nonreducing, then the organic compounds required for the emergence of life must have been brought in by interplanetary dust particles, comets, and meteorites. A significant percentage of meteoritic amino acids and nucleobases could survive the high temperatures associated with frictional heating during atmospheric entry and become part of the "primitive broth."

The Emergence of Biological Order: "Intelligent Design" or Evolution?

The synthesis of life's chemical constituents by nonenzymatic processes under laboratory conditions does not necessarily imply that these constituents were either essential for the origin of life or available in the primitive environment. However, these experiments do show that a number of naturalistic pathways for the emergence of the first life are possible. The significance of prebiotic simulation experiments is further enhanced by the occurrence of a large array of protein and nonprotein amino acids, carboxylic acids, purines, pyrimidines, hydrocarbons, and other molecules in the 4.6-billion-year-old Murchison meteorite (which fell on Australia in 1969), a carbonaceous chondrite that provides direct information about the chemistry of the early solar system and also yields evidence of liquid water (Becker et al. 2002). The presence of these compounds in the meteorite makes it plausible, but does not prove, that a similar synthesis took place on the primitive earth.

The fact that a number of chemical constituents of *contemporary* forms of life can be synthesized nonenzymatically under laboratory conditions does not necessarily imply by itself that they were also essential for the *origin* of life or that they were available in the primitive environment. Laboratory experiments suggest that the "prebiotic soup" must have been a bewildering organic chemical wonderland, but it could not include all the compounds or molecular structures found today even in the most ancient extant forms of life.

How, then, is it possible to imagine that the finely tuned, exquisitely organized cells evolved from such a complex mixture? In 1871, St. George Mivart, a Catholic convert and former student of Thomas Huxley (Ruse 1997), wrote *The Genesis of Species*, in which he questioned Darwin's mechanism in the emergence of evolutionary innovations. Prefiguring many modern anti-evolutionsts, he wrote: "'Natural Selection' is incompetent to account for the incipient stages of useful structures" (cf. Young 1992). By the same token, "intelligent design" advocates have argued that the intricate and complex molecular structures of subcellular components could not have begun through the evolutionary stepwise preservation and accumulation of successive variations. Their ultimate origin, they conclude, must be the work of an intelligent agent working outside the natural processes of biological variation, survival, and reproduction.

Theirs is a premature conclusion. The organic monomers of abiotic origin described in the previous sections would have accumulated in the primitive environment, providing the raw material for subsequent reactions. As shown by numerous experiments, clays, metal cations, organic compounds bearing highly reactive derivatives of HCN (such as cyanamide, dicyanamide, and cyanogen), or imidazole derivatives may have catalyzed polymerization reactions (Wills and Bada 2000). Selective absorption of molecules onto mineral surfaces has been shown to produce a chain of up to fifty-three nucleotides (Ferris et al. 1996), and other processes such as evaporation of tidal lagoons (Wills and Bada 2000) and eutectic freezing of dilute aqueous solutions (Kanavarioti et al. 2001) could have also assisted in concentrating organic precursors. (Compounds form more easily in concentrated solutions than in dilute ones.)

It is easy to imagine that as polymerized molecules became larger and more complex, some of them began to fold into configurations that could bind and interact with other molecules, thus expanding the list of primitive catalysts that could promote nonenzymatic reactions. Some of these catalytic reactions, especially those involving hydrogen-bond formation, may have assisted in making polymerization more efficient. As the variety of polymeric combinations increased, some compounds could have developed the ability to catalyze their own imperfect self-replication and that of related molecules—as some RNA molecules have now been shown to do (Uhlenbeck 1987). This could have marked the first molecular entities capable of multiplication, heredity, and variation, and thus the origin of both life and evolution (Bada and Lazcano 2002).

The above scheme is necessarily speculative, but its intrinsic heuristic value cannot be overemphasized. The possibility of simple prebiotic catalytic polymers is supported by the discovery of the rapid, highly specific cleavage reaction catalyzed by a small, synthetic 19-nucleotide RNA molecule under physiological conditions (Uhlenbeck 1987), and by the numerous examples of *in vitro* evolution experiments in which manifold catalytic activities—including peptide-bond formation and template-directed polymerization of ribonucleotides—are achieved when starting from random mixtures of RNA (Joyce 2002). It is possible that RNA itself was preceded by simpler genetic macromolecules lacking only the familiar 3',5' phosphodiester backbones of nucleic acids (Bada and Lazcano 2002). Nevertheless, experiments with ribozymes appear to support the possibility that random mixtures of catalytic and replicative macromolecules were available in the primitive earth and provide an excellent laboratory model for understanding the evolutionary transition from the nonliving to the living.

All known organisms share the same essential features of genome replication, gene expression, basic anabolic reactions, and membrane-associated ATPase-mediated energy production. The molecular details of these universal processes not only provide direct evidence of the monophyletic origin of all extant forms of life but also imply that the sets of genes encoding the components of these complex traits were fixed a long time ago—that is, major changes in them are very strongly selected against and are lethal. It is true that no ancient incipient stages or evolutionary intermediates of these molecular structures have been detected, but the existence of graded intermediates can be deduced—rendering unnecessary the supernatural origin advocated by both old-fashioned and contemporary creationists.

For instance, the fact that RNA molecules are capable of performing by themselves all the reactions involved in peptide-bond formation suggests that protein biosynthesis evolved in an RNA world (Zhang and Cech 1998)—that the first ribosome lacked proteins and was formed only by RNA. This possibility is supported by crystallographic data that have shown that the ribosome catalytic site where peptide-bond formation takes place is composed solely of RNA (Nissen et al. 2000). Clues to the genetic organization of primitive forms of translation are also provided by paralogous genes, which are sequences that diverge not through speciation but after a duplication event. For instance, the presence in all known cells of pairs of homologous genes encoding two elongation factors—enzymes that assist in protein biosynthesis—provides evidence of the existence of a more primitive, less-regulated version of protein synthesis that took place with only one elongation factor. Furthermore, the experimental evidence of successful *in vitro* translation systems with modified cationic concentrations that lack both elongation factors and other proteinic components (Gavrilova et al. 1976; Spirin 1986) strongly supports the possibility of an older ancestral protein-synthesis apparatus prior to the emergence of elongation factors.

The same is true of other enzymes. Detection of high levels of genetic redundancy in all sequenced genomes implies not only that duplication has played a major role in the accretion of the complex genomes found in extant cells but also that prior to the early duplication events revealed by the large protein families, simpler living systems existed that lacked the large sets of enzymes and the sophisticated regulatory abilities of contemporary organisms. If the existence of such functional intermediates can be deduced, then it may be concluded that the features of biochemical processes can be the outcome of natural selection acting from the start—a process in which no designer is required. The irreducible biochemical complexity claimed by Michael Behe and others does not exist.

Conclusions

We probably will never know in full detail how the emergence of life occurred. As reviewed here, we have manifold historical records that allow us to reconstruct, with different degrees of precision, the evolutionary processes that underlie these events. Unfortunately, the evidence required to understand the prebiotic environment and the nature of the events that led to the first life forms is scant and difficult to understand. It is not surprising, therefore, that the study of the origin of life is burdened with controversy and endless discussions (Shapiro 1986). Such disagreements do not indicate, as creationists erroneously imply, that a supernatural explanation is the ultimate solution. With few exceptions (see, for instance, Thaxton et al. 1984), acknowledgments of such differences in opinion have been recognized by the scientific community as intellectual challenges and have in most cases led to fruitful clarifying debates. Evidence of scientific ignorance is not evidence for creation, nor does disagreement among scientists in this or any other scientific field demonstrate that the scientific models are invalid.

If we were to accept the supernatural or extranatural proposals of anti-evolutionists, it would provide little useful information to help us understand the history and diversity of life, and it would put an end to all research into the matter. By contrast, mainstream scientific hypotheses on the origin of life—which have been developed within the framework of an evolutionary analysis—have led to a wealth of experimental results and the development of a coherent historical narrative linking many different disciplines and raising major philosophical issues. It is true that there is a huge gap in the current descriptions of the evolutionary transition between the prebiotic synthesis of biochemical compounds and the last common ancestor of all extant living beings, but attempts to reduce this gap have facilitated a more precise definition of what should be understood as the beginning of life. A central issue in contemporary origin-of-life research is to understand the abiotic synthesis of an ancestral genetic polymer endowed with catalytic activity, and its further evolution to an RNA world. We face major unsolved problems in understanding the origin of life, but it is no minor scientific achievement that this important period in the history of life is not completely shrouded in mystery. We have many questions to answer, but, as the Greek poet Konstantinos Kavafis once pointed out, Odysseus should be grateful not because he was able to return home but because of what he learned on his way back to Ithaca. It is the journey that matters.

References

Bada, J. L, and A. Lazcano. 2002. Some like it hot, but not the first biomolecules. *Science* 296: 1982–83.

Becker, L., J. Blank, J. Brucato, L. Colangeli, S. Derenne, D. Despois, A. Dutrey, P. Ehrenfreund, H. Fraaije, W. Irvine, A. Lazcano, T. Owen, and F. Robert. 2002. Astrophysical and astrochemical insights into the origin of life. *Reports on Progress in Physics* 65: R1–R56.

Behe, M. J. 1996. *Darwin's black box: The biochemical challenge to evolution.* New York: Free Press.

Brasier, M., O. R. Green, A. P. Jephcoat, A. K. Kleppe, M. J. van Kranendonk, J. F. Lindsay, A. Steele, and N. V. Grassineau. 2002. Questioning the evidence for earth's earliest fossils. *Nature* 416: 76–79.

Clark, R. W. 1986. *The survival of Charles Darwin: The biography of a man and an idea.* New York: Avon Books.

Dickerson, R. E. 1978. Chemical evolution and the origin of life. *Scientific American* 239: 70–86.

Eigen, M. 1992. *Steps towards life: A perspective on evolution.* Oxford: Oxford University Press.

Ferris, J. P., A. R. Hill, R. Liu, and L. E. Orgel. 1996. Synthesis of long prebiotic oligomers on mineral surfaces. *Nature* 381: 59–61.

Gavrilova, L. P., O. E. Kostiashkina, V. E. Koteliansky, N. M. Rutkevitch, and A. S. Spirin. 1976. Factor-free (non-enzymic) and factor-dependent systems of translation of polyuridylic acid by *Escherichia coli* ribosomes. *Journal of Molecular Biology* 101: 537–52.

Gesteland, R. F., and J. F. Atkins, eds. 1993. *The RNA world: The nature of modern RNA suggests a prebiotic RNA world.* Cold Spring Harbor, NY: Cold Spring Harbor Laboratory Press.

Gilbert, J. 1997. *Redeeming culture: American religion in an age of science.* Chicago: University of Chicago Press.

Guerrier-Takada, C., K. Gardiner, T. Marsh, N. Pace, and S. Altman. 1983. The RNA moiety of ribonuclease P is the catalytic subunit of the enzyme. *Cell* 35: 849–57.

Habgood, J. 1982. Myths of religion, myths of science. *Nature* 300: 118.

Joyce, G. F. 2002. The antiquity of RNA-based evolution. *Nature* 418: 214–21.

Kamminga, H. 1991. The origin of life on earth: Theory, history, and method. *Uroboros* 1: 95–110.

Kanavarioti, A., P. A. Monnard, and D. W. Deamer. 2001. Eutectic phases in ice facilitate nonenzymatic nucleic acid synthesis. *Astrobiology* 1: 271–81.

Kruger, K., P. J. Grabowski, A. J. Zaug, J. Sands, D. E. Gottschling, and T. R. Cech. 1982. Self-splicing RNA: Autoexcision and autocyclization of the ribosomal RNA intervening sequence of Tetrahymena. *Cell* 31: 147–57.

Lazcano, A. 2001. Origin of life. In *Palaeobiology II*, ed. D. E. G. Briggs and P. R. Crowther, 3–8. London: Blackwell Science.

Maden, B. E. H. 1995. No soup for starters? Autotrophy and origins of metabolism. *Trends in Biochemical Sciences* 20: 337–41.

Matsumura, M. 1995. *Voices for evolution*, 2nd ed. Berkeley, CA: National Center for Science Education.

Miller, S. L. 1953. Production of amino acids under possible primitive earth conditions. *Science* 117: 528.

Miller, S. L., and A. Lazcano. 2002. Formation of the building blocks of life. In *Life's origin: The beginnings of biological evolution*, ed. J. W. Schopf, 78–112. Berkeley: University of California Press.

Nisbet, E.G., and N. H. Sleep. 2001. The habitat and nature of early life. *Nature* 409: 1083–91.

Nissen, P., J. Hansen, N. Ban, P. B. Moore, and T. A. Steitz. 2000. The structural basis of ribosome activity in peptide bond synthesis. *Science* 289: 920–30.

Numbers, R. L. 1998. *Darwinism comes to America.* Cambridge, MA: Harvard University Press.

Oparin, A. I. 1938. *The origin of life.* New York: Macmillan.

Oró, J. 1961. Mechanism of synthesis of adenine from hydrogen cyanide under possible primitive earth conditions. *Nature* 191: 1193–94.

Pennock, R. T. 1999. *Tower of Babel: The evidence against the new creationism.* Cambridge, MA: MIT Press.

Ruse, M. 1997. John Paul II and evolution. *Quarterly Review of Biology* 72: 391–95.

Schopf, J. W. 1993. Microfossils of the early Archaean Apex chert: New evidence for the antiquity of life. *Science* 260: 640–46.

Schopf, J. W., A. B. Kudryavtsev, D. G. Agresti, T. J. Wdowiak, and A. D. Czaja. 2002. Laser-Raman imagery of earth's earliest fossils. *Nature* 416: 73–76.

Shapiro, R. 1986. *Origins: A skeptic's guide to the creation of life on earth.* New York: Bantam Books.

Sleep, N. H., K. J. Zahnle, J. F. Kasting, and H. J. Morowitz. 1989. Annihilation of ecosystems by large asteroid impacts on the early earth. *Nature* 342: 139–42.

Spirin, A. S. 1986. *Ribosome structure and protein synthesis.* Menlo Park, CA: Benjamin/Cummings.

Thaxton, C. B., W. L. Bradley, and R. L. Olsen. 1984. *The mystery of life's origin: Reassessing current theories.* New York: Philosophical Library.

Uhlenbeck, O. C. 1987. A small catalytic oligoribonucleotide. *Nature* 328: 596–600.

Van Kranendonk, M. J. 2002. The flourishing of early life on earth at hydrothermal vents: Geological evidence from the 3.49–3.43 Ga Warrawoona Group, Pilbara Craton, Western Australia. *Abstracts of the IAU Symposium 213 Bioastronomy 2002: Life among the Stars (Australian Centre for Astrobiology, Hamilton Island, Great Barrier Reef, Australia, July 8–12)*: p. 33.

Van Wyhe, J., ed. 2002. *The writings of Charles Darwin on the Web.* http://pages.britishlibrary.net/charles.darwin/ (last accessed July 13, 2006).

Van Zullen, M. A., A. Lepland, and G. Arrhenius. 2002. Reassessing the evidence for the earliest traces of life. *Nature* 418: 627–30.

Wächtershäuser, G. 1988. Before enzymes and templates: Theory of surface metabolism. *Microbiological Reviews* 52: 452–84.

———. 1992. Groundworks for an evolutionary biochemistry: The iron-sulphur world. *Progress in Biophysics and Molecular Biology* 58: 85–201.

Wetherill, G. W. 1990. Formation of the earth. *Annual Review of Earth and Planetary Sciences* 18: 205–56.

Wills, C., and J. Bada. 2000. *The spark of life: Darwin and the primeval soup.* Cambridge, MA: Perseus.

Young, D. 1992. *The discovery of evolution.* Cambridge: Cambridge University Press.

Zhang B., and T. R. Cech. 1998. Peptidyl-transferase ribozymes: Trans reactions, structural characterization and ribosomal RNA-like features. *Chemistry & Biology* 5: 539–53.

"Transitional Forms" versus Transitional Features

Kevin Padian & Kenneth D. Angielczyk

Introduction

Many creationists say that they accept microevolution, or change within species, but not macroevolution, defined as the transmutation of forms or the evolution of major clades and body plans. The study of changes in evolution on the macro level relies largely on the evidence of extinct organisms known only from the fossil record and from the phylogenetic relationships of those organisms, usually represented by cladograms. Increasingly, new mechanisms from evolutionary developmental biology provide the explanations of these patterns (e.g., Hall 1998). Typically, creationists disdain evolutionary explanations of transitions between what often seem like "unbridgeable" gaps between living groups, or between fossil and extant species. They commonly claim that appropriate intermediate forms, or "missing links," have not been found, hence there is no evidence for the transition. Here we show that (1) the concept of "missing link" is not only an archaic expression but also an outmoded approach to studying macroevolution; (2) the focus shifted decades ago from finding "transitional taxa" to finding "transitional features" shared by closely related forms with common ancestry; and (3) many such examples are well known today, and they illuminate not only the transition of features but also the functional and ecological contexts of those transitions.

Charles Darwin was suspicious of suggested "intermediate forms" in the fossil record; famously, he never used *Archaeopteryx* —an organism that combined features typical of both birds and reptiles—as a transitional form between reptiles and birds, despite the urgings of Thomas Huxley (Desmond 1982). The reason for this is not because Darwin did not accept transitions between major groups of organisms (he did), but rather because that was not the argument he was trying to make. Natural selection was a mechanism for evolutionary change, not an argument about transitional forms. *Archaeopteryx* and its extinct relatives were not well enough understood for Darwin or anyone else at the time to speculate about the action of natural selection on their features. Even as Huxley was crowing about the transitional characteristics of *Archaeopteryx*, Richard Owen was remonstrating that it had nothing of the sort, and that it was only an antiquated bird with feathers and hence warm-blooded, unlike any reptile could possibly be (Desmond 1982).

If, as Darwin argued, traditional schemes of classification emphasize the results of selective extinction (Padian 1999), what would make it easier to recognize and place so-called transitional forms? We suggest four points:

1. The fossil record is not complete, nor should we expect it to be. Nevertheless, there are ways to analyze how complete it is, within reason, for particular cases. Moreover, it is an adequate source of historical information even if it is incomplete. Critics of evolution should not be allowed to gain any advantage from this alleged fatal deficiency, unless they are prepared to admit that the recorded history of humans—including that of their own cultures, nationalities, and religions—is equally incompletely preserved.
2. The rise of phylogenetic systematics in the 1960s, pioneered by the entomologist Willi Hennig, changed the focus of classification to arranging according to genealogical order, rather than to enumerating and grouping according to notions of similarity and difference (which themselves frustrate the search for transition). Phylogenetic

systematics also changed the focus from *lineal* to *collateral* ancestors. We will seldom be able to identify the first, but we appear to have plenty of the second, and they are nearly as useful in revealing information about transitions between forms.

3. Shifting the focus from transitional *forms* to transitional *features* is much easier with cladistic analysis. Synapomorphies (shared derived characters) found in two closely related organisms are by definition hypothesized to have been present in their common ancestor, whether or not we have a record of that ancestor (even if we could recognize it). Hence, our hypotheses about evolutionary changes are more explicit, and the cladograms on which they are based are corroborated by other features.
4. Phylogenetic analysis also helps to reveal the sequence of changes that forged new adaptations from pre-existing structures. Because cladograms are based on a variety of characters, mostly unrelated to any single adaptive complex in question, they can test ideas about how adaptations evolved in particular lineages and organ systems, and so make the understanding of major functional change more explicit.

The Completeness of the Fossil Record

Want to start trouble? Ask the guy sitting on the next stool if he can produce proof of his unbroken patrilineal ancestry for the last four hundred years. Failing your challenge, the legitimacy of his birth is to be brought into question. At this insinuation, tables are overturned, convivial beverages spilled, and bottles fly. No fair, claims the gentle reader. This goes beyond illogic to impoliteness, because not only are you placing on the other patron an unreasonable burden of proof, you also are questioning his integrity if he fails. But isn't that what creationists do when they claim that our picture of evolution in the fossil record must be fraudulent because there are gaps in the fossil record?

When creationists attack the fossil record because it is incomplete, what they are really questioning is whether it is adequate to answer specific questions. In other words, they believe that because there are some gaps in the fossil record, it cannot answer the question of whether evolution occurred. Adequacy is a relative principle, however; it depends on the questions that we are asking. The fossil record is not 100 percent complete, and our knowledge of it will never be comprehensive, but neither will our knowledge be complete for the living world, as C. R. C. Paul (1998) points out. Yet incomplete knowledge does not stop biologists from approaching problems of evolutionary and ecological importance, nor historians from inferring human actions and motives; and it should not stop paleontologists from pursuing those problems for which the fossil record is perfectly adequate. It is unreasonable to expect no gaps in the fossil record and illogical to use the fossil record as a basis to assert ignorance of evolutionary patterns and processes.

To put the problem into perspective, consider the record of recent humans. Do we have a written archive of the name and domicile of every human living today? Are we even sure that we have accounted for all the far-flung tribes, populations, and language groups across the globe? In fact, with salient exceptions—such as the extensive genealogical database of the Church of Jesus Christ of Latter-Day Saints—our known records of human populations, both in terms of censuses and relationships, are far from perfect or even adequate to address many questions about patterns and processes of population growth, stability, health, and history. For example, our written records are probably not adequate to answer the question of exactly how many people were named George in the year 1840. However, if we are interested in the ratio of men to women in France in 1840, the incomplete data that we can gather from government records, graveyards, family histories, and other sources will most likely be adequate to provide an approximate answer.

Paleontologists, biologists, ecologists, epidemiologists, and many other scientists routinely face incomplete data. As a result, these scientists regularly attempt to determine the completeness of their data in order to address whether they are adequate to answer particular questions. For example, Alroy (1999) sought to determine whether the known fossil record of mammals could be used to falsify claims of a radiation of modern mammal groups in the Late Cretaceous that has been overlooked by paleontologists (e.g., Kumar and Hedges 1998). In other words, is the mammalian record sufficiently well sampled to instill confidence that the apparent rapid radiation observed after the Cretaceous is real? Alroy statistically analyzed a very large database of fossil occurrences in North America and concluded that the record is indeed complete enough to make a Late Cretaceous radiation of mammals unlikely. In another recent paper, Kidwell (2001) showed that the relative abundances of taxa in fossil assemblages of bivalve mollusks accurately reflect the abundances of the taxa in the original living community. Thus, the fossil record of bivalves is adequate to address questions regarding the abundance of different taxa over time. In addition to these examples, the fossil record also is adequate to address a wide variety of questions and problems, including the construction of phylogenetic hypotheses, studying trends in the evolutionary history of different organisms and inferring ancestor-descendant relationships between different species (e.g., Paul 1992; Smith 1992; Foote 1996; Wagner 1996, 1999; Roopnarine 2001). But for the remainder of this paper, we will consider the basic question of "transitional forms"—and show how this focus has been replaced among scientists by a focus on transitional features.

Better Living Through Phylogeny

The goal of the classic "Linnaean system" of taxonomy was only classification—creating names for groups— not deciphering relationships. The goal of this system was to understand the plan of God by ordering His creations. Phylogenetic trees, which began to be crafted in an evolutionary sense after Darwin, were a natural outgrowth of grafting a sense of evolution onto the Linnaean classification system. Trees built on the Linnaean system of classification often had artistically drawn, wavy branching patterns, and dotted lines and question marks where unknown relationships ought to have been established but were not known—usually because of gaps in the fossil record where there were not enough fossil representatives. There is nothing wrong with the existence of gaps in knowledge or with developing hypotheses about the uncertain aspects of the pattern. The point to be made here is that, as many authors have noted, the reconstruction of phylogeny, especially with recourse to the fossil record, became *a search for ancestors.* But what, in this context, is an ancestor?

There are several ways to define ancestors; the etymology refers simply to those who have gone before. In the search for fossil forms that are ancestral to others, it is commonly assumed that such forms were the *actual individuals* from which living or later forms were descended. This definition is impossible to establish, unlikely on statistical grounds, and unnecessarily restrictive. Take, for example, the fossil specimen of *Homo* designated as KNM-ER 1470, a fairly complete skull better known in paleontological circles by its museum number than by its species, which is still in question (Wood 1991 provides arguments for its inclusion either in *H. habilis or H. rudolfensis*). It meets the standards for a generalized hominin cranium—generalized enough that many more derived *Homo* skulls could be drawn from it. But we could never find evidence that "1470," or "Lucy," or any other famous hominin[1] fossil was the *actual genetic ancestor* of today's humans, because we could never assess the genetic lineage with sufficient precision. Even if we had a full genetic complement for such a specimen—and found that it was consistent with what we might

1. [Eds.] There is today some inconsistency in the use of the terms hominid and hominin in the anthropological literature. Physical anthropologists are increasingly adopting the methods and rules of taxonomic classification espoused by cladists. Years ago, the family name "Hominidae" was used to describe humans and all extinct species in the human lineage after the divergence of humans and chimpanzees. "Pongidae" was used to describe orangutans, gorillas, and chimpanzees. Because gorillas and chimpanzees are more closely related to humans than orangutans, this usage makes the family Pongidae paraphyletic, and thus violates the rules of phylogenetic taxonomy (which requires monophyly). The solution—to include the chimpanzees and gorillas (or the orangutans, chimpanzees and gorillas) along with humans in the family Hominidae, and to erect the tribe Hominini to exclusively describe the human lineage—has become increasingly popular. We follow Folinsbee and Begun (2004) in adopting this usage, although we acknowledge that a number of physical anthropologists prefer the traditional nomenclature.

expect in a common ancestor of many living human groups—we could not rule out the possibility that this individual was infertile or that its direct line of descent had died out at some time. The most we could say is that it has no features that would bar it from direct ancestry of living forms.

Returning for a moment to the record of living humans, it is instructive to visit a graveyard to understand the concept of ancestry. A sufficiently old church graveyard usually contains some, but not all, of the parishioners who were members of the parish. Some members moved away, some were buried elsewhere, the graves of some became neglected and the headstones tumbled or were eroded. Not all those interred have direct living descendants; some died without issue. David Raup (1978) looked into work that had been done by various scholars on the extinction of human surnames, tracing back at least to Malthus's 1826 *Essay on Population.* Malthus had noted that between 1583 and 1783, three-quarters of the prominent families who lived in Berne, Switzerland, became extinct. (There were, of course, no data for *hoi polloi.*) Think of famous U.S. presidents. Washington and Lincoln are common names, but does either president have many direct descendants who are still alive? And Thomas Jefferson had no living direct patrilineal descendants until geneticists reexamined the DNA profiles of the patrilineal descendants of Sally Hemings's son Eston (Foster et al. 1998)—which just goes to show that in history, as in paleontology, the more you look, the more you find.

The discovery of more living descendants of Jefferson was not a question of filling gaps in the historical record but of amending the methods of analyzing genealogy through the use of biomolecular similarities. In the analogous problem of evolutionary transitions and the adequacy of the fossil record, we have as much to learn by establishing reasonable phylogenies as by finding more fossils that close phylogenetic and stratigraphic gaps. To do this, we need testable methods, and this is one reason why cladistic analysis has gained a large share of its popularity. If creationists wish to be taken seriously when they dispute evolutionary trees, they will have to learn to develop and analyze them using contemporary methods.

Even if the denizens of the graveyard have no direct living descendants, they are by some measure the ancestors of the living. Anthropologists distinguish in this sense between *lineal* (direct) ancestors and *collateral* (side-branch) ancestors, and it is useful to borrow this concept to discuss real and apparent gaps in the fossil record. Collateral ancestors can still tell us much about the features, habits, and other characteristics of ancestors whose records may be lost but who would still be similar in most respects to those whose records we do have. Your grandfather is your lineal ancestor, whereas your great-uncle is a collateral ancestor; but were their lives and times necessarily much different? This is as true in paleontology as in anthropology. Even the member of a taxon who retains the most "primitive" characteristics and lacks the most derived ones does not have to be the *direct* ancestor of the more derived ones to be informative; we can accept it as a collateral ancestor, and learn from it a great deal about the features of the actual (though hypothetical) unknown direct ancestors. However, we need to consider the most effective methods for approaching this kind of analysis.

From Transitional Forms to Transitional Features and the Evolution of Complex Adaptations

The "transitional forms" problem is insoluble when dealing with anyone who does not want to accept that there could be transitional forms. Even for those willing to accept (at least in principle) the possibility of transitional forms, confusion often arises because many people are accustomed to thinking of the evolution of life in the terms of a ladderlike progression, with a different animal on each rung. In the case of vertebrate evolution, for example, they might envisage a fish on the bottom rung, a salamander on the next, then a lizard, a mouse, and finally a human on the top rung. Using a "ladder-thinking" approach rather than a "tree-thinking" one, it is not hard to understand why they might have difficulty imagining transitions between the different forms.

Although a ladderlike image of evolution remains common in the popular media, scientists have long realized that such a concept is simplistic and inaccurate. Instead of resembling a ladder, the evolution of life is more similar to a branching bush. Each branch on the bush represents a distinct lineage of organisms. Places where two or more branches diverge from a single point on the bush indicate that the lineages represented by the branches must have shared a common ancestor at a particular point in their history. To return to our example above, the branch including fish is joined at its base to the branch including tetrapods because both lineages share a common ancestor. On the tetrapod lineage, the salamander lineage would diverge first, followed by the lizard, the mouse, and the human. Thus, our focus in reconstructing the history of life shifts from trying to imagine how a mouse could transform into a human to discovering which features mice and humans inherited from their common ancestor and which features the lineages evolved after they diverged.

Phylogenetic analysis plays an important role in this process because it emphasizes inferring whether *features* are ancestral (or general) or derived, instead of searching for ancestral taxa. Shared derived features (synapomorphies) are the currency of phylogenetic reconstruction. If a synapomorphy is found in two or more related organisms, it is inferred to have been present in their common ancestor. (It could, of course, be independently evolved in each, and this question can be approached by adding more characters and taxa into the analysis.) So, rather than looking for fossils of lineal ancestors, we are now looking for synapomorphies that link collateral ancestors.

A phylogenetic approach is also critical to inferring how complex adaptations evolved. *Adaptation* is one of the most difficult concepts in biology to define, because the term has been used in so many senses and because so many kinds of evidence can be brought to bear on it. Rose and Lauder's (1996) excellent compilation of essays on the topic provides a good measure of the diversity of approaches and definitions. The method we will use here has been detailed elsewhere (e.g., Padian 1987) and is very similar to those applied to less inclusive phylogenetic levels (e.g., Mishler 1988; Brandon 1990). First, the adaptation must be defined, and the groups that possess it must be identified. Then, phylogenetic analysis must establish the relationships of the organisms that belong to the group in question and determine the closest relatives of the group. Using the phylogeny, it is possible to dissect the sequence of appearance of the components of the adaptation. From this sequence, hypotheses about the utility of such structures and associated functions at various stages in the phylogeny may be generated and tested. It is appropriate to use any and all lines of evidence that might be useful and testable.

The Transition to Birds and Flying

It is well established that birds evolved from small maniraptoran theropod dinosaurs (reviews in Dingus and Rowe 1998; Padian and Chiappe 1998a, 1998b). Troodontids and dromaeosaurs are thought to be the closest relatives of birds (figure 1), but neither is currently thought to include the direct ancestors of birds.

Most of the features that are traditionally thought to characterize birds did not evolve simultaneously with the animals that we would recognize as the first birds—e.g., the Late Jurassic *Archaeopteryx* (Dingus and Rowe 1998; Padian and Chiappe 1998a, 1998b). Many features are already present in other maniraptorans and even more general groupings of the theropod dinosaurs, such as thin-walled bones; a reduced tail; a calcified breastbone; a wishbone; a shoulder girdle with a slim, straplike scapula and an elongated coracoid braced to the sternum; a long forearm with a hand reduced to the first three fingers; a pelvis with a retroverted pubis and a shortened ischium; and a hindlimb with elongated tibia and metatarsals, straplike fibula, hingelike ankle with a triangular ascending process, and a functionally three-toed foot with the first toe pendant from the second metatarsal. The distributions of these features, as currently understood, are shown in figure 1; they make it possible to dissect out the adaptation of flight as seen in the first birds and to ques-

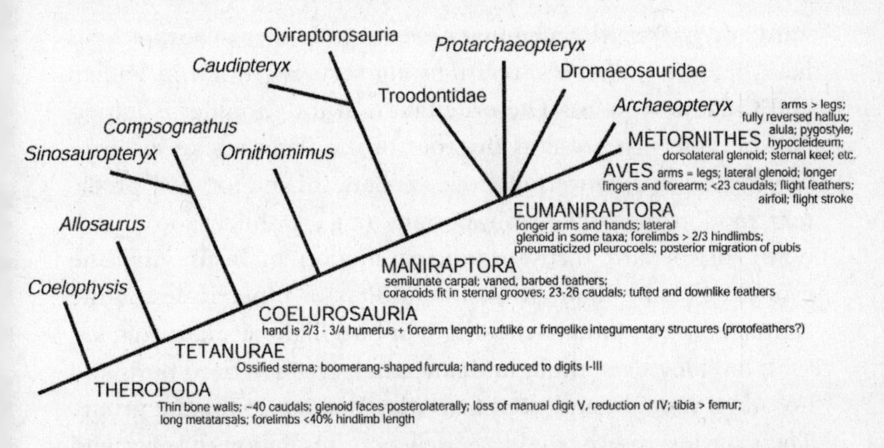

Figure 1. Cladogram of the immediate relatives of birds, showing evolution of some of their characteristic features. The names of particular theropod taxa are less important in the context than the characters that they share progressively with birds, showing how the avian body plan was assembled stepwise, with most of its components initially used for other purposes.

tion the original roles of some of the components of this complex adaptation (Padian and Chiappe 1998b).

Theropod dinosaurs were carnivorous. The long necks and fleet-looking hindlimbs of even the earliest forms suggest an ability to dart after prey, and remains of their meals have been found in the stomachs of even some of the most basal (*Coelophysis*) and smallest (*Compsognathus*) theropods. In theropods closer to birds, the arms and especially the hands became proportionally longer, suggesting a renewed emphasis on the forelimbs for predation. In these animals, the wishbone and breastbone first calcify, suggesting development of the musculature necessary to bring the arms forward and inward, as for grabbing prey (Padian and Chiappe 1998b).

The shoulder socket in typical theropod dinosaurs, as in other dinosaurs, faces downward and backward, but in *Archaeopteryx* it faces outward; in living birds, it faces outward and backward. Jenkins (1993) showed that in the dromaeosaurs such as *Deinonychus*, which have long been implicated as among the dinosaurs closest to birds, the shoulder socket still faced downward and backward. Since that time, several other maniraptorans have been found with outward-facing sockets (review in Padian and Chiappe 1998b). The orientation of the shoulder is important because it represents the root of the flight stroke in birds. This stroke was converted from an outward-and-forward predatory motion in basal maniraptorans (e.g., Gishlick 2001) to a down-and-forward thrust-producing motion in birds. And the evolution of the flight stroke, regardless of other issues, is the central consideration in the origin of bird flight (Padian 1987).

It has long been thought that feathers characterized birds and no other animals, and therefore were diagnostic for the group. Their obvious use in flight, as well as in insulation, display, and other functions, has long prompted questions about the original use of feathers and from what structures they were derived. Although there are still few answers to all these questions, it is clear that feathers with shafts, vanes, and barbs were already present in a variety of nonavian theropods that did not fly, and hence feathers did not evolve *for* flight (Padian et al. 2001). A tufted, filamentous integumentary covering is found in at least one basal coelurosaur, *Sinosauropteryx* (Chen et al. 1998), suggesting that insulation (and perhaps coloration) was important in the origin of such structures, which may be precursors to true feathers. Therefore, not only structures but functions may be "transitional," in the sense that they can have multiple purposes. A structure that originally keeps an animal warm can develop a color pattern that advertises or hides its owner, and the same structure, if its components develop features that stiffen and interlock its filaments, can contribute to a workable airfoil. This "co-opting" of existing structures and functions to new ones was termed *exaptation* by Gould and Vrba (1982), and it is quite probably how most functional changes take place in evolution.

Is the famous *Archaeopteryx*, then, a "transitional fossil"? Despite intensive scrutiny by many scholars, it appears to have no features that bar it from being an ancestor of more derived birds. As noted above, that is the most that can be said about any fossil proposed as a potential lineal ancestor. But we can say this because *Archaeopteryx* embodies some of the most basal features of the animals that we traditionally recognize as birds—namely, a wing comprising feathers that are long enough and have the structural integrity necessary to accomplish flight. And, although we cannot demonstrate that *Archaeopteryx* flew, we can show that it had the equipment and was capable of producing the motion needed to execute the flight stroke (e.g., Jenkins 1993). Virtually all the other features of *Archaeopteryx* that have been associated with birds, as noted above, were already present in its theropod ancestors. After *Archaeopteryx*, the wings lengthened further; many bones of the hand, pelvis, and foot fused; the skeleton became more pneumatic; the tail shortened; teeth were lost; and other typically avian features appeared, such as the alula (thumb-wing) and the perching foot (Dingus and Rowe 1998; Padian and Chiappe 1998a, 1998b).

So it can be said that, structurally, the features of *Archaeopteryx* are almost unexceptionally transitional between those of nonavian theropods and more derived birds—as far as we know now. Functionally, assuming that *Archaeopteryx* could fly, at least somewhat (which is generally agreed upon), it was transitional between nonflying and more proficiently flying forms; but because it retained the free three fingers of nonavian theropods, it can be assumed that the hand was also used for grasping, clambering, or other functions that were lost when the fusion of carpometacarpal elements strongly restricted motion in more derived birds.

The point that we emphasize here is that the "bird" and "reptile" features of *Archaeopteryx*, or any such animal, can be explained in quite orderly fashion with reference to their distribution on a cladogram such as figure 1. This distribution makes sense because it shows when each feature arose and was passed down, often in modified form, to the descendants of the first animals that had it.

Why Did the Whale Cross the Beach?

"Creation scientist" Duane Gish used to regale sympathetic audiences in the 1980s by mocking the evolutionary account of the origin of whales from terrestrial mammals. He represented to his listeners that evolutionists said that whales evolved from relatives of hoofed mammals, which he noted included the cows. He would then show a cartoon slide of Jersey cows, bells and all, outfitted with mermaidlike tails, and would ridicule what he alleged would be the necessary intermediate forms between terrestrial mammals and whales.

This origin story for whales was not new, just imprecise. Darwin suggested in the first edition of *On the Origin of Species* that a bearlike carnivore, successful in fishing excursions from the shore, might have spent more and more time in the water until ultimately aquatic features leading to a whalelike form were selected. Darwin had no specific fossils or transitional forms in mind, but his scenario was unusually speculative for his restrained prose, and criticism caused him to remove the passage from later editions. (It is interesting to note that the pinnipeds—seals, sea lions, and walruses—appear to be closely related to bears, weasels, and raccoons, so Darwin's hypothesis may be fairly accurate for this group of marine mammals.) By the early 1900s, a loose group of archaic mammals called condylarths had been identified as a possible group from which whales had evolved. Most of these animals had dentitions that suggested a herbivorous diet, and they were thought to have "given rise" to both major groups of hoofed mammals, the artiodactyls and perissodactyls. So Gish's confusion, which he persisted in transmitting to his audiences, can be traced. Ironically, recent discoveries in paleontology and molecular biology indicate that the use of a cow in his caricature of whale evolution was closer to the truth than most researchers at the time would have guessed.

For much of twentieth century, the history of whales was enigmatic. *Protocetus atavus*, known from the Eocene of Egypt, was generally accepted as a good model for basal cetaceans because of its primitive skull and tooth morphologies, but even it had many derived character states that obscured its relationship with terrestrial mammals (Thewissen 1998). It was not until the 1970s and 1980s that fossils documenting the transition to an aquatic lifestyle began to surface. Since then, many early cetacean fossils have been discovered, and our picture of whale evolution has become much clearer. Again the fossil record does not provide a series of lineal ancestors but instead has preserved a sequence of sister taxa to modern whales—collateral ancestors. Although each of these groups has its own unique specializations, each can be used to infer what ancestral forms were like at different points in the evolutionary history of whales. This is because each fossil whale shares some characteristics with others that mark the accumulation of features present in living whales.

The identity of the closest relatives of whales among living and fossil mammals has been the source of much controversy (e.g., see reviews by Gatesy 1998; Gatesy and O'Leary 2001; Thewissen and Williams 2002), but recently discovered fossils and nearly all molecular data suggest that whales are artiodactyls, or even-toed ungulates (e.g., Gatesy 1998; Shedlock et al. 2000; Gatesy and O'Leary 2001; Gingerich et al. 2001). Living artiodactyls include pigs, hippos, cows, deer, sheep, camels, and giraffes, and some lines of evidence suggest that hippos are the closest living relatives of whales, although they likely evolved an aquatic way of life independently.

The pakicetids, represented here by *Pakicetus*, are considered the most basal lineage within Cetacea (e.g., Gingerich et al. 1983; Thewissen et al. 1994; Uhen 1998, 1999; Geisler 2001; Uhen and Gingerich 2001), and some of their features seem transitional between fully aquatic whales and their terrestrial ancestors. For example, the ear region of the cetacean skull is extensively modified to facilitate directional hearing under water. However, in *Pakicetus*, the ear has been characterized as intermediate between those of terrestrial and fully aquatic mammals (Gingerich et al. 1983; Thewissen and Hussain 1993; Luo 1998; Nummela et al. 2004), and it could probably function in both environments. The known postcranial remains of pakicetids do not show modifications for an aquatic way of life, but instead are specialized for cursorial locomotion on land (Thewissen et al. 2001). Sedimentological (Aslan and Thewissen 1997; Williams 1998) and oxygen isotope ratio data (Thewissen et al. 1996a; Roe et al. 1998) suggest that pakicetids lived in or near fresh-water rivers, not marine or estuarine environments. The skull of *Pakicetus* has dorsally facing orbits placed near the midline of the skull (Thewissen 1998; Thewissen et al. 2001), implying that some members of the group may have hunted prey located above them, but it is not known whether this feature characterizes all members of the clade. Based on this information, the earliest whales appear to have lived in an inland environment and were terrestrial or perhaps semiaquatic. They may have been ambush predators, and their hearing system may have been modified to accommodate their forays in different environments.

Ambulocetus represents an amphibious stage in the cetacean transition to an aquatic lifestyle. It has a relatively long neck and tail, robust vertebrae, forelimbs fixed in a crouched position, and hindlimbs with very large feet (Thewissen et al. 1994, 1996b). Functional studies have concluded that *Ambulocetus* was capable of effective terrestrial locomotion and probably swam much like an otter (Thewissen et al. 1994; 1996b; Thewissen and Fish 1997; Buchholt 1998; Madar 1998). Ambulocetids are known from tidal deposits rich in marine invertebrates and plants (Williams 1998), but isotopic evidence indicates that, unlike modern cetaceans, they probably still needed to drink fresh water (Thewissen et al. 1996a; Roe et al. 1998). Thewissen et al. (1996b) have suggested that *Ambulocetus* was an ambush predator much like the crocodile, capable of subduing large, struggling prey items.

The remingtonocetids also appear to fit a crocodilian ecological analogy, although in this case the crocodilians in question are fish-eating forms like the living gharial. Like these crocodiles, remingtonocetids have long, narrow skulls (e.g., see figures in Bajpai and Thewissen 1998) capable of rapid lateral movements to capture fish. The morphology of the auditory region of the skull suggests that underwater hearing was well developed, although remingtonocetids probably also were sensitive to low-frequency airborne sounds (Bajpai and Thewissen 1998; Nummela et al. 2004). The eyes appear to have been relatively small (Thewissen 1998). Postcranial remains of remingtonocetids are rare, but the best-known taxon, *Kutchicetus,* has short, stocky limbs and a long, muscular back and tail (Thewissen and Williams 2002). They seem capable of terrestrial locomotion, and their swimming style was probably similar to that of the ambulocetids, although the tail may have played a somewhat greater role in remingtonocetid swimming (Buchholt 1998; Madar 1998). Remingtonocetids are also known from tidal and coastal sediments (Williams 1998), and isotopic data suggest that different members of the clade had differing degrees of tolerance of salt water (Roe et al. 1998).

The next node on the cladogram represents the divergence between *Rhodocetus,* a basal "protocetid," and the lineage leading to living whales. Here again, new features are associated with an increasingly marine lifestyle. Perhaps the most important of these are modifications to the vertebral column that represent the beginning of the transition to the swimming style seen in living whales. The neck vertebrae of *Rhodocetus* are shorter relative to their height than are those of earlier forms such as *Ambulocetus* (Gingerich et al. 1994), decreasing the flexibility of the neck and improving the stability of the front end of the body during swimming (Buchholt 1998). This feature is also seen in many other groups of aquatic vertebrates (e.g., sea cows), and is carried to extremes in living whales. *Rhodocetus* also lost the fused sacral, or hip, vertebrae that characterize terrestrial mammals (Gingerich et al. 1994), increasing the flexibility of the vertebral column. The hands and feet of *Rhodocetus* were relatively large and likely webbed in life (Gingerich et al. 2001). It appears that *Rhodocetus* could use its webbed appendages for paddling, although it probably also was capable of a limited degree of dorso-ventral undulation, similar to modern whales (Buchholt 1998; Gingerich et al. 2001; Gingerich 2003). Even though its limbs are clearly modified beyond those of its terrestrial ancestors, the hip bones of *Rhodocetus* are fused to the vertebral column (Gingerich et al. 1994), suggesting that *Rhodocetus* could still support its weight with its limbs on land. The auditory system shows further modifications for underwater hearing (Luo 1998), and the eyes are positioned laterally on the skull, much as in living whales (Thewissen 1998). *Rhodocetus* is known from deep continental-shelf deposits (Gingerich et al. 1994), and the stable isotopic signature of the closely related *Indocetus* indicates tolerance of salt water (Thewissen et al. 1996a; Roe et al. 1998). In general, *Rhodocetus* is probably a good model for what the most basal fully aquatic whales were like.

Dorudon, a basilosaurid known from Egypt (Uhen 1998), continues many of the trends begun in earlier taxa and appears to be fully committed to an aquatic lifestyle. The ear region of the skull is quite similar to those of living whales, suggesting well-developed underwater hearing abilities (Luo 1998; Nummela et al. 2004). The scapula is fan-shaped, as is common in aquatic mammals, and the forelimb is flattened into a flipper resembling those of living whales (Uhen 1998). However, *Dorudon* still retains some flexibility at the elbow joint and has not increased the number of phalanges, unlike later cetaceans. The neck vertebrae of *Dorudon* are shortened, and the body proportions of *Dorudon* and the closely related *Zygorhiza* show that they had a swimming style very similar to that of extant whales (Buchholt 1998; Gingerich 2003). The terminal caudal vertebrae closely resemble those of living cetaceans and suggest the presence of a tail fluke. The hindlimbs of *Dorudon* are not well known, but those of other basilosaurids are very small and limited in their range of movement (Gingerich et al. 1990). The hip bones are no longer fused to the vertebral column (Gingerich et al. 1990; Uhen 1998). The extensive modification of both the fore- and hindlimbs of *Dorudon* strongly suggests that, like other basilosaurids, it was incapable of terrestrial locomotion. *Dorudon* is probably a very good model for the ancestor of living cetaceans, but it still lacks a number of features, such as the telescoped skull bones, that are found in whales today.

In considering the evolution of whales, then, it is important to remember that each form found in the fossil record is not necessarily a lineal ancestor of later whales; in fact, most of the ones we know at present from the fossil record seem to have their own

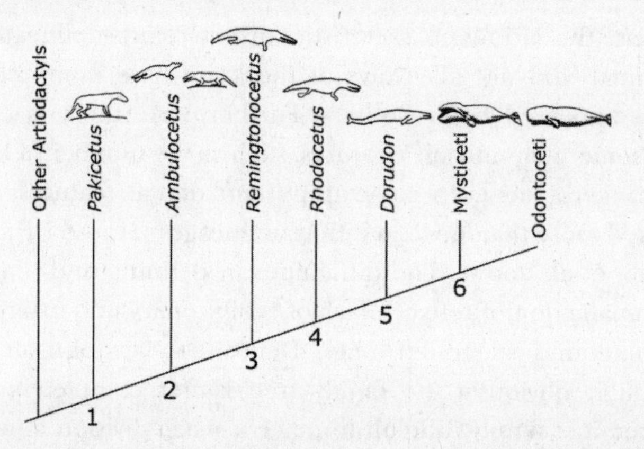

Figure 2. Cladogram outlining the relationships of Cetacea. Synapomorphies for the numbered nodes include: (1) reduced or absent paraconid cups on lower molars; partially reoriented ear ossicles; heavily ossified auditory bulla and ear ossicles; (2) enlarged mandibular foramen; absence of metaconid cusp on lower molars; (3) very large mandibular foramen; presence of a supraorbital shield; (4) reduction of hindlimb; unfused sacral vertebrae; four or fewer sacral vertebrae; laterally placed orbits; (5) hip bones not fused to vertebral column; very reduced hind limbs; upper third molar absent; (6) absence of deciduous teeth. Cladogram modified from Uhen (1998, 1999). Synapomorphies based on Luo (1998); O'Leary (1998); Thewissen (1998); Uhen (1998, 1999). Reconstructions modified from Romer (1959) and Madar (1998), and not to scale.

particular specializations. Again, it is the characters, not the animals, that are transitional (figure 2). We don't need to turn *Ambulocetus* into a sperm whale, any more than we need to start with Duane Gish's absurd Jersey cows. Instead, we can watch the phylogenetic progression of whalelike features accumulate in the lineage that led to the whales of today (Sutera 2001).

How to Recognize a Mammal

The diagnosis given by Linnaeus (1758) when he coined the term *Mammalia* can still be used to differentiate between living mammals and other tetrapods. However, when we begin to take into account many of the early fossil relatives of mammals, things become much more confusing; it becomes harder to draw a clear distinction between what is a mammal and what is not. This is because the various characteristics that so clearly delineate extant mammals did not all evolve at the same time; some character states appeared before others. Furthermore, there is evidence that some "mammalian" features, such as the number of bones in the fingers and even some important dental features, actually evolved more than once in different lineages (Hopson 1995; Luo, Cifelli, et al. 2001). The difficulties in defining and diagnosing Mammalia do not reflect a lack of evolutionary understanding, as is sometimes suggested (e.g., Denton 1985; Johnson 1993), although obviously the family tree is not completely known. Rather, it is simply difficult to make a sharp division in a continuum of organisms, because the division reflects the arbitrary placement of a name, not the understanding of genealogy. Readers interested in more detailed overviews of the early evolution of mammals and the sequence of acquisition of their characteristic features are referred to reviews by Hopson (1991, 1994), Sidor and Hopson (1998), Rubidge and Sidor (2001), and Kielan-Jaworowska et al. (2004).

Some of the problems concerning mammals and their extinct relatives are purely semantic. They have to do with definitions and diagnoses of clades. Rowe (1988) and de Queiroz (1994) provide useful discussions of the two terms in the context of mammalian phylogeny. In biological terms, the *definition* of a taxon describes its limits of membership and ancestry—in other words, who belongs to the taxon. Ancestry is a characteristic that all evolving organisms share, and differences in patterns of ancestry allow us to divide organisms into hierarchically arranged groups that are defined by genealogy. For example, the clade Mammalia can be defined as the monotremes, marsupials, and placentals, and all the descendants of their most recent common ancestor. A *diagnosis*, in contrast, gives unique characters that are useful for distinguishing one group from another. In a phylogenetic system of classification, synapomorphies are used in diagnoses, but because the characters may continue to evolve after they appear, diagnostic characters cannot be used to *define* a clade (in the sense above). For example, Mammalia might be diagnosed by the presence of hair, but whales have secondarily lost this character state, and we may discover evidence suggesting that some animals outside of monotremes, marsupials, and placentals had hair. We recognize whales as mammals because they share a genealogical relationship with other groups of mammals.

A second problem in terminology is simply what to call the fossil relatives of mammals. They are frequently referred to in the popular and scientific literature as "mammal-like reptiles," but this label is misleading and does not reflect our understanding of the relationships between mammals and reptiles. Mammals and the "mammal-like reptiles" are all members of the clade Synapsida and are characterized by having a single opening on the side of the skull through which jaw musculature passes. Synapsids that do not have the synapomorphies that diagnose mammals are simply called non-mammalian synapsids. Early nonmammalian synapsids are superficially similar to some reptiles because they retain many primitive characteristics inherited from the common ancestor of synapsids and reptiles, but synapsids (and mammals) did not evolve from reptiles. Critics of evolution, such as Denton and Johnson, never bother to understand or clarify this fact, because it is more to their purpose to suggest that scientists are vainly chasing nonexistent transitions between "reptiles" and "mammals" than to show their audiences that the groups of mammals and their relatives nestle nicely within a hierarchy of successively more inclusive phylogenetic groups.

There are, to be sure, several ways to define Mammalia (figure 3). We could just use the living forms and their most recent common ancestors (crown-group definition); we could go farther back in the tree, including several extinct groups such as docodonts and triconodonts and stop there (node-group definition); or we could call Mammalia all animals that are closer to today's mammals than to, say, dicynodont therapsids (stem-group definition). The definition is arbitrary; it is nomenclatural housekeeping. What matters is that people understand what is being communicated. What is not at issue is the phylogeny. There is much left to be discovered about mammal origins, and as new

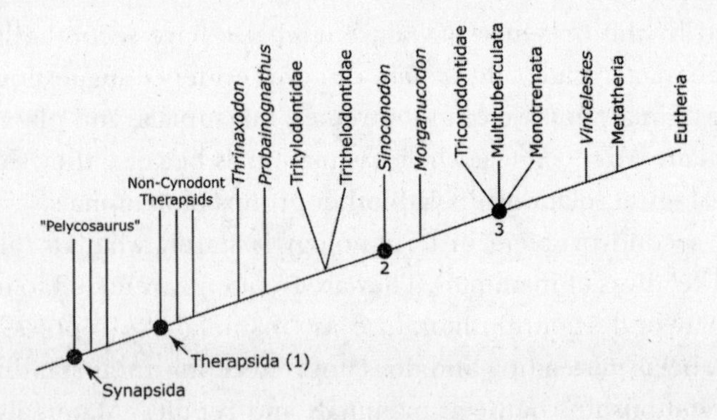

Figure 3. Simplified cladogram of synapsid relationships showing different definitions of Mammalia: (1) Mammalia of Van Valen (1960); (2) Mammalia of Kemp (1982); Hopson (1994); Luo, Kielan-Jaworowska, et al. (2002); Kielan-Jaworowska et al. (2004); (3) Mammalia of Rowe (1988); McKenna and Bell (1997); Luo, Crompton, et al. (2001); Wang et al. (2001). Cladogram based on phylogenies of Wible et al. (1995); Luo, Crompton et al. (2001); and Wang et al. (2001). The placement of the Triconodontidae varies among these phylogenetic hypotheses and the group is not always included within Mammalia under definition (3).

forms emerge, discussion continues about which are actually closest to one or another mammalian or near-mammalian group. These are separate problems, however.

Neo-Creationist Treatments of "Transitional Forms"

Traditional creationist writings have treated the problem of "transitional forms" in fairly consistent ways. They recur to Darwin's (1859) discussions that worry about how the absence of a complete sequence of both living and fossil forms poses so grave an objection to his theory. They bring up his discussion in the first edition of the *Origin* of how a bear was observed swimming with his mouth open, and his musing that such an animal's behavior in time might have triggered selection for more and more whalelike features. They point out how difficult it is to build an eye or a wing out of nothing, and how maladaptive an unfinished eye or half a wing would be.

Of course, as Darwin pointed out, there is nothing necessarily maladaptive about a structure that is less finished than the vertebrate eye. A nerve ending that is irritable by light, the photosensitive eye on the tentacle of a snail, the compound eye of an insect, and the complex one-lensed eye of a squid or a vertebrate are all useful to their possessors. The animals in question do not have to be regarded as ancestors of each other, nor do their optic organs have to form a chain of intermediates, in order to demonstrate that assembly of complex organs from simpler ones is possible. In the same way, half a wing, as noted above, can be perfectly functional, if not for flight; and if *Caudipteryx* and *Protarchaeopteryx* are any indication, the first wings and feathers were not at all for flight (Padian et al. 2001). It is not necessary to postulate that the first animals that possessed these wings and feathers could fly as well as a bird can today. At first these structures performed different functions entirely, and flight evolved in increments.

The neo-creationist literature has not really progressed beyond traditional anti-evolutionary arguments. Scientists continue to be misquoted or quoted out of context, suggesting that creationists spend a great deal of time combing the legitimate literature for sentences that can be snipped and used against the original purpose of the author. Neo-creationists, for example, have quoted the eminent mammalian paleontologist Michael Novacek as saying, "*Ambulocetus*, *Rhodocetus*, and other more aquatically specialized archaeocetes cannot be strung in procession from ancestor to descendant in a *scala naturae*." They infer from this that there is no fossil evidence for the evolution of whales. The excerpted sentence, which begins the last paragraph of a nearly full-page commentary, has been classically taken out of context. Novacek's (1994) quoted sentence means only to say that we do not regard these animals as successive *lineal* ancestors. This is because *Ambulocetus*, *Rhodhocetus*, *Pakicetus*, and other forms each have their own "autapomorphies" (distinguishing characteristics), as noted above, which they would have to lose in order to be considered lineal ancestors of other known forms. Novacek explains in his next sentence that these fossil forms show progressive specialization of features common to whales today: "Nonetheless, these fossils are real data on the early evolutionary experiments of whales." In previous paragraphs, he pointed out that archaic whales first evolved cetacean features of the middle ear, muzzle, skull roof, and teeth; then an amphibious habit with dorsoventral flexion of the body providing locomotion in the water, aided by paddlelike hind feet (*Ambulocetus*); then shorter neck vertebrae, unfused hip vertebrae, and the reduced femur (*Rhodocetus*); and so on, as we have detailed above. Finally, Novacek writes, "They powerfully demonstrate transitions beyond the reach of data, whether molecular or morphological, derived from living organisms alone."

Readers can certainly decide for themselves why Novacek has been so egregiously misquoted. Neo-creationists reply, however, that the perspective on the "evidence" that Novacek provided is only one interpretation of what may have happened. It is of course true that one is not required to accept an evolutionary interpretation of any evidence. On the other hand, the neo-creationists do not provide any alternative interpretation apart from "special creation" or "intelligent design," which are not testable by empirical means (contra Dembski 1999). And it seems ethically questionable to represent to the public that a scientist meant to say something in support of an idea or inference that is clearly opposite to what he intended—which would be clear to readers were his quote not taken out of context.

The Australian physician Michael Denton's (1985) *Evolution: A Theory in Crisis* is a well-written treatise on why evolution cannot have occurred. But, like most traditional and neo-creationist works, it confuses the evidence that evolution occurred with the notion that it all occurred by natural selection (which Darwin would never have maintained) and that it involved largely random processes (the opposite of forces such as selection, a characterization that appalled Darwin). Denton's sections having to do with the problem of evolutionary transitions (chapter 8) were old-fashioned even when he wrote them, in some senses, because he was unaware both of new paleontological findings and of the clarification of many traditional problems by phylogenetic analyses. He admitted that many forms in the fossil record have transitional features, but he insisted that because each appears with its own derived features, none could be ancestral to others (the problem of confusing collateral and lineal ancestors and the difficulty of recognizing ancestors with certainty to begin with). Like most creationists, he seized on the absence of transitional forms, denying the plausibility that so many could be missing if evolution between major groups had actually occurred (the problem of confusing completeness with adequacy). We have already dealt with this problem using examples from human history and genealogy. Inasmuch as we have no living or extinct transitional forms between many major races or cultural-geographic groups of humans, would creationists have to admit that humans do not plausibly share a common ancestor?

Of *Archaeopteryx*, for example, Denton wrote (1985, 176):

> No doubt it can be argued that *Archaeopteryx* hints of a reptilian ancestry but surely hints do not provide a sufficient basis upon which to secure the concept of the continuity of nature. Moreover, there is no question that this archaic bird is not led up by a series of transitional forms from an ordinary terrestrial reptile through a number of gliding types with increasingly developed feathers until the avian condition is reached. A much more convincing intermediate would be something like Heilmann's imaginary "pro-avis" a supposed hypothetical ancestor of *Archaeopteryx* which glided through the trees assisted by partially developed feathers.

This interesting passage suggests that Denton would accept, at least in theory, the possibility of the evolution of major groups of organisms such as birds from other forms of life. Denton made no allusion, however, to the fact that even in the 1970s, it had been established that the most likely candidates for bird ancestry were theropod dinosaurs, as explained above. If his search for bird origins was focused on such animals, and if he knew what we know now about feathering in nonavian theropods, would he accept the transition? This is doubtful, because he spent some time (1985, pp. 204–10) discussing how difficult it is to evolve flight, especially from the ground up, and then proceeded to discuss the problems of evolving an avian lung, without mentioning any specific animals implicated in bird origins. It seems odd to discuss theoretical questions of evolution in the complete absence of animal models. It is even more puzzling why Denton would find a hypothetical animal, for which there is no evidence, a more plausible ancestor than any real animals could be. He seems to accept in advance that flight must begin from a gliding, arboreal ancestor, but this seems less plausible in light of current evidence (Padian and Chiappe 1998a, 1998b; Dial 2003). In the end, however, Denton would not really accept the possibility of transitions—not because they are unimaginable but because no amount of evidence or logic would be convincing to someone who has already decided that there cannot be transitions.

Retired criminology professor Phillip E. Johnson has won a wide audience among the scientifically illiterate of the United States during the past decade. Johnson has not studied or done research in any science, and his book *Darwin on Trial* (2nd ed., 1993) is a hash of selective quotations from secondary sources and coffeetable books that lacks any firsthand understanding of the evidence for evolution. Most of his writing on transitional forms appears to be derived from Denton and other creationist sources, because the phraseology and examples are the same (the old chestnut of the juvenile hoatzin's atavistic separate fingers and claws, for example, and the difficulties with evolving flight and avian lungs). Even then, Johnson has difficulty in denying that there are many transitional forms in the fossil record; he insists, however, that there are not enough, and he implies freely that this is because such transitions are impossible (1993, chapter 6). He cannot accept that "amphibians" evolved from "fishes" because the living coelacanth is not a direct ancestor of tetrapods. He cannot accept that "amphibians" gave rise to "reptiles," but he has no idea what basal tetrapods were like, so he is reduced to trying to derive living reptiles from frogs. Johnson has no choice but to accept that *Archaeopteryx* is an almost ideal intermediate between birds and nonbirds, but he will not accept that fossil whales with reduced legs are intermediate between terrestrial mammals and today's whales, though he cannot seem to imagine why the fossil whales would have small legs. He quotes Stephen Jay Gould and Douglas Futuyma on the origin of mammals, though neither is an authority, and he confuses mammals and reptiles, and monophyly and paraphyly. Johnson is forced to admit that Hopson (1987) successfully arranges a sequence from basal therapsids to the basal mammals, but he will not accept the evolutionary model (Johnson 1993: 191) because "Hopson does *not* present a genuine ancestral line." The confusion is again between lineal and collateral ancestry; Johnson's perception of phylogeny has not moved beyond the *scala naturae*; moreover, he conflates stratigraphic order with ancestry. Both false problems were dealt with above.

On more theoretical grounds, biochemist Michael Behe (1996) cannot accept evolutionary transitions, and his major reason seems to be the problem of "irreducible complexity," which reforms the old creationist notion that a complex structure with a complex function is useless when atomized into its parts, and the assembly of such parts in stepwise, always functional fashion is inconceivable. Fortunately, new discoveries, advances, and techniques tend to make today's "irreducible complexity" tomorrow's understood complexity. There are now many more examples of intermediate complexes of features in the sequences between whales and nonwhales, mammals and nonmammals, birds and nonbirds, tetrapods and nontetrapods, than we had twenty years ago, as well as legitimate, testable hypotheses about how these intermediate features functioned. If one is not willing to accept such intermediates, however, it does not matter how many are found.

In the late 1700s, the Cambridge cleric William Paley apotheosized an argument for divine design in Nature that, although not considered intellectually deep or philosophically respectable by many scholars, found broad success. Paley's (1802) *Natural Theology* was taught at Cambridge (Darwin read it and respected its logic, though he did not agree with its arguments, even in his pre-evolutionary days), and it is still in print today. Intellectual descendants of Paley now resurrect many of his arguments and cloak them in cryptoscientific language (Moreland 1994; Dembski 1998). This view assumes that if one cannot imagine how a complex structure with complex functions could have existed and worked in a simpler state, or if there are no obvious transitional forms involved, or if the probability that a certain phenomenon or pattern could be produced is apparently very small, then the situation must have been divinely created. (Does this force us to accept that lottery winners must be divinely selected because the probability of winning is so small?) Apart from the logical non sequitur that if one cannot *imagine* natural causes (and whose imagination is at stake?), one must resort to supernatural processes, there is a larger issue here.

Proponents of "intelligent design," "the design inference," "irreducible complexity," and their colleagues are often not particularly interested in changing how science is done—they are realistic about this—though they demand that scientists stop assuming that all natural phenomena are explainable in natural terms. They are often more interested in the goal of having their views represented in American classrooms—or at least in having the scientific view of evolution removed. This is a primary goal of books such as *Of Pandas and People* (Davis et al. 1989), and Johnson (1993) spends a good deal of time on the subject (see also his more recent [1997] *Defeating Darwinism by Opening Minds*). This goal is shared by the conservative Christian organization known as the American Scientific Affiliation, as well as by conservative political groups such as the Rutherford Institute.

Conclusions

The problems of recognizing "transitional forms" are many and have deep histories, but they are mostly pseudo-problems that can be alleviated by substituting tree-thinking for ladder-thinking, by focusing on transitional characteristics instead of transitional taxa, and by recognizing that although the fossil record is not complete, it is complete enough to answer many major evolutionary questions reasonably. In any case, the fossil record is not relatively worse than the historical record of human beings for most analogous questions. Notions of "intelligent design" (e.g., Moreland 1994) or "irreducible complexity" (Behe 1996) that allege that evolutionary theory cannot account for certain evolutionary transitions merely reflect the inability of their authors to explain these features in testable terms, and their misunderstanding of how these problems are approached and analyzed in a contemporary evolutionary framework. They demolish a straw man and set up an ectoplasmic illusion in its place.

Acknowledgments

We are indebted to Dr. Eugenie Scott and Yohannes Haile-Selassie for helpful discussions, and to John Hutchinson for reviewing parts of the manuscript. We are particularly grateful to the Paleontological Society for allowing us to revise, update, and reprint our contribution to *The Paleontological Society Papers*, Volume 5 (1999), edited by P. H. Kelley, J. R. Bryan, and T. A. Hansen.

References

Alroy, J. 1999. The fossil record of North American mammals: Evidence for a Paleocene evolutionary radiation. *Systematic Biology* 48: 107–18.

Aslan, A., and J. G. M. Thewissen. 1997. Preliminary evaluation of paleosols and implications for interpreting vertebrate fossil assemblages, Kuldana Formation, Northern Pakistan. *Paleovertebrata* 25: 261–77.

Bajpai, S., and J. G. M. Thewissen. 1998. Middle Eocene cetaceans from the Harudi and Subathu Formations of India. In *The emergence of whales: Evolutionary patterns in the origin of cetacea,* ed. J. G. M. Thewissen, 213–68. New York: Plenum Press.

Behe, M. J. 1996. *Darwin's black box: The biochemical challenge to evolution.* New York: Free Press.

Brandon, R. N. 1990. *Adaptation and environment.* Princeton, NJ: Princeton University Press.

Buchholt, E. A. 1998. Implications of vertebral morphology for locomotor evolution in early Cetacea. In *The emergence of whales: Evolutionary patterns in the origin of Cetacea,* ed. J. G. M. Thewissen, 325–52. New York: Plenum Press.

Chen, P.-J., Z. Dong, and S. Zhen. 1998. An exceptionally well preserved theropod dinosaur from the Yixian Formation of China. *Nature* 391: 147–52.

Darwin, C. 1859. *On the origin of species by means of natural selection.* London: John Murray.

Davis, P., D. H. Kenyon, and C. B. Thaxton. 1989. *Of pandas and people: The central question of biological origins.* Dallas: Haughton Publishing Co.

Dembski, W. A. 1998. *The design inference: Eliminating chance through small possibilities.* Cambridge: Cambridge University Press.

———. 1999. *Intelligent design: The bridge between science and theology.* Downers Grove, IL: InterVarsity Press.

Denton, M. 1985. *Evolution: A theory in crisis.* London: Burnett Books.

de Queiroz, K. 1994. Replacement of an essentialistic perspective on taxonomic definitions as exemplified by the definition of "Mammalia." *Systematic Biology* 43: 497–510.

Desmond, A. J. 1982. *Archetypes and ancestors.* Chicago: University of Chicago Press.

Dial, K. P. 2003. Wing-assisted incline running and the evolution of flight. *Science* 299: 402-4.

Dingus, L., and T. Rowe. 1998. *The mistaken extinction: Dinosaur extinction and the origin of birds.* New York: W. H. Freeman.

Donovan, S. K., and C. R. C. Paul, eds. 1998. *The adequacy of the fossil record.* New York: John Wiley.

Folinsbee, K. E., and D. R. Begun. 2004. Phylogenetic nomenclature of living and fossile catarrhines. First International Phylogenetic Nomenclature Meeting. Muséum national d'Histoire naturelle, Paris, July 6–9 (Abstracts, PhyloCode), p. 33. www.ohiou.edu/phylocode/IPNM.pdf (last accessed July 14, 2006).

Foote, M. 1996. On the probability of ancestors in the fossil record. *Paleobiology* 22: 141–51.

Foster, E. A., M. A. Jobling, P. G. Taylor, P. Donnelly, P. de Knijff, R. Mieremet, T. Zerjal, and C. Tyler-Smith. 1998. Jefferson fathered slave's last child. *Nature* 396: 27–28.

Gatesy, J. 1998. Molecular evidence for the phylogenetic affinities of Cetacea. In *The emergence of whales: Evolutionary patterns in the origin of Cetacea,* ed. J. G. M. Thewissen, 63–112. New York: Plenum Press.

Gatesy, J., and M. A. O'Leary. 2001. Deciphering whale origins with molecules and fossils. *Trends in Ecology and Evolution* 16: 562–70.

Geisler, J. H. 2001. New morphological evidence for the phylogeny of Artiodactyla, Cetacea, and Mesonychidae. *American Museum Novitates* 3344: 1–53.

Gingerich, P. D. 2003. Land-to-sea transition in early whales: Evolution of Eocene Archaeoceti (Cetacea) in relation to skeletal proportions and locomotion of living semi-aquatic mammals. *Paleobiology* 29: 429–54.

Gingerich, P. D., M. ul Haq, I. S. Zalmout, I. H. Khan, and M. S. Malkani. 2001. Origin of whales from early artiodactyls: Hands and feet of Eocene Protocetidae from Pakistan. *Science* 293: 2239–42.

Gingerich, P. D., S. M. Raza, M. Arif, M. Anwar, and X. Zhou. 1994. New whale from the Eocene of Pakistan and the origin of cetacean swimming. *Nature* 368: 844–47.

Gingerich, P. D., B. H. Smith, and E. L. Simons. 1990. Hind limbs of Eocene *Basilosaurus*: Evidence of feet in whales. *Science* 229: 154–57.

Gingerich, P. D., N. A. Wells, D. E. Russell, and S. M. I. Shah. 1983. Origin of whales in epicontinental remnant seas: New evidence from the Early Eocene of Pakistan. *Science* 220: 403–6.

Gishlick, A. D. 2001. The function of the manus and forelimb of *Deinonychus antirrhopus* and its importance for the origin of avian flight. In *New perspectives on the origins and early evolution of birds: Proceedings of the international symposium in honor of John H. Ostrom*, ed. J. Gauthier and L. F. Gall, 301–18. New Haven, CT: Peabody Museum of Natural History.

Gould, S. J., and E. S. Vrba. 1982. Exaptation—a missing term in the science of form. *Paleobiology* 8: 4–15.

Hall, B. K. 1998. *Evolutionary Developmental Biology*, 2nd ed. London: Chapman & Hall.

Hopson, J. A. 1987. The mammal-like reptiles: A study of transitional fossils. *American Biology Teacher* 49 (1): 16–26.

———. 1991. Systematics of the nonmammalian Synapsida and implications for patterns of evolution in synapsids. In *Origins of the higher groups of Tetrapods: Controversy and consensus*, ed. H.-P. Schultze and L. Trueb, 635–93. Ithaca, NY: Comstock Publishing Associates.

———. 1994. Synapsid evolution and the radiation of non-eutherian mammals. In *Major features of vertebrate evolution: Short courses in paleontology*, ed. R. S. Spencer, 190–219. The Paleontological Society, no. 7.

———. 1995. Patterns of evolution in the manus and pes of non-mammalian therapsids. *Journal of Vertebrate Paleontology* 15: 615–39.

Jenkins, F. A., Jr. 1993. The evolution of the avian shoulder joint. *American Journal of Science* 293-A: 253–67.

Johnson, P. E. 1993. *Darwin on trial*, 2nd ed. Downers Grove, IL: InterVarsity Press.

———. 1997. *Defeating Darwinism by opening minds.* Downers Grove, IL: InterVarsity Press.

Kemp, T. S. 1982. *Mammal-like reptiles and the origin of mammals.* London: Academic Press.

Kidwell, S. M. 2001. Preservation of species abundance in marine death assemblages. *Nature* 294: 1091–94.

Kielan-Jaworowska, Z., R. L. Cifelli, and Z.-X. Luo. 2004. *Mammals from the Age of Dinosaurs: Origins, evolution, and structure.* New York: Columbia University Press.

Kumar, S., and S. B. Hedgesm. 1998. A molecular timescale for vertebrate evolution. *Nature* 392: 917–20.

Linnaeus, C. 1758. *Systema naturae per regna tria naturae, secundum classes, ordines, genera, species cum characteribus, differentiis, synonymis locis.* Editio decima, reformata. Vol. 1. Stockholm: Laurentii Salvii.

Luo, Z.-X. 1998. Homology and transformation of cetacean ectotympanic structures. In *The emergence of whales: Evolutionary patterns in the origin of Cetacea*, ed. J. G. M. Thewissen, 269–302. New York: Plenum Press.

Luo, Z.-X., R. L. Cifelli, and Z. Kielan-Jaworowska. 2001. Dual origin of tribosphenic mammals. *Nature* 409: 53–57.

Luo, Z.-X., A. W. Crompton, and A.-L. Sun. 2001. A new mammaliaform from the Early Jurassic and evolution of mammalian characteristics. *Science* 292: 1535–40.

Luo, Z.-X., Z. Kielan-Jaworowska, and R. L. Cifelli. 2002. In quest for a phylogeny of Mesozoic Mammals. *Acta Palaeontologica Polonica* 47: 1–78.

Madar, S. I. 1998. Structural adaptations of early archaeocete long bones. In *The emergence of whales: Evolutionary patterns in the origin of Cetacea,* ed. J. G. M. Thewissen, 353–78. New York: Plenum Press.

McKenna, M. C., and S. K. Bell. 1997. *Classification of mammals above the species level.* New York: Columbia University Press.

Mishler, B. D. 1988. Reproductive ecology of bryophytes. In *Plant reproductive ecology: Patterns and strategies,* ed. J. L. Doust and L. L. Doust, 285–306. Oxford: Oxford University Press.

Moreland, J. P., ed. 1994. *The creation hypothesis.* Downers Grove, IL: InterVarsity Press.

Novacek, M. J. 1994. Whales leave the beach. *Nature* 368: 807.

Nummela, S., J. G. M. Thewissen, S. Bajpai, S. T. Hussain, and L. Kumar. 2004. Eocene evolution of whale hearing. *Nature* 430: 776–78.

O'Leary, M. A. 1998. Phylogenetic and morphometric reassessment of the dental evidence for a mesonychian and cetacean clade. In *The emergence of whales: Evolutionary patterns in the origin of Cetacea,* ed. J. G. M. Thewissen, 133–61. New York: Plenum Press.

Padian, K. 1987. A comparative phylogenetic and functional approach to the origin of vertebrate flight. In *Recent advances in the study of bats,* ed. B. Fenton, P. A. Pacey, and J. M. V. Rayner, 3–22. Cambridge: Cambridge University Press.

———. 1999. Charles Darwin's views of classification in theory and in practice. *Systematic Biology* 48: 352–64.

Padian, K., and L. M. Chiappe. 1998a. The origin of birds and their flight. *Scientific American* (February): 28–37.

———. 1998b. The origin and early evolution of birds. *Biological Reviews* 73: 1–42.

Padian, K., Q. Ji. and S.-A. Ji. 2001. Feathered dinosaurs and the origin of flight. In *Mesozoic vertebrate life,* ed. K. Carpender and D. Tanke, 117–35, pls. 1–3. Bloomington: Indiana University Press.

Paley, W. 1802. *Natural theology.* London: R. Faulder.

Paul, C. R. C. 1992. The recognition of ancestors. *Historical Biology* 6: 239–50.

———. 1998. Adequacy, completeness and the fossil record. In *The adequacy of the fossil record,* ed. S. K. Donovan and C. R. C. Paul, 1–22. New York: John Wiley and Sons.

Raup, D. M. 1978. Approaches to the extinction problem. *Journal of Paleontology* 52: 517–23.

Roe, L. J., J. G. M. Thewissen, J. Quade, J. R. O'Neil, S. J. Bajpai, A. Sahni, and S. T. Hussain. 1998. Isotopic approaches to understanding the terrestrial-to-marine transition of the earliest cetaceans. In *The emergence of whales:*

Evolutionary patterns in the origin of Cetacea, ed. J. G. M. Thewissen, 399-422. New York: Plenum Press.

Romer, A. S. 1959. *The vertebrate story.* Chicago: University of Chicago Press.

Roopnarine, P. D. 2001. The description and classification of evolutionary mode: A computational approach. *Paleobiology* 27: 446–55.

Rose, M. R., and G. V. Lauder, eds.. 1996. *Adaptation.* San Diego: Academic Press.

Rowe, T. 1988. Definition, diagnosis, and origin of Mammalia. *Journal of Vertebrate Paleontology* 8: 241–64.

Rubidge, B. S., and C. A. Sidor. 2001. Evolutionary patterns among Permo-Triassic therapsids. *Annual Review of Ecology and Systematics* 32: 449–80.

Shedlock, A. M., M. C. Milinkovitch, and N. Okada. 2000. SINE evolution, missing data, and the origin of whales. *Systematic Biology* 49: 808-17.

Sidor, C. A., and J. A. Hopson. 1998. Ghost lineages and "mammalness": Assessing the temporal pattern of character acquisition in the Synapsida. *Paleobiology* 24: 254–73.

Smith, A. B. 1992. *Systematics and the fossil record.* Oxford: Blackwell Scientific.

Sutera, R. 2000. The origin of whales and the power of independent evidence. *Reports of the National Center for Science Education* 20 (5): 33–41.

Thewissen, J. G. M. 1998. Cetacean origins: Evolutionary turmoil during the invasion of the oceans. In *The emergence of whales: Evolutionary patterns in the origin of Cetacea,* ed. J. G. M. Thewissen, 451–64. New York: Plenum Press.

Thewissen, J. G. M., and F. E. Fish. 1997. Locomotor evolution in the earliest cetaceans: Functional model, modern analogues, and paleontological evidence. *Paleobiology* 23: 482–90.

Thewissen, J. G. M., and S. T. Hussain. 1993. Origin of underwater hearing in whales. *Nature* 361: 444–45.

Thewissen, J. G. M., S. T. Hussain, and M. Arif. 1994. Fossil evidence for the origin of aquatic locomotion in archaeocete whales. *Science* 263: 210–12.

Thewissen, J. G. M., S. I. Madar, and S. T. Hussain. 1996b. *Ambulocetus natans,* an Eocene cetacean (Mammalia) from Pakistan. *Courier Forschungsinstitut Senckenberg* 191: 1–86.

Thewissen, J. G. M., L. J. Roe, J. R. O'Neil, S. T. Hussain, A. Sahni, and S. Bajpai. 1996a. Evolution of cetacean osmoregulation. *Nature* 381: 379–80.

Thewissen, J. G. M., and E. M. Williams. 2002. The early radiations of Cetacea (Mammalia): Evolutionary pattern and developmental correlations. *Annual Review of Ecology and Systematics* 33: 73-90.

Thewissen, J. G. M., E. M. Williams, L. J. Roe, and S. T. Hussain. 2001. Skeletons of terrestrial cetaceans and the relationship of whales to artiodactyls. *Nature* 413: 277–81.

Uhen, M. D. 1998. Middle to Late Eocene basilosaurines and dorudontines. In *The emergence of whales: Evolutionary patterns in the origin of Cetacea,* ed. J. G. M. Thewissen, 29–62. New York: Plenum Press.

———. 1999. New species of protocetid archaeocete whale, *Eocetus wardii* (Mammalia: Cetacea) from the Middle Eocene of North Carolina. *Journal of Paleontology* 73: 512-28.

Uhen, M. D., and P. D. Gingerich. 2001. New genus of dorudontine archaeocete (Cetacea) from the Middle-to-Late Eocene of South Carolina. *Marine Mammal Science* 17: 1–34.

Van Valen, L. 1960. Therapsids as mammals. *Evolution* 14: 304–13.

Wagner, P. J. 1996. Contrasting the underlying patterns of active trends in morphological evolution. *Evolution* 50: 990–1007.

———. 1999. The utility of fossil data in phylogenetic analyses: A likelihood example using Ordovician-Silurian species of the Lophospiridae (Gastropoda: Murchinsoniina). *American Malacological Bulletin* 15: 1–31.

Wang, Y., Y. Hu, J. Meng, and C. Li. 2001. An ossified Meckel's cartilage in two Cretaceous mammals and the origin of the mammalian middle ear. *Science* 294: 357–61.

Wible, J. R., G. W. Rougier, M. J. Novacek, M. C. McKenna, and D. Dashzeveg. 1995. Mammalian petrosal from the Early Cretaceous of Mongolia: Implications for the evolution of the ear region and mammaliamorph interrelationships. *American Museum Novitates* 3149: 1–19.

Williams, E. M. 1998. Synopsis of the earliest cetaceans: Pakicetidae, Ambulocetidae, Remingtonocetidae, and Protocetidae. In *The emergence of whales: Evolutionary patterns in the origin of Cetacea*, ed. J. G. M. Thewissen, 1–28. New York: Plenum Press.

Wood, B. 1991. *Koobi Fora research project, volume 4: Hominid cranial remains.* Oxford: Clarendon Press.

Biological Complexity

Robert Dorit

In the lyrical concluding passage of *On the Origin of Species*, Darwin entreats:

> It is interesting to contemplate an entangled bank, clothed with many plants of many kinds, with birds singing on the bushes, with various insects flitting about, and with worms crawling through the damp earth, and to reflect that these elaborately constructed forms, so different from each other, and dependent on each other in so complex a manner, have all been produced by laws acting around us. (Darwin 1964, 409)

This passage hints at the immense complexity of the natural world, emerging as a result of the operation of time, chance, and natural selection. As he wrote, Darwin knew full well just how difficult it would be for his Victorian contemporaries to see the richness and adaptedness of the living world as the consequence of simple material forces. Today, almost a century and a half later, creationists still resist the idea. For them, complexity is the Waterloo of evolutionary theory, the place where the materialist explanation of the history of life finally fails. Here, for instance is Michael Behe, with a folksy statement invoking a designer to explain biochemical complexity:

> The conclusion of intelligent design flows naturally from the data itself [*sic*]—not from sacred books or sectarian beliefs. Inferring that biochemical systems were designed by an intelligent agent is a humdrum process that requires no new principles of logic or science. It comes simply from the hard work that biochemistry has done over the past forty years, combined with consideration of the way in which we reach conclusions of design every day. (Behe 1998, 193)

And here is Phillip Johnson, pushing the same agenda, this time leavened with an appeal to fairness:

> The question . . . is whether Darwinism is wrong in principle in assuming that marvelously complex structures like the human body, or even the bacterial cell, can be built up by an unguided material process. . . . If mutation and selection cannot accomplish wonders of creativity, then science is on the wrong track and needs to be brought back to reality. (Johnson 1993)

That anti-evolutionists resist a material explanation for biological complexity should come as no surprise: accepting it requires one to abandon cherished notions of purpose, progress, and human preeminence. But, their objections notwithstanding, we are closer to understanding Darwin's tangled bank than ever before. Our task in this chapter will be to deal with the profound confusion that surrounds creationist interpretations of complexity—a confusion emerging at least in part from the multiple, overlapping meanings and definitions of the term. Once the meanings of *complexity* are spelled out and a proper taxonomy of the concept is established, the understanding of biological complexity emerges both as one of the great triumphs of evolutionary theory and as one of its most exciting challenges.

What exactly do we mean when we speak of *complexity*? In the vernacular, complexity has come to mean roughly the same as "very complicated," with the implicit suggestion that any account of complexity is likely to be suspiciously convoluted, if not altogether unintelligible. Active critics of a materialist explanation for the living world have often capitalized on the many meanings of the term *complexity*, using them interchangeably and shifting the meaning to suit the argument. At times, they seem to be echoing the flawed logic of William Paley's analogy, arguing that design must imply both designer and purpose. As Paley would poignantly state at the beginning of his *Natural Theology: or, Evidences of the Existence and Attributes of the Deity, Collected from the Appearances of Nature*, published fifty-six years before *On the Origin of Species*:

> . . . suppose I had found a *watch* upon the ground, and it should be inquired how the watch happened to be in that place; . . . when we come to inspect the watch, we perceive . . . that its several parts are framed and put together for a purpose, *e. g.* that they are so formed and adjusted as to produce motion, and that motion so regulated as to point out the hour of the day; that, if the different parts had been differently shaped from what they are, of a different size from what they are, or placed after any other manner, or in any other order, than that in which they are placed, either no motion at all would have been carried on in the machine, or none which would have answered the use that is now served by it. . . . This mechanism being observed . . . , the inference, we think, is inevitable, that the watch must have had a maker. (Paley 1802)

More recently, as we will see below, anti-evolutionists have put forth more sophisticated definitions of *complexity*. These range from the convoluted Bayesian notion of "complex specified information" (abbreviated to the cooler-sounding "CSI" [Dembski, 1998]) to the supposedly novel notion of "irreducible complexity" (Behe 1998), which echoes—albeit in molecular terms—Paley's argument above. While it is beyond the scope of this article to present and dismantle the various and shifting meanings of "complexity" employed by anti-evolutionists (for a detailed treatment, see Pennock 1999), there is always something to be gained by trying, as did Dr. Seuss's faithful elephant, Horton, to say what you mean and mean what you say. In the sections that follow, we disentangle the multiple meanings of *complexity* as they coexist in our increasingly sophisticated evolutionary understanding of the living world.

At least four distinct meanings of *complexity* appear relevant to our rational understanding of the evolutionary process. These are: (1) complexity of cause; (2) complexity of outcome; (3) complexity of structure; and (4) complexity of organization.

The first two categories—*complexity of cause* and *complexity of outcome*—relate principally to the *mechanism(s)* underlying organic change: mutation, development, adaptation, and so on. The latter two categories—*complexity of structure* and *complexity of organization*—refer to the *products* of the evolutionary process: metabolic networks, cellular structures, tissues, populations, and/or ecosystems. In the sections below, we will examine each of these meanings in more detail and explore the challenge that they present to a coherent and rich theory of material explanation.

Complexity of Cause

The task of evolutionary biology consists in providing a material, historically based explanation for current (or extinct) biological forms. For methodological reasons, this task often involves the decomposition of organisms (or ecosystems, for that matter) into simpler component parts, which can then be analyzed in greater detail. This approach, broadly known as *reductionism*, has proven enormously successful. Incautiously used, however, reductionism has led to a dangerously oversimplified shorthand in evolutionary biology—an interpretation in which organisms are decomposed into features and features are accounted for by single evolutionary causes. Thus, a bird becomes no more than a set of individual features in search of explanation: beak, bone density, feathers, wings. Each of these, in turn, is explained by appealing to a primary evolutionary force: wings are designed for flight, bone density in birds is reduced to provide strength with less weight, and so on.

While these explanations capture something crucial about the evolution of birds, they deliberately avoid what all biologists know to be true: Organisms are more than collections of independent features, and features are rarely shaped by single causes. Thus, for instance, when we speak of the hemoglobin of the bar-headed goose (a species of goose that routinely migrates over the Himalayas, reaching altitudes above 29,000 feet) as having evolved to "enable the capture of oxygen at high altitudes," we cannot really mean that the only force shaping the evolution of this protein is the low-oxygen tension present at high altitudes. Similarly, we would not want to claim that the extraordinary migration of the bar-headed goose is made possible solely by the small number of amino-acid changes that characterize its hemoglobin: modifications of the circulatory, muscular, and respiratory systems are all part of the adaptive nexus that permits routine high-altitude migration. In effect, when evolutionary biologists speak of this molecule as an adaptation to high-altitude flight, we are using shorthand. The full (and unwieldy) statement would go something like this:

> The hemoglobin of the bar-headed goose has evolved in response to multiple evolutionary pressures, among them the need to bind oxygen; the requirement that the bond between oxygen and hemoglobin be reversible as a function of tissue needs; the stability of the component parts of hemoglobin; and their ability to interact to form a stable multimeric protein. The evolution of this protein is also constrained by the history of hemoglobin, a protein with a deep evolutionary past whose counterparts (homologs) exist in plants, suggesting a protein that has been evolving for more than 1.5 billion years. In addition, there are particular environments that exert additional demands on hemoglobin, including the ability to bind oxygen at very low external oxygen pressures. We have identified a small number of amino-acid changes in the hemoglobin of the bar-headed goose (relative to the sequence of hemoglobin in a closely related, lower-flying goose species) that appear to confer this additional functional ability. We have also carried out the necessary experiments to confirm that in fact the observed amino-acid changes alter the oxygen-binding capacity of the bar-headed goose hemoglobin.

In the face of such detail, it should come as no surprise that we use a more efficient shorthand, focusing on the particular feature of hemoglobin that appears as an adaptation to high-altitude flight. But, in the hands of creationists, the shorthand description substitutes for the full account: hemoglobin evolves as response to single selection pressures. In this formulation, high-altitude flight is the "problem" and bar-headed goose hemoglobin the "solution." While such a description may seem harmless, it in fact results in a deep misunderstanding of the evolutionary process: that any one feature of an organism arises in response to a single evolutionary pressure. Every aspect of the living world is the result of myriad forces, often varying in time and space. These forces frequently act in opposition to each other, and biological forms embody the balance point of these opposing selective forces. Chance events —the unpredictability of mutation, the inevitable loss of advantageous traits through the sampling process, the inexorable nature of extinction—also play into the evolutionary process. Finally, a functioning organism requires integration: all of the features that we choose to study in isolation are in fact embedded in a complicated network of interactions. Every step of metabolism, every component of cell structure, every part of the living machine is shaped by its ability to interact with other components. In short, although the mechanisms of evolution are deceptively easy to grasp, the causes of evolution are, even for the simplest of traits, jarringly complex.

Of course, what is sauce for the goose must be sauce for the gander. Evolutionary biologists, too, need to acknowledge the complexity of causes that underlie the evolution of biological features. Armchair evolutionary biology—a beguiling two-step process that consists of pointing at some feature of the living world and making up a plausible adaptive story to account for its existence—should not be confused with the real thing. Biological adaptation is a strong claim and demands strong proof. To return to our hemoglobin example: My colleagues have shown (not speculated, *shown*) that the amino-acid changes that are unique to bar-headed goose hemoglobin (Perutz 1983; Jessen et al. 1991) do in fact confer higher affinity for oxygen. Furthermore, they have examined the hemoglobin of other high-altitude migrators and fliers and found it to bind oxygen more readily than does the hemoglobin of lower-altitude birds (particularly evident in the hemoglobin of Ruppell's griffon vulture, seen flying more than six miles above the earth). The results of these comparative studies are satisfying: In some cases, the same amino-acid positions have undergone similar replacements in several species; in others, a different set of replacements converge to produce similar structural changes in the hemoglobin. The case linking the amino-acid changes we observe with the increased-oxygen affinity of these hemoglobins is exceptionally tight. Finally, evolutionary biologists working on this problem remind us that the hemoglobin changes are but a part of a much larger suite of changes that accompany high-altitude flight, including changes in lung structure and function, in the circulatory system, and in the synthesis of red blood cells (Faracci 1991). Evolutionary explanation is seldom simple, but it is precisely this rigor that protects evolution from the simple-minded dismissals that anti-evolutionists deploy.

Complexity of Outcome

This second meaning of complexity in the context of evolutionary theory centers on the apparent discrepancy between the rich, layered, noisy character of the living world and the simplicity of the mechanisms that give rise to it. Anti-evolutionists frequently express their utter amazement at this aspect of evolution: How could the mindless agency of differential survival and reproduction result in so many exquisite and elaborate forms? The disbelief is frequently expressed in folksy appeals to common sense ("C'mon now, could this possibly be true?") or logically flawed metaphors trying to masquerade as argument (a million monkeys typing away on a million typewriters could not come up with *Hamlet*).[1]

Notwithstanding the agenda that underlies their critiques of evolution, the creationists' discomfort stems from a common but incorrect assumption about the relationship between cause and

1. See, for instance, www.nutters.org/docs/monkeys (last accessed May 11, 2006).

effect in the material world—namely, that complex outcomes can only derive from complex causes. The flip side of the same fallacy would have us believe that simple mechanisms can only give rise to simple effects.

Appealing as that construction seems, a good deal of science in the last twenty years has been devoted to undercutting the notion that simple mechanisms preclude complex outcomes. This phenomenon is known as "emergent complexity." In fields as disparate as computer science, mathematics, physics, ecology, genomics, and developmental biology, for example, the notion of emergent complexity has altered our perspective in dramatic ways (Holland 1998).

One of the most intuitive examples of this phenomenon of emergence comes from the study of ant colonies. Even a casual observer is struck by the organization of an ant colony. Their remarkable specialization and division of labor have enabled these social insects to successfully colonize a vast array of habitats. The architecture of ant nests, the efficiency of foraging, the defense of the colony are all exquisite, highly complex, and adaptive outcomes. Yet the architects of this complexity are simple organisms whose brains harbor fewer than 10,000 neurons. An individual ant obviously does not carry a mental picture of the tunnel system designed to keep the temperature of the nest constant, nor the instructions on how to build such tunnels. Nonetheless, ant colonies consistently, successfully, and reproducibly generate complex nest architectures—even adapting the details to the particular heat and humidity conditions of specific habitats (Gordon 1999; Pereira and Gordon 2001).

What lies at the core of this apparent paradox is the concept of *emergence*: Complex outcomes arise from the interaction of much simpler units. Given a small set of rule-based behaviors, a system for incorporating information from the environment, and a sufficient number of interactions, complexity not only can, but will, inevitably arise. The notion of emergence presents a straightforward solution to creationist incredulity: Complex outcomes really do arise from simple rules. The ant colony provides a concrete example, with the design arising from the constant, coordinated, but not goal-directed action of tens of thousands of simpler units. Ironically, the notion of emergence is already foreshadowed in the Bible (Proverbs 6:6), when the authors urge:

> Go to the ant, thou sluggard; consider her ways, and be wise: which having no guide, overseer, or ruler, provideth her meat in the summer, and gathereth her food in the harvest.

The concept of emergent properties has become the focus of many disparate scientific disciplines. Growing interest in chaos theory stems in part from the increasing realization that even the simplest of processes (and the mathematical functions meant to capture such processes) can become complex and unpredictable in certain regions of parameter space. Similarly, simple and predictable forces may generate processes and patterns that appear unintelligibly complex and intractable. This conceptual revolution in our understanding of disorder and its rules has been driven in part by the advent of computers, which enable us to simulate processes in detail across a wide range of parameter values. These computational models have consistently revealed the counterintuitive result that so puzzles creationists: Complexity arises, for the simplest of reasons.

Complexity of Structure

From the outset, critics of evolutionary explanation seized on Darwin's somewhat disingenuous assertion, "If it could be demonstrated that any complex organ existed, which could not have possibly been formed by numerous, successive, slight modifications, my theory would absolutely break down" (1964, 189). Darwin knew that biological features involving a large number of interacting parts appeared to pose a very particular challenge to the notion of evolution by natural selection. But in the very next line, seldom quoted, he goes on to say, "But I can find out no such case" (1964, 189).

Creationist critics, alas, have failed to grasp Darwin's rhetorical device. Instead, *plus catholiques que le pape,* they declare complex structures to be definitive evidence of a designing intelligence. Their argument is elusive, but it appears to rest primarily on the notion of "irreducible complexity": biological features that provide some obvious "advantage" to the bearer, but where the absence of a single component in the structure would render it a meaningless, useless agglomeration of parts. How, then, they ask, could natural selection possibly have built this complex structure, given that it only becomes useful when all of the parts are in place? They illustrate their argument by using the analogy of a mousetrap: The full device works to catch unsuspecting mice, but remove the spring or the trigger or the hinge and it becomes no more than a cheese platter. The mousetrap works because it was designed (by James Henry Atkinson, a British inventor who came up with the "Little Nipper" around 1897). So, too, they argue, must it be with biologically complex structures: they work because they are designed by some ineffable intelligence, and they cannot have arisen by the interplay of chance and history.

This meaning of *complexity* has arisen of late as the preferred target of creationist anxiety, and it forms the basis of their most recent attempts to undermine evolutionary logic. In an effort to make this objection sound particularly contemporary, creationist arguments have recently focused on molecular examples of presumed "irreducible complexity": the clotting cascade in vertebrates, flagellar motors, and the like.

While the advent of molecular methods has certainly supplied us with an embarrassment of riches when it comes to examples of the complex and the baroque in biology, the "irreducible complexity" chestnut is the old nineteenth-century argument from design: no design without a designer. And, like the argument from design, it fails because it rests on some profound misunderstandings of the evolutionary process. The notion that a complex structure, or a bit of cellular machinery, or a biochemical pathway cannot have evolved, but must have been designed unless a "direct gradual route" from simple origins to finished product can be found is, at best, a caricature of the task of evolutionary biology. If anything, the accumulation of molecular data makes it clear that existing genes and their products are constantly coopted and adapted to new functions, and their current role is often a poor guide to their original *raison d'être.*

The eye provides a particularly good example of intelligible complexity at both the structural and the molecular level. Darwin fretted about the challenge that the eye, "an organ of extreme perfection," presented to the theory of evolution by natural selection. He would write:

> To suppose that the eye, with all its inimitable contrivances for adjusting the focus to different distances, for admitting different amounts of light, and for the correction of spherical and chromatic aberration, could have been formed by natural selection seems, I freely confess, absurd in the highest possible degree. (1964, 186)

Darwin's concern is again merely rhetorical. As he often does in the *Origin,* he is posing the problem in order to disarm it. He goes on to discuss a trajectory for possible evolution of the eye in some detail, using a variety of examples to illustrate exactly how the operation of natural selection might have resulted in a complex organ for the transmission of light and image. Ironically, Darwin would have been delighted by the latest developments in our understanding of the ontogeny and evolution of the eye. How he would have enjoyed, for instance, the discovery of a family of homeobox-containing genes (*eyeless* and its homologs) involved in the development of the eye in insects and vertebrates (Quiring et al. 1994; Halder et al. 1995).

These genes are part of a large and fascinating family whose members all encode transcription factors—proteins that can directly bind to specific targets in DNA and prevent (or facilitate) the transcription of other genes. Work over the past ten years suggests that the presence of the *eyeless* (ey) protein triggers the developmental pathway that generates eye structures in the fruit fly *Drosophila.* This gene can be experimentally expressed in tissues where it would normally be turned off—such as the antennae—and there too induce the formation of eye structures (where eyes would normally never form). Homologs of *eyeless* have now been found in cephalopods (Tomarev et al. 1997). The vertebrate eye, the cephalopod eye, and the insect eye are clearly very different in their development and organization, and they have always been considered a stunning example of multiple origins and convergence of complex structures. The discovery of *eyeless* and its homologs points to a common developmental thread to this instance of convergence in an apparently "irreducibly complex" organ.

And there is more: Eyeless homologs have now been found in organisms without any discernible eye, such as cnidarians (jellyfish and corals) (Groger et al. 2000; Miller et al. 2000). This discovery suggests an even earlier role—possibly the regulation of photopigment expression—for genes that later become integral parts of the developmental pathway of these organs "of extreme perfection." It also underscores the surprising connections that link preexisting bits of molecular machinery to more recent complex innovations. We could not have predicted that a homeobox gene would illustrate a link in the developmental pathways of cephalopod, vertebrate, and insect eyes, nor would we have assumed that it would link creatures capable of sight to those that can only sense light in the most rudimentary fashion (Gehring 2002). But there is no metaphysics here, and the inability to conceive of complex features evolving from simpler beginnings is a failure of imagination and not of evolutionary logic.

The eye supplies a second lesson refuting "irreducible complexity," this time concerning the evolution of lens proteins. The problem appears daunting at first: How can a structure such as the lens—which must be stable, transparent, and nonrefractive—possibly have evolved? Without a light-transmitting lens, the complex eye fails. The resolution of this puzzle is both simple and surprising: The proteins of the lens are a motley collection of preexisting housekeeping proteins recruited into doing double duty as lens crystallins. Would we have predicted that epsilon crystallin—the protein that provides stability, minimal refractivity, and light diffraction to the lens of the bird eye—would be the same protein that catalyzes the conversion of lactate to pyruvate in glycolysis?

Or that the same light-transmitting function in the squid eye would be performed by a close cousin of glutathione-S-transferase, a protein normally involved in detoxifying certain classes of mutagens (Piatigorsky and Wistow 1989; Piatigorsky et al. 1994)? These surprising discoveries illustrate a common theme in evolutionary reconstruction: The process relies on the use and reuse of preexisting parts. The evolution of the complex eye did not involve the simultaneous construction, from scratch, of all the parts required to capture, focus, and perceive light. Instead, the eye emerges from the constant experimentation of mutation and the constant sorting of available variants.

In certain cases, such as the evolution of lens crystallins, the details of this evolutionary tinkering are well understood. A series of studies has uncovered the details of this molecular recruitment. We now know, for instance, that the proteins recruited for the lens tend to be small, globular proteins that can be tightly packed (facilitating light transmission). In addition, they tend to be stable proteins, a useful feature for a structure like the lens that is not vascularized and hence cannot renew its protein constituents. The mechanisms of recruitment are as varied as the recruits themselves (Piatigorsky 2003). In certain cases, the recruited gene is duplicated and paired with a set of lens-specific control regions. In other instances, the machinery that keeps a household gene quiescent in the developing lens is reversed and the protein is expressed, but now for a very different reason. The pattern is startling and unexpected, and a healthy reminder of the opportunistic nature of evolutionary change.

When anti-evolution critics demand evidence of a long linear series of intermediates linking "simple" and "complex" structures, they not only misunderstand the evolutionary process, they vastly underestimate the number of possible trajectories that connect the past to the present. We would never have predicted *a priori* that a glycolytic enzyme would end up as part of the light-transmitting matrix of the eye. Yet it has, in a humbling reminder of the number and variety of evolutionary experiments underway at any given time.

That said, it is important for us as evolutionary biologists to understand the constraints, contingencies, and necessity that underlie the evolution of complex structures. But that is a welcome challenge, and not a threat to the coherence of evolutionary explanation. In fact, for many complex features of organisms, the mission is close to accomplished.

Complexity of Organization

The web of interrelationships that characterizes every functioning ecosystem on this earth strikes casual observers and professional biologists alike. This interdependence of living things, the variety of associations—from parasitism to obligate symbiosis—captures yet another meaning of complexity in biological systems. Here again, the apparent fit between organisms seems to suggest some higher intelligence at work, some supervisory gardener bringing harmony and color to the garden. Darwin understood that this Romantic vision of nature as a benevolent Eden was undercut by the material explanation of natural selection. He was careful to define the *struggle for existence*—the engine of evolution by natural selection—as a metaphor that need not imply violence or destruction:

> I use the term Struggle for Existence in a large and metaphorical sense, including dependence of one being on another, and including (which is more important) not only the life of the individual, but success in leaving progeny. (1964, 62)

Still, he knew that a fully benevolent vision of nature was an imposed illusion:

> We behold the face of nature bright with gladness, we often see superabundance of food; we do not see, or we forget, that the birds which are idly singing round us mostly live on insects or seeds, and are thus constantly destroying life. . . . (1964, 62)

He argued against the Romantic vision because a more objective view of the natural world undercut the notion of a designer, or else (even more scandalous) forced the acceptance of a fundamentally cruel or uncaring designer. This problem of defining intention in the natural world might seem quaint to our twenty-first century sensibilities, but it was a major issue in the nineteenth century. Naturalists and theologians worried, for instance, about the meaning behind *parasitism*. Victorian society was agitated by the discovery of parasitic wasps, which laid their eggs in the bodies of caterpillars that they had previously paralyzed (but did not kill), thus allowing the young wasp larvae to hatch and feed on the helpless caterpillar, essentially eating their way out of it. Extensive writings and sermons addressed "the Ichneumonid problem": What benevolent designer would create an ecological relationship that involved so much ostensible cruelty and suffering? In a letter written some years after the publication of *The Origin*, Darwin would write:

> I own that I cannot see as plainly as others do, and as I should wish to do, evidence of design and beneficence on all sides of us. There seems to me too much misery in the world. I cannot persuade myself that a beneficent and omnipotent God would have designedly created the Ichneumonidae with the express intention of their feeding within the living bodies of Caterpillars, or that a cat should play with mice. (1993)

Some years later, a similar sentiment would be echoed by John Stuart Mill: "If there are any marks at all of special design in creation, one of the things most evidently designed is that a large proportion of all animals should pass their existence in tormenting and devouring other animals" (1987, 26).

The notion of a benevolent and charitable nature—from which humans are to draw moral lessons—endures in the more sentimental strains of the contemporary environmental movement. Perhaps unsurprisingly, a similar designer-driven conception of ecosystem complexity pervades creationist discussions. It seems unimaginable to them that the complex organization of even the simplest ecosystem—producers and consumers, competitors and parasites, mutualists and symbionts, could have arisen through the agency of chance and natural selection. Recent experiments, both in the laboratory and in the field, suggest instead that the emergence of complex ecological interrelationships is virtually certain and requires only the availability of usable resources and the existence of a species pool from which ecological players can be drafted. Newly emergent volcanic islands, or environments denuded by catastrophic events (fire, hurricane), provide clear examples of the self-assembling nature of complex ecosystems.

In a contingent but still predictable succession, available habitats are invaded by rapid colonizers, which are in turn displaced by more efficient competitors. The arrival of every new participant in an ecosystem in turn creates novel ecological opportunities for parasites, symbionts, and predators. Ecosystem complexity increases (up to a point), entrained by this positive-feedback loop. In a complex process of trial and error, new participants try to gain a foothold in an increasingly crowded environment. Some species cannot successfully insert into the maturing ecosystem, others are displaced by more effective competitors. A similar triage may be occurring at the ecosystem level: Unstable networks of ecological interactions give way to more robust networks of interaction (Schneider and Kay 1995; Levin et al. 2001). The result, in the absence of a designer, is a complex ecosystem in which little appears to go to waste.

A similar result emerges from *in vitro* experiments where populations of viruses or bacteria are allowed to vary and are subjected to the forces of chance and selection. In almost every case, these systems result in the emergence of one or more unexpected and unselected-for ecological interactions: parasitism, mutualism, or symbiosis. Even in the simplest systems—this time involving populations not of organisms but of molecules (catalytic RNA)—a similar result obtains: Parasites, commensals, and symbionts arise and carve out space in the experimental ecosystem (Hanczyc and Dorit 1998). In short, complexity of interactions is not only unsurprising in evolving systems, it is completely predicted.

Epilogue

The problem of complexity is likely to pervade scientific debate in the twenty-first century. Understanding how outputs can be more varied and complex than the inputs that produce them is a profound and stimulating question that exists almost everywhere we look. When we seek to explain how 30,000 genes in the human genome can give rise to a human being, how a handful of interacting genes involved in floral development can generate the beauty and diversity of floral shapes, how the richness of the Amazonian ecosystem is self-sustaining in the absence of human interference, or how a simple set of equations can produce the complicated and unique fractal patterns that adorn so many of our computer desktops, we are seeking to understand complexity. In a narrow and unexpected sense, anti-evolutionists are right: Complexity is a profoundly stimulating problem for the sciences. Where they are wrong—deeply, misguidedly, anti-intellectually wrong—is in their conviction that because we do not have all the answers, we must be on the wrong track. Difficult questions—not pat answers—are what get us up in the morning. Anything else is simply not worth the trouble.

References

Behe, M. J. 1998. *Darwin's black box.* New York: Free Press.

Darwin, C. R. 1859 (reprinted 1964). *On the Origin of Species,* facsimile of 1st ed. Cambridge, MA: Harvard University Press.

Darwin, C. R., writing to Asa Gray, May 22, 1860 (reprinted 1993). In *The Correspondence of Charles Darwin, Vol. 8,* ed. F. Burkhardt. Cambridge: Cambridge University Press.

Dembski, W. A. 1998. *The design inference: Eliminating chance through small probabilities* (Cambridge Studies in Probability, Induction and Decision Theory). Cambridge: Cambridge University Press.

Duncan, M. K., J. I. Haynes 2nd, A. Cvekl, and J. Piatigorsky. 1998. Dual roles for Pax-6: A transcriptional repressor of lens fiber cell-specific beta-crystallin genes. *Journal of Molecular Cell Biology* 18 (9): 5579–86.

Faracci, F. M. 1991. Adaptations to hypoxia in birds: How to fly high. *Annual Review of Physiology* 53: 59–70.

Gehring, W. J. 2002. The genetic control of eye development and its implications for the evolution of the various eye-types. *International Journal of Developmental Biology* 46 (1): 65–73.

Gordon, D. M. 1999. Interaction patterns and task allocation in ant colonies. In *Information Processing in Social Insects*, ed. C. Detrain, J. M. Pasteels, and J.-L. Deneubourg, 51–76. Basel, Switzerland: Birkhauser Verlag.

Groger, H., P. Callaerts, W. J. Gehring, and V. Schmid. 2000. Characterization and expression analysis of an ancestor-type *Pax* gene in the hydrozoan jellyfish *Podocoryne carnea*. *Mechanisms of Development* 94 (1–2): 157–69.

Halder, G., P. Callaerts, and W. J. Gehring. 1995. Induction of ectopic eyes by targeted expression of the eyeless gene in *Drosophila*. *Science* 267 (5205): 1788–92.

Hanczyc, M. M., and R. L. Dorit. 1998. Experimental evolution of complexity: In vitro emergence of intermolecular ribozyme interactions. *RNA* 4: 268–75.

Holland, J. H. 1998. *Emergence: From chaos to order.* Oxford: Oxford University Press.

Jessen, T. H., R. E. Weber, G. Fermi, J. Tame, and G. Braunitzer. 1991. Adaptation of bird hemoglobins to high altitudes: Demonstration of molecular mechanism by protein engineering. *Proceedings of the National Academy of Sciences USA* 88 (15): 6519–22.

Johnson, P. E. 1993. Creator or blind watchmaker? *First Things* 29: 8–14.

Levin, S. A., J. Dushoff, and J. E. Keymer. 2001. Community assembly and the emergence of ecosystem pattern. *Scientia Marina* 65 (suppl. 2): 171–79.

Mill, J. S. 1874 (reprinted 1987). On nature. In *Three essays on religion: Nature, the utility of religion and theism.* Amherst, NY: Prometheus Books.

Miller, D. J., D. C. Hayward, J. S. Reece-Hoyes, I. Scholten, J. Catmull, W. J. Gehring, P. Callaerts, J. E. Larsen, and E. E. Ball. 2000. *Pax* gene diversity in the basal cnidarian *Acropora millepora* (Cnidaria, Anthozoa): Implications for the evolution of the *Pax* gene family. *Proceedings of the National Academy of Sciences USA* 97 (9): 4475–80.

Paley, W. 1802. *Natural theology: or, Evidences of the existence and attributes of the deity, collected from the appearances of nature.* Albany, NY: Printed for Daniel & Samuel Whiting.

Pennock, R. T. 1999. *Tower of Babel: The evidence against the new creationism.* Cambridge, MA: Bradford Books.

Pereira, H., and D. M. Gordon. 2001. A trade-off in task allocation between sensitivity to the environment and response time. *Journal of Theoretical Biology* 208: 165–84.

Perutz, M. F. 1983. Species adaptation in a protein molecule. *Molecular Biology and Evolution* 1 (1): 1–28.

Piatigorsky, J. 2003. Crystallin genes: Specialization by changes in gene regulation may precede gene duplication. *Journal of Structural and Functional Genomics* 3 (1–4): 131–37.

Piatigorsky, J., M. Kantorow, R. Gopal-Srivastava, and S. I. Tomarev. 1994. Recruitment of enzymes and stress proteins as lens crystallins. *EXS* 71: 241–50.

Piatigorsky, J., and G. J. Wistow. 1989. Enzyme/crystallins: Gene sharing as an evolutionary strategy. *Cell* 57 (2): 197–99.

Quiring, R., U. Walldorf, U. Kloter, and W. J. Gehring. 1994. Homology of the eyeless gene of *Drosophila* to the small eye gene in mice and Aniridia in humans. *Science* 265 (5173): 785–89.

Schneider, E. D., and J. J. Kay. 1995. Order from disorder: The thermodynamics of complexity in biology. In *What is life? The next fifty years: Reflections on the future of biology*, ed. M. P. Murphy and L. A. J. O'Neill, 161–72. New York: Cambridge University Press.

Tomarev, S. I., P. Callaerts, L. Kos, R. Zinovieva, G. Halder, W. J. Gehring, and J. Piatigorsky. 1997. Squid *Pax*-6 and eye development. *Proceedings of the National Academy of Sciences USA* 94: 2421–26.

Logic and Math Turn to Smoke and Mirrors: William Dembski's "Design Inference"

Wesley R. Elsberry

Introduction

"THE DESIGN INFERENCE" (TDI) IS ADVERTISED as an argument for finding the property of specified complexity (SC) in certain events. According to its originator, William A. Dembski, one can infer from such complexity whether or not such events must have been caused by an intelligent agent. This argument and the ideas that support it provide the intellectual foundation of the modern "intelligent design" movement. In this essay, I will describe the general arguments that comprise "intelligent design" and review specific problems with the ideas of the design inference, specified complexity, and various claims about the testability of "intelligent design."

The basis of "intelligent design" argumentation is a straightforward denial of the sufficiency of evolutionary processes to account for the history and diversity of living organisms and of natural processes to account for the origin of life. Dembski's argument is an argument by elimination of alternatives; one of the alternatives he claims to eliminate is natural selection. Dembski views Michael Behe's (1996) concept of irreducible complexity (IC) as an application of specified complexity to molecular biology (Dembski 2002). Dembski's claim is that whenever one identifies a system as having the IC property, one can also show that the same system will have the SC property. Dembski's sole support for this claim is that his (erroneous) analysis of the *E. coli* flagellum shows that this IC system has the SC property.

Dembski's Specified Complexity

William Dembski is a scholar with advanced degrees in divinity (Princeton Theological Seminary), mathematics (University of Chicago), and philosophy (University of Illinois at Chicago). He is currently research professor in philosophy at Southwestern Baptist Theology Seminary in Fort Worth, Texas. He is also a senior fellow of the Discovery Institute's Center for Science and Culture. His concept of specified complexity is developed in *The Design Inference* (Dembski 1998a) and significantly modified in chapter 2 of *No Free Lunch* (Dembski 2002). I will attempt to summarize Dembski's major points and arguments in non-technical terms. After that, I will review a number of criticisms that Dembski's concepts and arguments have received.

Dembski's concept of specified complexity comes with a fair amount of its own jargon (Dembski 1997), but it is rife with redefinitions of common terms and phrases, both scientific and vernacular. It is important to keep in mind the differences between common usage and the meanings that Dembski ascribes to these terms. He identifies *specified complexity* as a property of an event and claims that the presence of specified complexity is a reliable indication that an intelligent agency caused the event in question. Dembski has used a number of phrases interchangeably for this concept, including "complexity–specification criterion" and "complex–specified information." The property of specified complexity is a compound one, linking complexity (defined by Dembski as the improbability of an event given a chance causal hypothesis) and specification (a match to an independently given pattern). When both conditions are satisfied, he considers the event to be explicable only as being due to design.

According to *The Design Inference*—considered seminal by "intelligent design" advocates—a specification is an independently given pattern that allows us confidently to reject all relevant causal hypotheses for an event that are based on chance. Dembski's methods for testing hypotheses derive from work by the British mathematician Sir Ronald A. Fisher, probably best known in the United States for his work on applied mathematical models of population genetics and natural selection (Fisher 1930). Fisher uses a statistical approach to hypothesis testing. Any experiment will have data that follow some sort of distribution. If the observed measurements lie in the *rejection region*—far from what would be expected if the results were due to chance—then one may reject the null hypothesis (that the results are, in fact, due to chance) as inadequate. For data that are distributed in the classic bell curve, for example, observed values that lie in the tails of the curve either far to the left or far to the right of the center of the curve would encourage us to reject the null hypothesis and instead prefer a hypothesis that something other than chance explains the observed data values.

A key problem in experimentation and statistical hypothesis testing is whether one can legitimately choose the rejection regions (where observed data lead to the rejection of the chance hypothesis) after the fact—that is, after one already has the results of the experiment. Standard procedure in scientific experiments calls for planning data collection and statistical tests in advance of actually collecting the data, in order to avoid "cherry-picking"—the deprecated practice of examining a data set after the fact and then figuring out what statistical test will produce a *significant* result of the desired form. It is a form of pattern recognition or construction, since some significant pattern can be found in any data set, given enough after-the-fact examination.

Dembski often uses an example of an archer who releases an arrow at a large wall. If the arrow lands in the middle of an already-painted bull's-eye on the wall, we may conclude that it is far less likely that the archer is incompetent, but lucky, than that the archer is skilled and intended to place the arrow there. The already-painted bull's-eye, Dembski tells us, is a *specification*. If the archer instead shot an arrow willy-nilly at the wall, then painted a bull's-eye around where the arrow happened to stick, Dembski would consider the pattern as a *fabrication* and not anything that could legitimately guide us to making a design inference.

Dembski cites four necessary conditions that distinguish *specifications* (or these useful patterns) from *fabrications*, which give us no cause to infer design because they are little more than descriptions of the outcomes of events. These specifications, according to Dembski, exist independently of the events to which they are applied. The first condition is *conditional independence*, that knowing about the specification does not alter our estimate of the probability of occurrence of the event in question. The second is *tractability*, the idea that a problem is capable of solution given the resources at hand. The third is *detachability*, another way in which one assesses whether the specification is independent of the event being analyzed. The fourth condition is what Dembski calls *small probability*. This means basically that the specification being examined has a low probability of being observed in the type of event under study. These conditions have been examined by a number of critics whose judgment uniformly is that these do not set aside or satisfy concerns about analyses of data sets made after the fact.

Dembski needs to establish his concept of specification as a causal pattern that avoids the criticisms of "cherry-picking," thereby allowing us to reject other competing hypotheses as explanations for already-observed patterns. This is why he spends so much time arguing for the validity of his procedure for establishing a specification as legitimate for rejecting chance: His is a pattern formed after the occurrence of the event being classified. Without this methodological sleight of hand, Dembski would be unable to apply his concept of specified complexity to phenomena in the field of biology and assert his claim that the explanatory filter allows us to determine, even after the fact, that a pattern is due to specified complexity and to determine that a particular event exhibits specified complexity.

We should notice, though, that Dembski's example of the archer works against him in this project. The distinction between the specification of an already-painted bull's-eye that the archer hits corresponds to the usual experimental procedure of determining rejection regions and tests prior to data collection, and the fabrication (or illegitimate painted-around-the-arrow bull's-eye), corresponds precisely to the concerns scientists have about any project based upon after-the-fact analysis that we know by experience is prone to "cherry-picking."

The Explanatory Filter

To determine whether an event has the property of specified complexity, Dembski deploys what he calls the "explanatory filter" (Dembski 1996). This is a stepwise sequence of actions whose logic Dembski expounds as the basis of a design inference (Dembski 1998a). An event is analyzed in Dembski's explanatory filter/design inference (EF/DI). This approach marks a break from William Paley's discussion of artifacts. In Dembski's framework, an event is "any actual or possible occurrence in space and time" (1998a, 72). This gives enormous flexibility to his application of specified complexity. One need not analyze an artifact but rather the *event* that produced such an artifact. (Elsewhere, Dembski has argued against casting events as processes [Dembski 1999, 78].) He evaluates an event according to two criteria: its complexity (i.e., its improbability of occurrence given a causal hypothesis) and whether a specification can be given for it.

Dembski's explanatory filter/design inference is a classification scheme for events with three decision nodes (see figure 1).

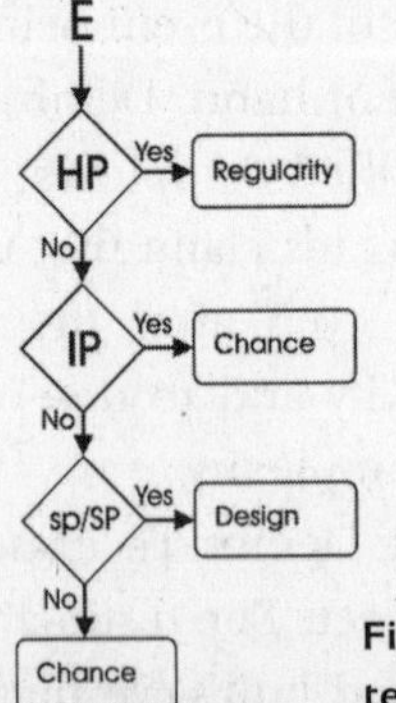

Figure 1. William Dembski's explanatory filter for inferring design from biological data.

In step 1, if an event is deemed to have high probability, it is classified as due to a regularity—i.e., the event can be explained through law-like physical processes. (Perakh [2004] notes that, in actuality, we determine that something has high probability because it is caused by a law-like process.) An as-yet-unclassified event then moves on to the second decision node. If it has intermediate probability, it is classified as due to chance, which is a sufficient explanation for many events. Thus-far-unclassified events (which have low probability) then move on to the third decision node. If the event has low probability and *also* conforms to a specification, it is classified as due to design; if it has low probability and is unspecified, it is classified as due to chance. Dembski describes this as the "law of small probability"—that specified events of small probability do not occur by chance. Throughout *The Design Inference,* he claims that the deductive argument of the explanatory filter leads conclusively to certain events' being classed as due to design. The catch is that Dembski is using his own definition of design—where design is simply the residue that remains after chance and regularity are eliminated (Dembski 1998a).

Dembski makes a number of claims based upon his EF/DI. These include: His EF/DI is reliable in the sense that it does not make false-positive attributions of design (Dembski 1996, 1998a, 1998b, 2002, 2004); biologists cannot be concerned that further knowledge might overturn a finding of design for biological systems (1998b, 2001c, 2002, 2004); algorithms, functions, and natural law are in principle incapable of "generating" events with specified complexity (1997, 2002); "intelligent design" is rendered a fully scientific research program because of the EF/DI (1998b, 2001c, 2002, 2004); specified complexity is falsifiable/refutable (2001b, 2002, 2004); and specified complexity is testable (2001b, 2002, 2004).

Dembski's concept of specified complexity and the claims enumerated above have elicited a number of criticisms. These range from doubts about the basic validity of Dembski's arguments to outcries that his logical apparatus is inapplicable to particular problems. These critiques of specified complexity come from several different topical areas, including epistemology, logic, mathematics, and pragmatics. Dembski's responses—when he makes them—tend not to allay the criticism. To keep things relatively simple here, I will concentrate on reviewing the criticisms based on epistemology and bring up other concerns as they become relevant.

The Importance of Epistemology

Epistemology is the study of the nature of knowledge. It is concerned with how we can reliably acquire knowledge or at least recognize when what we think we know may be unreliable. It is within epistemology that we encounter discussion of deduction, induction, and abduction, among other things. Consideration of specified complexity involves epistemology because it is claimed that SC gives us a reliable indication of an intelligent agent being causally responsible for a particular event.

Let us be clear about what is at issue here: Dembski has made some large claims concerning his EF/DI. Not only is the EF/DI supposed to deliver a reliable indication of causation by an intelligent agent, but it also is supposed to make unjustifiable the worry that biologists have had in making a design inference for biological systems (Dembski 1998b, 2002, 2004). In the past, such inferences have been overturned when further knowledge became available, but Dembski claims that his EF/DI overcomes such considerations. It does so because, he says, the EF/DI never gives a false-positive indication of design when natural processes actually caused the event. Further, once an event is found to be due to design via the EF/DI, no additional information can cause the classification to change to a nondesign category (Dembski 1998b).

To recognize specified complexity, we apply the EF/DI. But we do so with limited knowledge. This is an important point. The EF/DI is only useful if it allows us to make reliable design inferences under limited knowledge. Dembski points out only two ways in which his EF/DI could be wrong: It can fail to find design when design actually is present (a false negative), or it can find design for an event whose causal history is one of natural processes (a false positive). Dembski's stand on this issue is that his EF/DI may produce a large proportion of false negatives without harm, but that it never produces a false-positive attribution of design. Thus, he dismisses those who have been reluctant to conclude design in the past:

> Yet without such a method for distinguishing the two, how can we be sure that our ascriptions of design are reliable? It is this worry of falsely ascribing something to design (here construed as creation) only to have it overturned later that has prevented design from entering science proper.
>
> This worry, though perhaps justified in the past, can no longer be sustained. There does in fact exist a rigorous criterion for discriminating intelligently caused from unintelligently caused objects. Many special sciences (for example, forensic science, artificial intelligence, cryptography, archeology, and the Search for Extraterrestrial Intelligence) already use this criterion, though in a pre-theoretic form. I call it the complexity-specification criterion. When intelligent agents act, they leave behind a characteristic trademark or signature that I define as specified complexity. The complexity specification criterion detects design by identifying this trademark of designed objects. (Dembski 2002, 6)

Elsewhere, though, Dembski responds to criticism with a different view of his criterion:

> John Wilkins and Wesley Elsberry attempt to offer a general argument for why the filter is not a reliable indicator of design. Central to their argument is that if we incorrectly characterize the natural necessities and chance processes that might have been operating to account for a phenomenon, we may omit an undirected natural cause that renders the phenomenon likely and that thereby adequately accounts for it in terms other than design. Granted, this is a danger for the Explanatory Filter. But it is a danger endemic to all of scientific inquiry. (Dembski 2002, 14)

This is a fairly forthright admission that false positives may be generated by the EF/DI. Dembski simply seems not to see that this undercuts his claims that we need not worry about falsely ascribing design anymore. If the EF/DI can produce false positives, we certainly are justified in worrying about ascribing design, since such ascriptions may, contrary to Dembski's claims, be overturned as we acquire more knowledge about a phenomenon. It is not the case, as Dembski elsewhere asserts, that Wilkins and I want "risk-free" science (Dembski 2004). What we want is "truth in advertising." The strong claim that the EF/DI produces no false positives continues to be asserted by Dembski even after his concession that false positives are possible (Dembski 2002, 2004). Human history records a large number of false-positive attributions of design when further knowledge revealed them to be eminently false. Dembski has not shown that his EF/DI attenuates any of the worry that should accompany making a design inference.

The critical question to ask is, "When do we have enough knowledge to use the EF/DI such that it works reliably?" Dembski never addresses this question directly. Indirectly, he hints that we can have too little knowledge to apply the EF/DI. For example, in his discussion of the Oklo natural nuclear reactors of Gabon, in Africa, he says that not enough is known to generate the probability estimates needed for application of the EF/DI (Dembski 2002, 27). He does not, however, develop any guidelines for when to apply—and when not to apply—his EF/DI. The general tenor of his writings indicates that his EF/DI may be applied at any time, and specifically that the knowledge of biology current in 1998 was sufficient for its application to any biological system (Dembski 1998b). But his EF/DI lacks a "don't know" option (Wilkins and Elsberry 2001). Without explicit consideration of the sufficiency of one's background knowledge within the EF/DI, and with inconsistent indications of what constitutes sufficient knowledge, applying the EF/DI appears to be a risky venture, even in the few instances that Dembski provides.

When background knowledge is incomplete, Dembski's EF/DI fails as a reliable classifier of an event. Changes in background knowledge easily change classifications for an event, especially those events previously classed as due to design. John Wilkins and I came to the conclusion that the only circumstance in which one is safe in using the EF/DI is when one already knows the causal history of an event—in which case, Dembski's EF/DI is utterly superfluous (Wilkins and Elsberry 2001).

Jeffrey Shallit has noted that Dembski fails to give credit to alternative mathematical approaches relevant to making ordinary design inferences (Shallit 2002). One of these is the "universal probability distribution" (Kirchherr et al. 1997). Walter Kirchherr and coworkers describe a tractable method of testing randomness of bit strings based upon Kolmogorov complexity. Applying the "universal probability distribution" is simplicity itself when compared with Dembski's apparatus of the EF/DI. The inference made by Kirchherr, however, is that bit strings conforming to a pattern are due to a simple computational process. This procedure has been explicitly adapted for providing an alternative to Dembski's EF/DI; Shallit and I term this alternative *specified anti-information* (SAI) (Elsberry and Shallit 2003b). The apparent success of Dembski's EF/DI in these examples can be entirely explained by reference to such non-Dembskian principles, and Dembski's further conclusions of obligate intelligent agency are unsupported in light of these alternative explanations.

The Design Inference and Arguing from Ignorance

Dembski objects to critics' characterizing his EF/DI as an argument from ignorance (Dembski 2000). In response to such claims, he responds that arguments from ignorance have a logical structure of "not X, therefore Y," and that applying his EF/DI in analysis requires considerable knowledge. It is curious that Dembski himself lays out the "not X, therefore Y" structure of a fallacious argument as if his EF/DI were in some respect different. It is clear from *The Design Inference* and *No Free Lunch*, however, that Dembski's EF/DI is a deductive eliminative argument that demands acceptance of design when regularity and chance hypotheses of causation have been excluded. His formulation of design from *The Design Inference* (1998a, 54) which has not been retracted, reads as follows:

$$\text{des}(E) = \text{def} \sim \text{reg}(E) \;\&\; (\forall H) \sim \text{ch}(E|H) \quad (1)$$

Simply stated, this equation says that an event is due to design if it is not due to regularity and also is not due to chance, for all relevant chance hypotheses. This matches the framework Dembski himself stipulates as constituting an argument from ignorance. Design is here defined as the negation of regularity and chance; "not X, therefore Y" can be expanded as "not (regularity or chance), therefore design."

As for the considerable knowledge that Dembski cites as being necessary for deploying his EF/DI, it is easy to see that most of his knowledge is used to demonstrate "not (regularity or chance)," and none of it is applied to establishing the validity of "therefore design." In the absence of knowledge—which we may take here as empirical evidence of the natural world—any two propositions X and Y might be considered to be possibly true. Empirical evidence may show that particular consequences of proposition X, but not proposition Y, are false—in which case, we are justified in taking proposition X as having been falsified. But this does nothing to tell us whether Y is true or false. The level of effort (or "considerable knowledge") that goes into showing X to be false has no bearing upon whether Y is true or not (Sober 2004).

Calculations of Blood, Sweat, Toil, and Tears

Dembski criticizes Murray Gell-Mann's theory of "effective complexity" for having "resisted detailed applications to real world problems" (Dembski 2002, 133), yet this criticism is much more aptly aimed at Dembski's own apparatus for determining whether an event has "specified complexity." Dembski proposed his EF/DI in his 1996 doctoral dissertation, which was published as *The Design Inference* in 1998. In the years since that time, Dembski has provided only four examples of application of his EF/DI that either partly or completely demonstrate his logical and mathematical apparatus at work (Elsberry and Shallit 2003b). A number of other events or phenomena are claimed by Dembski to represent specified complexity, but these other cases rely only on a bare assertion without showing the calculated results (Elsberry and Shallit 2003b). No other researcher has published any example in which Dembski's EF/DI is applied to any event whatsoever. Given that few examples have been provided, even by Dembski, over a number of years, one may suspect that the application of the EF/DI in a rigorous and complete way is both difficult and time-consuming.

For some time, critics requested that Dembski produce the details of his calculations for finding specified complexity in biological examples, as his 1998 "Science and Design" article implied had already been done (Dembski 1998b). He deferred fulfilling this reasonable request until the 2002 publication of *No Free Lunch*. Dembski's single-example calculation, based on the *Escherichia coli* (*E. coli*) flagellum, has attracted criticism for its failure to implement his EF/DI fully, its failure to consider the probability that the system might have arisen through a process of natural selection, and its reliance upon made-up and sometimes false-to-fact parameters (Elsberry and Shallit 2003b). Howard Van Till gave a particularly thorough analysis and critique of the proffered *E. coli* flagellum example (Van Till 2002).

A standard measure of the importance and utility of scientific concepts is the citation of significant work by other researchers. Dembski's dissertation and monograph have been cited primarily by critics. Dembski himself brought to my attention a technical paper whose first sentence made a positive reference to *The Design Inference* (Chiu and Lui 2000). That the work thereafter completely ignores Dembski's EF/DI framework seemed not to deter him. Dembski's particular phrases—"specified complexity" and "complex specified information"—have similarly failed to become part of the technical jargon typically used by other workers, as some time spent with various literature search engines demonstrates. Despite hyperbolic statements from "intelligent design" advocates about "specified complexity" being increasingly accepted and used in the scholarly community, there appears to be no good evidence to support this.

Putting "Intelligent Design" and Specified Complexity to the Test

One of the most troubling problems I have found in Dembski's work is his consistent failure to give a testing methodology for determining whether his EF/DI actually delivers upon the promises that he makes for it. Despite his authorship of essays with such titles as "Is Intelligent Design Testable?" Dembski provides no way to check the validity of applying his EF/DI to the biological systems he likes to portray as evidence for the existence of an unembodied intelligent agent acting through biological history (Dembski 2001b, 2004). He gives four categories of possible tests: falsifiability/refutability, confirmation, prediction, and explanatory power.

For the category of *falsifiability/refutability*, Dembski tells us that specified complexity and irreducible complexity could be "falsified" or "refuted" by a suitable account of Darwinian evolution that would explain the origin of bacterial flagella (Dembski 2001b, 2004). This does not match the concept of falsifiability as Sir Karl Popper deployed it (1959). Popperian falsification applies when a statement that follows as a necessary implication from a theory is found to be false. (Popper considered refutability to be a synonym for falsifiability [Popper 1985, 1992].) Under those circumstances, we can reliably conclude that the theory that generated the false statement is itself false. But "intelligent design of bacterial flagella" does not follow as a necessary implication from either specified complexity or irreducible complexity, and thus finding a naturalistic mode of explanation for bacterial flagella would not provide a falsifying test of SC or IC. Other systems could then be portrayed as having the IC or SC properties without having placed those concepts at risk.

Dembski admitted in *No Free Lunch* that he was using the term *falsifiability* in a nonstandard fashion (2002, 357), and he substituted *refutability* for *falsifiability* in the corresponding section of *The Design Revolution* (2004, chapter 39). The change in label did nothing to improve the argument, which is as fallacious as it was before. "Irreducible complexity" is a far more malleable concept than Dembski would lead us to believe. Michael Behe has significantly altered his description of what properties of the human blood-clotting system render it irreducibly complex without any indication that this would place his theoretical concept at risk (Behe 2001).

It is harder to form an opinion of how malleable "specified complexity" may prove, since Dembski thus far has provided no nontrivial examples of the complete application of his EF/DI to a biological system. Despite his earlier claims to have made multiple calculations from various applications of his complexity-specification criterion to biological systems (Dembski 1998b), at the 2001 conference, "Interpreting Evolution," held at Haverford College, Dembski admitted that he had no calculations to show at that time—though he promised that such would appear in his then-forthcoming book, *No Free Lunch.*

That book contains a section entitled "Doing the Calculation," in which Dembski attempts a probability calculation for the assembly of an *E. coli* flagellum. Unfortunately, this calculation is woefully incomplete with respect to the "rigorous" framework proposed in Dembski's *The Design Inference* or even in the earlier sections of *No Free Lunch.* In particular, Dembski does nothing to show that a specification for the *E. coli* flagellum can be produced; he merely asserts that no biologist he knows of would deny that such a system is specified. This is an equivocation, for biologists use "specified" in a different way than Dembski has deployed it as a concept for his EF/DI (Elsberry and Shallit 2003b; Shallit and Elsberry 2004). Instead, Dembski urges a view that "intelligent design" has empirical *confirmation* (2001b, 2002, 2004). He even goes so far as to base his calculations on a fictional scenario (the prime numbers in the movie *Contact*). Dembski's example of SETI (search for extraterrestrial intelligence) researchers' inferring design comes from popular fiction, but one cannot employ or invent fictional cases that would fit one's viewpoint. Further, SETI research of the type seen in *Contact* falls under the category of ordinary design, not the rarefied design that Dembski seeks to underwrite (Wilkins and Elsberry 2001).

Dembski also seems confused about the specified complexity of biological systems. At some points, it is clear that he is speaking hypothetically, but in others it seems that he is taking demonstration of specified complexity in biological systems as having been achieved and as evidence of the action of an intelligent agent. Neither point has been satisfied.

For another category of testability, *prediction*, Dembski admits that "intelligent design" provides no basis for principled statements about what will happen (2001b). This Dembski attributes to the essential character of intelligent designers being that of inventors. But here again Dembski is wrong, for while designers may (occasionally) be innovators, it is still possible to make predictions about innovation. A famous example is Moore's law of electronic technology, wherein Moore predicted that "the number of transistors per integrated circuit would double every eighteen months." In popular use, this is typically taken to mean that computational resources (speed and memory space) will double every eighteen months. Moore made his prediction in 1965, and it is still holding up quite well. Moore did not predict *how* those transistors could be crammed into ever-smaller spaces, but he did predict that the requisite innovations would happen, and he predicted quite closely the rate at which they would happen. Further, none of Dembski's proffered instances of specified complexity concern cases of innovation.

The last category of testability Dembski considers, *explanatory power*, is discussed only briefly in "Is Intelligent Design Testable?" (2001b) before he switches to another topic. Dembski asserts that explanatory power identifies the best explanation from among competing hypotheses, and that now "intelligent design" outperforms "Darwinism" in this regard, though he does admit that "Darwinists" may not reach the same conclusion (Dembski 2001b). His treatment of explanatory power relies upon a qualitative and subjective view, where the question of which of two or more hypotheses actually has greater explanatory power is dependent upon the views and commitments of each observer. By contrast, the fields of statistics and philosophy of causality have already provided quantitative interpretations of explanatory power. In statistics, explanatory power corresponds to goodness of fit in regression, where the causal factor under analysis is said to explain a certain proportion of the variance of a dependent variable. The statistic used for this is R^2. This metric is appropriate when quantitative values can be given for both the causal factor (as the independent variable) and the dependent variable.

Judea Pearl offers two quantitative measures of explanatory power, cast in terms of probabilities, specifically to deal with general causes and singular causes (Pearl 2000, 221–22). In both of these interpretations, we obtain a quantitative result between 0 and 1 inclusive. While the kind of analysis Dembski and ID advocates explore via the EF/DI may not be amenable to quantification via regression, the metrics advanced by Pearl appear to be excellent matches. Dembski may be loath to deploy them, because in situations for which we have sufficient evidence to determine the causal history of biological features, we find that these features are due exclusively to ordinary evolutionary processes.

Dembski then poses another question, again exemplified by a hypothetical situation invented for the purpose (Dembski 2001b). He posits a situation where he designs and implements a "stinger" for a bacterium, and, having inserted this device into bacteria, wipes out all traces of the development and implementation he performed. Dembski claims that the origin of the stinger would remain mysterious under a Darwinian framework of inquiry but could be easily explained under "intelligent design." The question, Dembski asserts, is not whether this "stinger" actually exists, but rather which hypothesis can be applied to a larger number of phenomena.

The metric proposed by Dembski appears to be an exceedingly poor match to the problem. The "number of questions" covered by a concept is an ill-defined indicator. This has been known since the time of John Locke, who noted that an almost infinite number of trivial relations could be generated concerning any phenomenon of interest (Locke 1690). It seems useless to note that one concept can be applied to a greater number of questions than another when there is no constraint placed upon the significance of the questions proposed. Further, Dembski's preferred interpretation leaves "intelligent design" conjectures open to the criticism that they "explain" too much. In Dembski's EF/DI, a designer of sufficient capability is postulated to exist in the appropriate time and place to "explain" any event that is found to have specified complexity, without any independent evidence of the existence of any such entity.

Dembski objects strenuously to the assumption that natural processes are sufficient to the tasks posed by certain events, yet he does not hesitate to assume credulously that an intelligent agent actually existed who could do the job. Unconstrained postulates of this sort do nothing to satisfy us that an event has been explained, rather than explained away. Of the four categories of possible tests of a concept, specified complexity is stipulated not to be applicable to one category, is supported only by hypotheticals and dubious argument in two others, and is erroneously said to meet the criteria of the fourth. This does not bode well for the foundational concept underlying what has been billed as a valid scientific research program.

Testing the Explanatory Filter/Design Inference

While "intelligent design" advocates have been unable to show how their "theory" might undergo testing, critics have had no difficulty in proposing challenges and methods for actually putting the concepts at risk. Dembski has argued by induction from confirming cases to a generalization that specified complexity is applicable to certain biological systems—specifically those identified as being irreducibly complex by Behe. According to Dembski, whenever his EF/DI finds design and we know the causal story of that event, it was due actually to design. Thus, Dembski concludes, if his EF/DI finds design in irreducibly complex biological systems, we are warranted in saying that an intelligent agent must be the cause. However, Dembski has not proposed any sort of test here. Selective application of the EF/DI to cases where an intelligent agent is known to have acted cannot possibly place the EF/DI at risk. One way to do so is to expose the EF/DI to cases where it is conceded that a complex system arose through natural (and putatively nonintelligent) processes. I have challenged Dembski to utilize the Krebs citric-acid cycle and the evolutionary transition leading to the impedance-matching apparatus of the mammalian middle ear as two appropriate cases (Elsberry 2001). He has yet to take up that challenge.

In order to test the EF/DI, we must examine cases where we have both (1) evidence concerning the ontogeny of the event and (2) a willingness to exclude hypotheses of an intelligent designer. Another case was suggested by Rebecca Flietstra in an exchange with Michael Behe: a "fairy ring"—a circular arrangement of mushrooms explained in folklore as due to fairies, and explained by modern biology as a concomitant of exhaustion of food resources by an outward-spreading subterranean fungus (Flietstra 1998). With an increase in background knowledge about this biological phenomenon, scientists concluded that the pattern could be explained on the basis of a causal relationship between food resources and the life cycle of these fungi; no intelligent agency was necessary.

To analyze a "fairy ring" or some other situation with the EF/DI, we must first find an appropriate test. For example, when changes are made to programs in software engineering, a common method of reducing the number of bugs introduced into the program is called "regression testing." In a regression test, previously run cases are presented to the new program, and the output is examined to see if there are any differences in how the cases are handled. Regression-testing Dembski's EF/DI as a new program for distinguishing intelligent causation consists of presenting cases where design was once inferred but further information now causes the case to be attributed to regularity or chance. The critical difference is in the amount of knowledge presented to the EF/DI. Under a medieval knowledge set, would Dembski's EF/DI put fairies in the fairy rings?

Another example provided by Dembski himself is the conclusion by Sir Isaac Newton that the will of God sustained regular planetary motion (Dembski 1998a). If Newton were provided a copy of Dembski's *The Design Inference*, would he still have made the same attribution? I believe that after application of Dembski's EF/DI, Newton would have still concluded that stabilized planetary motion must be due to design. Yet, we now know conclusively that such planetary motion is due not to God but rather to physical laws, many of which were first discovered by Newton! If the EF/DI is found to be incapable of handling the historical examples correctly, it seems obvious that it should not be relied upon for correctly handling modern cases where our knowledge is yet fragmentary and incomplete.

In 2003, Jeffrey Shallit and I proposed a series of challenges for "intelligent design" advocates to show that the concept of complex specified information (CSI) was well founded, applicable to real-world problems, and produced the results claimed by the ID advocates (Elsberry and Shallit 2003a). These included: publication of a rigorous definition in a peer-reviewed journal of information theory or statistics; publication of the fully worked-out calculations for all the claimed examples of CSI (or a retraction for each that failed the claim); and use of CSI for a number of real-world problems taken from a variety of fields. If Dembski's framework is to live up to the claims made for it, our challenges—or something very similar—will have to be undertaken.

These considerations do not explain why no "intelligent design" advocate has yet to present a method for testing (in the sense of putting the tested concept at risk) either specified complexity or irreducible complexity. I argue that Dembski's EF/DI is capable in principle of being tested for reliability, since I have proposed no fewer than two separate means of testing the concept. In my judgment, if or when it is tested, it will fail to provide a reliable indication of the action of an intelligent agent. This is so because having been designed is a historical property, and not an inherent property of an artifact or event (Shallit and Elsberry 2004).

Taking Back the Marker of Intelligent Agency

Dembski bills his EF/DI as a reliable marker of intelligent agency (Dembski 1998a; 1998b, 2002, 2004). I want to explore briefly what it would mean to have a reliable marker of causation by an intelligent agent. This marker would be a property of events such that it is never found within the class of events caused by processes that are non-intelligently directed. There is an asymmetry in the epistemology for such proposed markers: We can disconfirm such markers by finding events or artifacts that have the property and that we know were formed without intelligent agency, but we can never confirm with certainty that a proposed property does yield a reliable marker of "intelligent design." The status of a proposed marker of intelligent agency (MIA) is critically dependent upon the state of our knowledge concerning the class of events that have the property specified.

I will illustrate the point with an example. If we have an observer whose personal knowledge of objects is extremely limited, we could propose that the property of "red color" is a reliable MIA. To bolster this impression, we could present a series of "confirming" cases to our observer, including a red-colored flashlight, a red fire hydrant, and a red stop sign. A large number of such cases could be enumerated—all of which we could use to demonstrate that "red color" is a good MIA. And we could make the argument that since "red color" has been shown to be a reliable MIA for those confirming cases, by induction we should infer design when we see "red color" in biological examples, such as squirrelfish, roses, and poison dart frogs. But, given a larger knowledge set, our observer will point out that "red color" is also observed in some clays, rust, garnet, and ruby, and for these objects no one seriously proposes that natural processes are insufficient. Claims of reliable MIAs are arguments from ignorance, in that our ignorance of causative factors for a class of events is taken as evidence of the reliability of the MIA. This view mistakenly asserts that confirmation of an MIA is possible via induction.

Conclusions

The "intelligent design" movement relies upon William Dembski's concept of specified complexity as its philosophical underpinning and its basis for making a claim of having a legitimate scientific program in place. Dembski's concept of specified complexity, however, has been critiqued in detail and found wanting in most of its technical aspects. Alternative formulations of Dembski's explanatory filter and his concept of specification have been proposed that correct several deficiencies in Dembski's approaches but that still do not lead to an inference of design by an intelligent agent.

Acknowledgments

I am grateful to John Wilkins, Jeffrey Shallit, Glenn Branch, and Nick Matzke for discussion of concepts in this essay and useful comments.

References

Anonymous. 2000. Press release. U.S. Newswire (May 8). www.usnewswire.com/topnews/Current_Releases/0508-102.html (last accessed May 9, 2000).

Behe, M. J. 1996. *Darwin's black box.* New York: Free Press.

———. 2001. Order and design. Presentation at "Interpreting Evolution," sponsored by Center for Theology and the Natural Sciences and American Association for the Advancement of Science and hosted by the Philadelphia Center for Religion and Science at Haverford College, PA, June 17.

Chiu, D. K. Y., and T. H. Lui. 2000. Integrated use of multiple interdependent patterns for biomolecular sequence analysis. *International Journal of Fuzzy Systems* 25.

Dembski, W. A. 1996. The explanatory filter: A three-part filter for understanding how to separate and identify cause from intelligent design. www.origins.org/articles/dembski_explanfilter.html (last accessed June 14, 2004).

———. 1997. Intelligent design as a theory of information. *Perspectives on Science and Christian Faith* 49: 180–90.

———. 1998a. *The design inference: Eliminating chance through small probabilities.* New York: Cambridge University Press.

———. 1998b. Science and design. *First Things* 86 (October): 21–27.

———. 1999. *Intelligent design.* Downers Grove, IL: InterVarsity Press.

———. 2000. Intelligent design coming clean. *Metaviews* 98. www.anti-evolution.org/people/dembski_wa/wad_mv_20001118.txt (last accessed August 17, 2003).

———. 2001a. Signs of intelligence: A primer on the discernment of intelligent design. In *Signs of Intelligence,* ed. W. A. Dembski and J. M. Kushiner. Grand Rapids, MI: Brazos Press.

———. 2001b. Is intelligent design testable? *Metanexus Views* 004. www.metanexus.net/metanexus_online/show_article.asp?2667 (last accessed June 14, 2004).

———. 2001c. Darwin's unpaid debt. Lecture at the University of California at San Diego, April 25, 2001.

———. 2002. *No free lunch: Why specified complexity cannot be purchased without intelligence.* Lanham, MD: Rowman and Littlefield.

———. 2004. *The design revolution.* Downers Grove, IL: InterVarsity Press.

Elsberry, W. R. 2001. Order and design: Philosophical issues. Presentation at "Interpreting Evolution," sponsored by Center for Theology and the Natural Sciences and American Association for the Advancement of Science and hosted by the Philadelphia Center for Religion and Science at Haverford College, PA, June 17.

Elsberry, W. R., and J. O. Shallit. 2003. Eight challenges for intelligent design advocates. *Reports of the National Center for Science Education* 23 (5–6): 23–24.

Evans, S. 2001. Doubting Darwinism through creative license. *Reports of the National Center for Science Education* 21 (5–6): 22–3.

Fisher, R. A. 1930. *The genetical theory of natural selection.* Oxford: Clarendon Press.

Flietstra. R. J. 1998. The design debate. *Christianity Today* 4 (5): 34

Kirchherr, W., M. Li, and P. Vitányi. 1997. The miraculous universal distribution. *Mathematical Intelligencer* 19 (4): 7–15.

Locke, J. 1690. *An essay concerning human understanding.* Printed for Tho. Basset and sold by Edw. Mory, London.

Pearl, J. 2000. *Causality: Models, reasoning, and inference.* New York: Cambridge University Press.

Perakh, M. 2004. *Unintelligent design.* Amherst, NY: Prometheus Books.

Popper, K. R. 1959. *The logic of scientific discovery.* New York: Harper and Row.

———. 1985. Metaphysics and criticizability. In *Popper Selections*, ed. D. Miller. Princeton, NJ: Princeton University Press.

———. 1992. *Conjectures and refutations*, 5th ed. London: Routledge.

Shallit, J. O. 2002. Review of William Dembski, *No free lunch: Why specified complexity cannot be purchased without intelligence. BioSystems* 66: 93–99.

Shallit, J. O., and W. R. Elsberry. 2004. Playing games with probability: Dembski's complex specified information. In *Why intelligent design fails*, ed. M. Young and T. Edis, 121–38. New Brunswick, NJ: Rutgers University Press.

Sober, E. 2004. The design argument. In *The Blackwell Guide to the Philosophy of Religion*, ed. W. E. Mann. Oxford: Blackwell Synergy.

Van Till, H. J. 2002. *E. coli* at the no free lunchroom: Bacterial flagella and Dembski's case for intelligent design. www.aaas.org/spp/dser/ evolution/ perspectives/vantillecoli.pdf (last accessed June 14, 2004).

Wilkins, J. S., and W. R. Elsberry. 2001. The advantages of theft over toil: The design inference and arguing from ignorance. *Biology and Philosophy* 16: 711–24.

Human Emergence: Natural Process or Divine Creation?

C. Loring Brace

Human Uniqueness

Humans may be the most remarkable creatures in the entire world. Of course, one of the reasons we say that is that we ourselves are human, and self-esteem is part of the human condition. Still, when we look at the enormous realm of learned behavior of which humans are capable, such a statement seems fully justified. No other creature displays anything more than the beginning rudiments of such a capacity.

The essential key to this extraordinary human distinction is language. The ability to verbalize the learning of previous generations and of contemporaries who have experienced events not immediately shared by others is unique. No other living community can even begin to encode experience in symbolic form and transmit it to contemporaries and subsequent generations as humans do. This process enables us to cope successfully with circumstances that we have never directly encountered. The mastery of language is of equal value to all human groups, and while we recognize that certain individuals are more gifted in that regard than others, there is no reason to suspect that there is any average difference in such capability between any one group and another. The ability to master a language is of equal importance to the survival of a factory worker, a farmer, a fisherman, a hunter, a lawyer, a mechanic, and virtually any other human occupation that comes to mind.

All kinds of anatomical, physiological, and genetic evidence supports the generalization that humans are more closely related to chimpanzees than to any other nonhuman animal. Human and chimpanzee DNA may differ by as little as 2 percent (Britten 1986), yet it is obvious that a furry quadruped is profoundly different from our own species of hairless bipeds. In the nonhuman world, chimpanzees are the epitome of cleverness. They can be taught to do a great many of the things that people can do, and they can learn to respond to simple verbal instructions. Learned behaviors of different sorts are transmitted from generation to generation in the wild chimpanzees—one of the markers of actually having culture (Whiten et al. 1999; Whiten and Boesch 2001).

Despite this attribute, they have not taken even the first tentative steps toward the development of linguistic capabilities. We can infer that their cleverness is based on the same ancestral abilities that formed the foundation from which human capabilities were derived. However, language does not fossilize, and the most we can do is surmise that those capabilities emerged from the modifications of features from a population of common ancestors that produced both humans and chimpanzees. It is these morphological changes that we can actually see in the fossil record of our human lineage over the last seven million years. In all but two features, the earliest of these fossils look far more like chimpanzees than they do like humans. Their canine teeth, while larger than those of modern humans, do not project beyond the occlusal —or contact— surfaces of the other teeth, as they do in chimpanzees, and their skulls show changes related to upright walking. The fossil evidence shows an incremental transformation through time from near-chimpanzeelike form to the fully human. In somewhat more recent fossils—but still well before full human form emerges—we see an incremental increase in sophistication in the manufacture of stone tools. Based on this anatomical and archaeological evidence, we can infer that behavioral features also had undergone a similar increase in sophistication in this period. Certainly nothing in the fossil and archaeological evidence shows a sudden appearance of the modern human form. Rather, the evidence points to the incremental emergence of the human by natural processes from a common ancestry with chimpanzees.

Africa and the Emergence of Early Hominins

The gap between linguistic and behavioral capabilities of humans and those of the cleverest members of the nonhuman world—the great apes—is profound. African apes, especially chimpanzees, not only are the most intelligent of nonhuman animals, they also are genetically the most similar to humans. The difference in bodily form, while less profound, is nevertheless quite considerable. Apes are fur-covered and humans are not. The canine teeth are long and extend past the surfaces of the other teeth (especially in males) and can serve as formidable weapons—which is not the case with humans. An ape's brain size is approximately one-third that of a human of comparable body size.

There are also significant differences between humans and apes in the anatomy of the hands and feet, although it is clear that these features of humans are closer to that of the great apes than to those of any other creature in the world. The arms and shoulders of apes facilitate suspension from overhead sources of support, such as tree branches. Although humans are also capable of such feats, they are outstripped by the apes in this mode of locomotion. Apes also surpass humans in their ease of quadrupedal locomotion. On the other hand, humans are obligatory bipeds for whom quadrupedal locomotion is not an effective alternative, and this locomotor specialization is reflected in morphological differences between humans and apes in the skull, spine, hips, knees, ankles, and feet.

Because these differences between humans and apes in the pattern of anatomical features reflect modifications derived from a similar structural foundation, it has long been tempting to use the form of a chimpanzee to suggest what the ancestor of humans might have looked like. Of course, the scholars who have made such comparisons have regularly cautioned that chimpanzees have been responding to the forces of natural selection for just as long as humans have, and they may be just as different from *their* remote ancestors. Even so, there is reason to infer that the chimpanzee habitat and way of life has changed little over the past six million years or more, while the human one has undergone a radical change. Because the chimpanzee habitat has changed so little during this important stage for the emergence of humans, it is reasonable to study living chimpanzees as the best approximation of the general features of the population that was ancestral to both chimpanzees and modern humans. The anatomy of both modern chimps and modern humans began with the morphology of this ancestral population and was subsequently modified in response to different lifeways and environmental conditions in the intervening millions of years.

So far, we have only been considering comparisons in the complexity of contemporary living organisms. Starting in the 1920s, fossil discoveries, initially in South Africa, began to fill in the morphological gap between chimpanzee and human form. Raymond Dart's *Australopithecus africanus* of 1925 was so apelike in form that the eminent English anatomist Sir Arthur Keith stated flatly, "The skull is that of a young anthropoid ape" (Dart 1925; Keith 1925, 11). Ten and twenty years later, with the discovery of more specimens from other South African sites—particularly individuals with the trunk, pelvis, and legs preserved—it became apparent that while the brain of the "australopithecines" (as they had come to be called) was of essentially chimpanzee size, the creatures were terrestrial bipeds—even though the upper body retained many of the climbing capabilities of the apes (McHenry and Corruccini 1978; Vrba 1979). Not only were legs relatively short in proportion to the body, but the arms were relatively long. The rib cage was somewhat cone-shaped, and the shoulders were oriented to facilitate arm-hanging. Yet the lengthened lumbar region, the flare of the pelvis, and the structure of the knees and heels clearly were those of bipeds.

Those first South African fossil hominins confirmed Darwin's inference that Africa must have been the locale where the crucial initial steps of human evolution had taken place (Darwin 1871). Darwin had based his surmise on the comparison of the spectrum of form in living species analogous to humans. As Darwin noted, Africa contained the greatest variety of living members of the primates, the zoological order to which humans belong. That spectrum of variation runs from relatively primitive primates such as the lemurs of Madagascar through the lorises, diverse kinds of monkeys, and the great apes—the gorillas and chimpanzees of the African mainland. Furthermore, Darwin correctly observed that the African apes were more closely related to humans than were any other nonhuman animals. This was the kind of linkage that embryologists and comparative morphologists had made to show the progressive series of organic changes running from *Amphioxus* to fish to amphibians, reptiles, and mammals.

The portentous finds of Raymond Dart and colleagues in South Africa were the first trickles of what has become a growing stream of discoveries of australopithecines and what would appear to be *their* ancestors in Africa. At first it was assumed that the australopithecines were of Middle Pleistocene age and therefore approximately half a million years old. That would have made them too recent to have served as human ancestors, since Lower Pleistocene hominins with twice the brain size—the famous *Homo erectus* specimens—had already been found in Java in the last decade of the nineteenth century (Dubois 1894). However, the dating techniques used to assess the African fossils were relatively crude, and, as it happens, greatly underestimated their actual antiquity.

The 1959 discovery of another kind of australopithecine by that extraordinary husband-and-wife team Louis and Mary Leakey, at Tanzania's Olduvai Gorge, produced a quantum of change in the assessment of the African fossil evidence. The Leakeys named their discovery *Zinjanthropus boisei* (Leakey 1959). The genus name has not been accepted in the paleoanthropological community, although we still refer to the specimen informally as *Zinj*. The South African finds had already provided evidence that more than one kind of *Australopithecus* had existed at the same time, but *Zinj* was clearly yet another species. The molar teeth were too enormous to belong to any other hominin group, and they suggested a dietary specialization that separated the group to which it belonged from the other known australopithecines (Tobias 1967). But at least as significant was the fact that radiometric techniques had been developed that allowed *Zinj* to be given an absolute age—an age nearly three times as great as that which had initially been assumed. At 1.4 million years, it was clearly of Lower and not of Middle Pleistocene age (Leakey et al. 1961).

The date seemed almost breathtakingly old when it was announced, but, as subsequent work has demonstrated, *Zinj* actually represented the latest survival of an australopithecine. Since 1959, australopithecines have been found in many parts of Africa: from as far west as Chad, at the south-central edge of the Sahara; eastward to southern Ethiopia, just beyond the southeastern end of the Sahara; northern Kenya, near the shores of Lake Turkana; western Kenya; northern Tanzania; Mali; and down to the northeastern part of South Africa. All those locations plot the distribution of the African savanna and its interface with adjacent river-flanking gallery forests as that habitat existed from the early Middle Pliocene (four million years ago) right up to the present (Cane and Molnar 2001). Earlier hominins and probably ancestors of the australopithecines going all the way back to the Late Miocene evidently lived in more closed woodlands, although their mode of locomotion was terrestrial bipedalism (Leakey and Walker 1997; Brunet et al. 2002).

The dates of those australopithecines range from the Lower Pleistocene of *Zinj* back to the early Pliocene. And their more woodland-dwelling predecessors go back to the Late Miocene, nearly seven million years ago (Brunet et al. 2002). With a time range of maybe four to five million years, the fossils indicate that they occupied a wide area spanning about 3,500 miles. Starting some 2.5 million years ago, the australopithecines were making and using stone tools of the sort first identified by Louis Leakey at Olduvai Gorge. For that reason, they are referred to as Oldowan-type tools (Leakey 1966, 1971). Unlike the bones of their makers, stone tools are imperishable and provide evidence for an even wider distribution of those makers. The presence of these ancient stone tools would indicate that australopithecines apparently existed throughout the entire African savanna and from the Atlantic coast in the west, across the Mediterranean coast in the north, and all the way south to the tip of the African continent (Isaac 1976).

Given this range, there is real reason to question whether all the hominin fossils that have been found even belong in the same genus, let alone the same species. Since the discoveries have been made by a considerable number of field workers, it is hardly surprising that there has been a tendency to give each new find a new species (if not a new genus) name. The argument has been made that humans have classified other primates such as forest monkeys more minutely, producing a great species diversity even though their skeletal and dental anatomy is much alike (Tattersall 1993). The counterargument is that, for most of their existence, hominins were never tree-living creatures. To be sure, the earliest hominins clearly used trees as refuges and may well have slept in them, but as terrestrial bipeds, they moved from one place to another on foot on the ground. For that reason, they were never isolated from each other the way tree-living monkeys were when forest patches produced prolonged separation among populations.

Periodically in the past, during the cooler and drier periods that coincided with episodes of glaciation in the temperate zone, the now-continuous extents of African forest were split into patches. When those forest patches re-coalesced as the climate became warmer and wetter, the separated arboreal monkey groups had come to differ genetically from each other through the random occurrence of mutations that could not be shared because of the groups' mutual isolation. Eventually, those genetic differences had accumulated to such an extent that the groups in which they had arisen could no longer interbreed, even when the inhabitants of those once-separated forest segments confronted each other anew. Speciation had taken place—produced not by adaptation to different ways of life but simply by the lapse of time without sharing genetic material (Tattersall 1993). A similar phenomenon is observable in tropical birds that inhabit areas of the same forests that had been separate tracts at some point in the past. Thus, there are many species of parrots that live the same kind of life in the same areas, using the same resources, yet have dramatically different coloration (Moreau 1966). The differences in color allowed birds to recognize inappropriate mating partners on sight. Such dramatic color distinction—or at least coat pattern difference—is just as striking in related, but specifically distinct, arboreal monkeys.

The savanna, however, was never cut up into patches as the rain forest had been (Kingston et al. 1993). It is true that there may well have been a time when the rain forest of the Congo basin spread east and coalesced with the east coast rain forest. This would have separated the great central region of the African continent into northern and southern sections, which could account for the speciation of some savanna animals. The springbok of South Africa, for example, looks and behaves much like the gazelles of East Africa, but it is nonetheless a distinct species. Still, there are other savanna and savanna–gallery-forest denizens that do not show specific differences now and evidently did not do so in the past. Lions are the best-known example, but others include leopards and the African wild dogs, *Lycaon pictus.*

While these are all members of the Order Carnivora rather than Primates, there is in fact a savanna-dwelling primate counterpart. Members of the genus *Papio*—the baboons—have a reproductive continuity that extends from Cairo in the north to Cape Town in the south. There are several differently named varieties of baboon, including the chacma baboon, the hamadryas baboon, and the olive baboon. While these populations have received separate specific names, they commonly exchange mates and genes where one population grades into another, even thought they look and behave in slightly different ways. They certainly all belong to a single genus, and arguably to a single polytypic species (Jolly 1993, 2000). One can infer that this must have been the case for regionally and temporally different groups of australopithecines in the Pliocene in excess of two million years ago.

On the other hand, some australopithecines differed from each other to such an extent that it is reasonable to suggest that they must have been distinct species. Contemporary with Dart's *Australopithecus africanus* was a group referred to as the "robust" australopithecines. Actually, their bodies were no more robust and no larger than those of what have been called the "gracile" australopithecines. The difference is to be seen in the size of the molar teeth and the muscles of mastication. This manifestation reaches its extreme in the molars of Leakey's *Zinj*, although, by that time, there were no gracile contemporaries left and the hominin with which they coexisted was an early member of our own genus—*Homo erectus*. It is clear from the fossil record that australopithecines have differentiated into a big-toothed form specializing in a diet of hard nuts, seeds, and tubers, and a smaller-toothed form with a more general diet. It is also true that their overall physical appearance changed noticeably through time.

The fossils of the earliest australopithecines at the beginning of the Pliocene and their Miocene predecessors are associated with fossils of forest-living monkeys; they were clearly not savanna dwellers (White et al. 1996; Brunet et al. 2002). Their continued connection to life in the trees is demonstrated by arm bones that showed muscle attachments characteristic of climbing apes and that diminished in more recent forms that were increasingly terrestrial but still climbing-adapted later forms (White et al. 1994, 1996). By the time we see fossils of the genus *Homo*, savannas are a major part of the East African landscape and shape the genus as it emerges from its australopithecine ancestors (Cerling 1992; Cachel and Harris 1995).

Furthermore, the anatomical details of the wrist in later specimens show that their immediate ancestors not only had been quadrupeds but also that they had supported their weight not by placing their hands palm down, but rather by using the second knuckles of each hand in the fashion of the starting stance of a lineman in American football (Richmond and Strait 2000). Primatologists and anthropologists refer to this as "knuckle-walking," and it is still characteristic of the mode of terrestrial progression in chimpanzees and gorillas. Through time, the australopithecines lost some of the chimpanzeelike features and began to display a form that foreshadows the first members of the genus *Homo.* Although it is still spotty and incomplete, the fossil record from the early to the late australopithecines shows the steps by which creatures with ancestors that looked like chimpanzees produced descendants that began to look like humans.

The hominin fossil record provides just as convincing a documentation of how humans arose by natural means from nonhuman antecedents as the paleontological record demonstrates how other groups of organisms arose from a succession of ancestral forms (Jarvik 1980), or of evidence for how a hoofed mammal returned to the sea and gave rise to the whales (Gingerich et al. 2001). Still, we have no comparable direct evidence of the development of the behavioral capabilities that most obviously distinguish humans from their nearest primate relatives. As mentioned before, language does not fossilize, and it had obviously been the key to human survival long, long before it was first written some 5,000 years ago (Lawler 2001). There are some aspects of human form, however, that allow us to suggest the steps by which the learning of previous generations could be transmitted to descendants without the requirement that they would be perpetually forced to reacquire the knowledge learned by their predecessors.

Bipedalism and the Reliance on Handheld Tools

There are two things that clearly distinguish the australopithecines and their evident ancestors from any other primate except those that belong in the genus *Homo.* First, the lower back, pelvis, knee joints, and the heels of their feet clearly show that they were habitual if short-legged bipeds. Second, whatever the other features of the jaws and teeth, the tips of the canines did not project beyond the level of the chewing surface of the rest of the components of the dental arch. There have been attempts to attribute bipedalism to prolonged standing to pick the fruits, nuts, and berries of bushes and shrubby trees growing at the savanna's edge (Hunt 1996). Others have suggested that the advantage lay in the lower energy requirements of bipedalism over quadrupedalism for low-speed movement over long distances (Wheeler 1993; Leonard and Robertson 1995). Another argument has stressed the lesser area of skin exposed to direct sunlight by a vertical rather than a horizontal body posture (Wheeler 1984). However, if one considers the possibility that bipedalism and nonprojecting canine teeth may be responses to the same set of circumstances, then another explanation becomes more plausible.

A well-conditioned human biped can run down an animal by trotting after it persistently for a day or two or three, but no human being can run fast enough to escape a predator such as a leopard or a pack of hunting dogs. A short-legged australopithecine between four and five feet tall would have had even less luck outrunning a carnivore. Nor could australopithecines—even when gathered together in a group—have defended themselves by biting, as do savanna baboons. Survival on the savannas and their adjacent open woodlands would have been impossible for slow-footed bipeds with no innate anatomical means of defense if there had been no other survival strategy.

Modern humans, however, even though they are the slowest runners for their body size in the animal world, traditionally are at less risk from predation than any other noncarnivorous animal. The reason is related not to any particular aspect of anatomy in and of itself but rather to the capability of wielding handheld defensive implements. A five-foot, eight-inch Tanzanian Masai weighing all of 150 pounds, carrying a six-foot spear that is too heavy to throw any appreciable distance, is at least a match for a quarter-ton of lion. He can jab the iron tip of his spear at the predator, and if the latter throws caution to the wind and charges, the Masai simply plants the butt end of the spear on the ground and aims the point at the oncoming lion—which by simple momentum then impales itself on the stationary spear. Needless to say, this is a far more effective way of dealing with a savanna predator than trying to bring one's mouth into position to bite it.

Do we know how far back hominins were using spears? Of course not. Iron spear points were obviously not available in prehistoroic times. We do know that hafted stone spear points were being used in Africa a quarter of a million years ago (McBrearty et al. 1996), although we do not know whether the spears on which they were mounted were designed for throwing or thrusting. Wooden spears 400,000 years old have been found in a Lower Paleolithic site in Germany (Thieme 1997), but it took special circumstances for wood to have been preserved that long, and there just is no evidence from older sites. We do know that australopithecines were making and using stone tools 2.5 million years ago (Semaw et al. 1997), and that those tools were in continual use thereafter. Although we have no direct evidence for this, the most cogent consideration is that those tools were being used for processing the food collected. Sheer physical survival, however, would have demanded some sort of assistance that could have compensated for the lack of speed afoot and an ineffective dental means of defense. An object carried to maintain a physical distance between the hominin and a would-be predator would have had more survival value than fearsome canine teeth or great speed of foot.

Of course, the hominin would have had to carry the object at all times, which would be more than a trifle awkward if all four appendages were normally used in locomotion. The advantage in carrying one's defensive implement whenever crossing open ground could well have been the selective force that led to habitual bipedalism. At the same time, whatever selective advantage had previously been operating to preserve canine teeth that projected above the surface of the rest of the dental arch would have been completely eliminated because of the obvious superiority of handheld implements. Traits that are not maintained by selective forces tend to be deemphasized in the course of time. This could well be why nonprojecting canine teeth always characterize the early bipeds (Brace 1967, 61 and all subsequent editions; Kortlandt 1980, 87, 104).

If the simultaneous occurrence of bipedalism and nonprojecting canine teeth indicates the reliance on handheld defensive weapons, then this suggests that the role of tools as essential for hominin survival goes back well over six million years to the Late Miocene, even though we have no direct evidence of the tools themselves. Such a conclusion suggests that the australopithecines and their predecessors had taken an important step beyond the degree of behavioral sophistication shown by chimpanzees. Different chimpanzee groups do pass on behavioral traditions from one generation to the next. In one area they use hammer stones to crack open nuts placed on a flat rock; other groups have developed the technique of poking a de-leafed twig into a termite mound, angering the insects into biting the intruding object. The chimpanzee then pulls out the twig and eats the termites that cling to it. Primatologists have isolated other instances of traditions' being passed down by chimpanzee groups, demonstrating that chimpanzees clearly have the rudiments of culture (Whiten and Boesch 2001). In none of those cases, however, is the tradition essential to the survival of the group in question. That, however, was not true for the australopithecines. A terrestrial biped, slow of foot and lacking defensively enlarged canine teeth, was absolutely dependent on a handheld deterrent in the form of a branch, perhaps one taken from a thorn tree or acacia at the edge of a savanna.

Stone Tools and a Change of Diet

After having spent some four million years relying on handheld tools of perishable material, some hominins by 2.5 million years ago had discovered the use of simple stone tools. These are associated with the butchered remains of large animals, such as the hippopotamus and the elephant. It would appear that the toolmakers

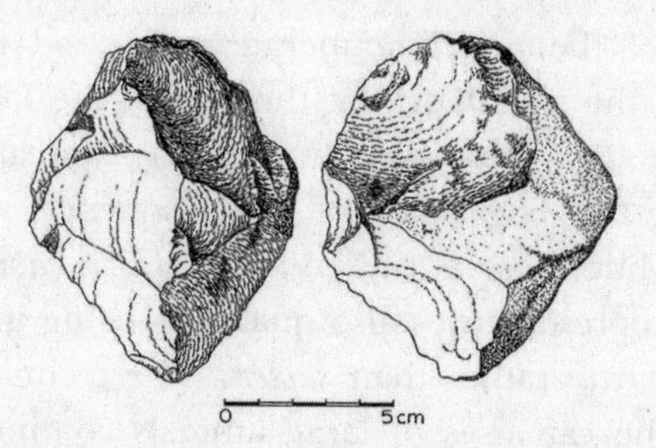

Figure 1. A 1.88-million-year-old Oldowan "tool" from FxJj 1, the Kay Behrensmeyer Site in the Koobi Fora area of northern Kenya. Actually, this is a core from which flakes were struck. The flakes were the functional tools, although the first archaeologists to collect them thought that the cores themselves were the tools. Drawn by Kay Clahassey from Figure 1.1013, Appendix 6AA, in Isaac and Isaac, eds. 1997,324.

were striking flakes from chunks of volcanic rock and using the flakes to get through the skin of the animals and to the meat underneath. This enabled them to function as scavengers (see figure 1). A dead elephant represents a lot of food for any creature able to process the carcass, and elephants eventually do die of old age. Lions and hyenas do not have the dental equipment to get through pachyderm skin, but it has been experimentally demonstrated that stone flakes chipped off cores—similar to those characteristic of the Oldowan toolmaking tradition—work quite well even when being used by the less muscular hands of modern human beings (Stanford et al. 1981; Toth and Schick 1983).

From the occurrences and distribution of Oldowan tools 2.5 million years ago, it is clear that at least one branch of the australopithecines was engaged in large-animal scavenging. This would have put them in competition with the saber-tooth cats. The latter had much smaller brains per unit of body size and less agility than the large hunting felines, and they may well have evolved not as hunters but specifically to focus on scavenging pachyderm carcasses. In any case, the ultimate disappearance of these two species coincides with the successful spread of hominin meat users in the Lower Pleistocene of Africa, the Middle

Pleistocene of the north temperate zone, and the spread of humans into the New World at the end of the Pleistocene and beginning of the Holocene (Brain 1981; Marean 1989). The hominins at the beginning of that sequence were australopithecines, while those at the end were *Homo sapiens.* The early Pleistocene representatives also qualify as being in genus *Homo,* but species *erectus* rather than *sapiens.* The techniques used for scavenging the carcasses of large animals continued for more than a million years without a break, while during this same time, the skeletal evidence shows a series of anatomical changes in the users from *Australopithecus* to succeeding versions of the genus *Homo.* The likelihood of unrelated hominins' inventing the same techniques independently and using them in sequence is comparable to the independent achievement of the same sequence of nucleotides in the homeobox genes of fruit flies, fish, frogs, and us. In both instances, to regard these cases as the results of the accumulation of difference from a common origin—i.e., evolution, or, as Darwin originally phrased it, "descent with modification" (Darwin 1859, 420)—requires far less speculation and reliance on wildly improbable coincidence.

From Scavenging to Hunting

Somewhere between the Late Pliocene of the 2.5-million-year-old Oldowan tools and the beginning of the Pleistocene just under two million years ago, the makers of those butchering tools had stopped just waiting for large animals to die before they cut them up for food. The evidence from Olduvai Gorge itself shows that at some time between the end of Bed I and the beginning of Bed II, the toolmakers had begun to hunt prime adult game animals. Evidently the defensive spears and stone tools of earlier times were now being used as offensive weapons (Kortlandt 1980, 87). At Olduvai and also east of Lake Turkana in northern Kenya, there is evidence that brain size had increased to beyond the chimpanzee range. Robust australopithecines such as *Zinj* still survived, but the fossil record shows that the larger-brained hominins were also becoming longer of leg and larger in body. At the same time, the pattern of the venous drainage of the blood from the brain was altered by the appearance of an increased number of emissary foramina, evidently associated with an increased need to reduce heat buildup (Falk 1990).

These somewhat disparate pieces of evidence suggest that the hominin population that was on the threshold of becoming *Homo* had begun to practice what has been called "persistence hunting"—that is, camping on the trail of an animal and trotting after it until it can go no farther (Carrier 1984). Chimpanzees kill and eat the newborn young of antelopes and bush pigs when they encounter them, but they do not trot persistently after potential prey, as the hominins evidently learned to do. Chimpanzees will use persistence in the trees where they hunt arboreal monkeys, evidently considering them a delicacy (Goodall 1967, 71; Stanford 1995). However, they do not go after older terrestrial quadrupeds, since those invariably can run faster than chimpanzees. Newborn antelopes and wildebeests soon become quite mobile, but even so, it takes some time before they can keep up their speed for significant distances. The early hominin predators could well have discovered that by trotting after a young gazelle that could run much faster for the first hundred yards or so, they could tire it out and eventually be rewarded by a meal. This sort of knowledge could accrue and develop so that the hunter, concentrating first on the newborn, could use that same persistence technique on slightly older youngsters, graduating eventually to full adults. By the end of the Pleistocene, the proliferation of technological ingenuity had reduced the importance of persistence hunting, though it was still regarded as a laborious but reliable means of acquiring large game among peoples as diverse as the South African !Kung, Australian aborigines, Tarahumara Indians in northern Mexico, and the Caribou Eskimo (Watanabe 1971).

The increasing reliance of successive hominin species on meat in the diet is reflected in the fossil record. The relatively large teeth of our australopithecine predecessors become significantly smaller over time. The reduction is particularly evident in the molars—a reflection of the change in diet engendered by increasing the meat component and reducing the amount of rough, uncooked vegetable matter that had been the hominin staple for so long. Meat does not need the amount of chewing necessary to make plant foods digestible, so selection maintaining large molars was eased, and molar size was duly reduced in time. Tooth size remained stable for nearly two million years after this shift until "modern" human form emerged from its nonmodern predecessors (Brace 1995a).

Around two million years ago, there were no more "gracile" australopithecines, and the hominins who were making the Oldowan tools invented by the australopithecines half a million years earlier showed a marked increase in brain size compared with that of their predecessors—although it was still only about half the size of the average living human (Gabunia et al. 2000). Since the tools they were making and the way of life they were pursuing represented an unbroken continuity of the tools and way of life of their predecessors, it should follow that they themselves were simply the descendants of those earlier australopithecines. The greater brain size, longer legs, larger bodies and smaller teeth all represent only minor modifications on the earlier body plan and do not require any sudden and unlikely leaps. There are a number of scraps of evidence from the late Pliocene suggesting that brain size was getting larger in creatures that otherwise looked like australopithecines, and there are a number of instances of reduced tooth size. Clearly, the changes that were to characterize the subsequent manifestation of human form were realized in this period, but fossil evidence for the sequence of change is tantalizingly incomplete. We do not have full-scale representatives of the transition, although there are increasing quantities of suggestive pieces.

Emergence of the Genus *Homo*

Starting with the beginning of the Pleistocene just under two million years ago, a *Zinj*-like australopithecine with huge teeth and a chimpanzee-size brain continued to survive, but the

Oldowan toolmaking hunters, with nearly twice the brain size, are recognized as members of our own genus, *Homo*. Although a segment of the business has become what one might call "name-happy," bestowing designations such as *habilis*, *ergaster*, *rudolfensis*, and *heidelbergensis* on various adaptively identical specimens, the most accepted designation for these early Pleistocene hunting hominins is *Homo erectus*. There will always be a name problem when the subject being considered is a gradually evolving continuum, but, whatever it is called, the continuity of the tool technology suggests that it is unlikely that more than one species of the genus *Homo* existed at a single time. The idea that stone toolmaking and hunting were independently discovered by a series of separate species is too improbable to warrant serious consideration. The same is also true for the suggestion that speciation occurred after the *erectus* level had been reached at the extremities of the range occupied by that species. As the stone tools show, the distribution of *H. erectus* was continuous. For that reason, there is no support for the suggestion that there ever was reproductive isolation of the sort necessary for a separate species of the genus *Homo* to have arisen, let alone three or four separate species.

Except for the size of the brain, the anatomy of *Homo erectus* from the neck on down, as revealed in the skeleton, looks startlingly like that of a living human, though there are some significant differences that have been discovered upon closer examination. The size of the vertebral canal in the thoracic vertebrae is smaller than it is in *Homo sapiens* individuals of comparable size, which suggests that *Homo erectus* lacked the innervation necessary for the breath control that makes articulate speech possible (MacLarnon 1993; MacLarnon and Hewitt 1999). In addition, a detailed examination of tooth-root formation and eruption patterns shows that individuals in the *Homo erectus* species matured at a rate that fell somewhere between those of modern chimpanzees and living humans. Based on a study of the same features in australopithecines, it was shown that the latter followed a maturation timetable similar to that of chimpanzees, while Neanderthals were the same as modern humans (Smith 1992, 1993, 1994). *Homo erectus*, then, reached adolescence later than a chimpanzee but earlier than *Homo sapiens*, which presumably implies that *H. erectus* was geared for a longer life span than that of a chimp but not as long as our own.

Out of Africa

As soon as *Homo erectus* populations became established in Africa and—if we take the archaeological evidence as indicative—widespread throughout that continent, they extended their range throughout the southern temperate zone and the tropics of Eurasia. Evidently, once *H. erectus* specimens had become successful members of the "large carnivore guild" as diurnal predators, there was nothing to stop them from filling that niche in the warm temperate and tropical grasslands of the entire Old World (Walker 1984). They had reached as far as Java (Indonesia) and the Yangtze basin of south central China between 1.9 million and 1.7 million years ago (Huang et al. 1995; Larick and Ciochon 1996); and there are *H. erectus* skulls and jaws with quantities of Oldowan tools dated at 1.7 million years ago at Dmanisi in the Republic of Georgia (Gabunia et al. 2000).

Approximately 1.5 million years ago, the famous "hand axe" had been invented, a teardrop-shaped, bifacially flaked tool that was evidently used in butchering large animal carcasses (Asfaw et al. 1992). Oldowan flakes had worked fine for skinning and removing meat, but the first hominin hunters evidently did not dismember the animals they processed (Shipman and Rose 1983; Toth and Schick 1983). The new technique, initially developed in Africa, was rapidly adopted throughout the area inhabited by *Homo erectus*, and bifaces or hand axes remained a part of their technology for more than a million years (see figure 2). Archaeologists tend to look at that long stretch as a period of relative cultural stagnation, but of course we must realize that the stone tools we find can only represent a very small fraction of Lower Pleistocene

Figure 2. A biface or "hand axe" of a kind called Acheulean from the site of St. Acheul in northwestern France. While this is a Middle Pleistocene example, the form is the same as that of specimens made as far back as 1.5 million years ago in Africa. Drawn by Mary L. Brace.

hominins' culture. The ones that persist through time are the ones that worked so well that there was no need to change their form, but we have no insight into the nature of the perishable bulk of their cultures.

During that stretch of time, brain size slowly increased so that, by somewhere between 200,000 and 300,000 years ago, modern levels of brain size had been reached. Tooth size, although smaller than australopithecine levels, was greater than that found in any living human population. Robustness and muscularity were also greater, on average, than among any living people. By that time also, the control of fire had been mastered and had spread to virtually all human groups (Straus 1989). In this same period, stone flakes were shaped to be hafted and mounted on shafts as projectile points, a technique that was first developed in East Africa and rapidly spread to other parts of the world (McBrearty et al. 1996; Schick 1998). Some have concluded that the inventors migrated out of Africa, taking that technique with them. Others have noted that the continuity of local toolmaking traditions indicates that the indigenous people elsewhere simply copied the techniques that were adopted by their neighbors. The low population densities of these innovators makes extensive migratory propagation of tool techniques less likely than copying and locally modifying the tools used by adjacent groups.

Local Style Differences, Universal Control of Fire, and Probability of Language

When the innovations of the control of fire and the production of hafted spear points appeared and spread quickly throughout much of the continuously inhabited world, we also see the first appearance of local differences in elements of technological style that do not really improve the chances for survival of the groups in which they occur. Local idiosyncrasies appear in the contours of flaked stone tools. When the distributions of such regional variants are plotted on the map, the result is eerily similar to the distribution of language families in the recent world (see figure 3). Although much of the technique of tool manufacture can be acquired by sheer imitation, the appearance of regional differences in style leads one to suspect that people were actually specifying that this is how "*we*" do it here, as opposed to how of "*they*" do it over there.

By itself, the first appearance of systematic regional distinctions in style does not provide conclusive evidence that people were actually talking about such things, but it does raise the suspicion that they had begun to deal with the world by reference to shared symbols. Simply, one begins to suspect that the basics of language as we know it had been established. Because of the continuity of the other features of anatomy and material culture throughout human history, the emergence of this basic linguistic ability could not have been a sudden occurrence. Although the eminent linguist Noam Chomsky has promoted the view that linguistic ability arose not by the cumulative modification of what was already there, but rather as a "true 'emergence' "—the result of a single "mutation" (Chomsky 1972, 70, 97) analogous to the "flick of a switch"—there is no such sudden jump or discontinuity in the fossil record that allows us to infer the emergence of linguistic abilities in humans (Stringer and Gamble 1993, 204).

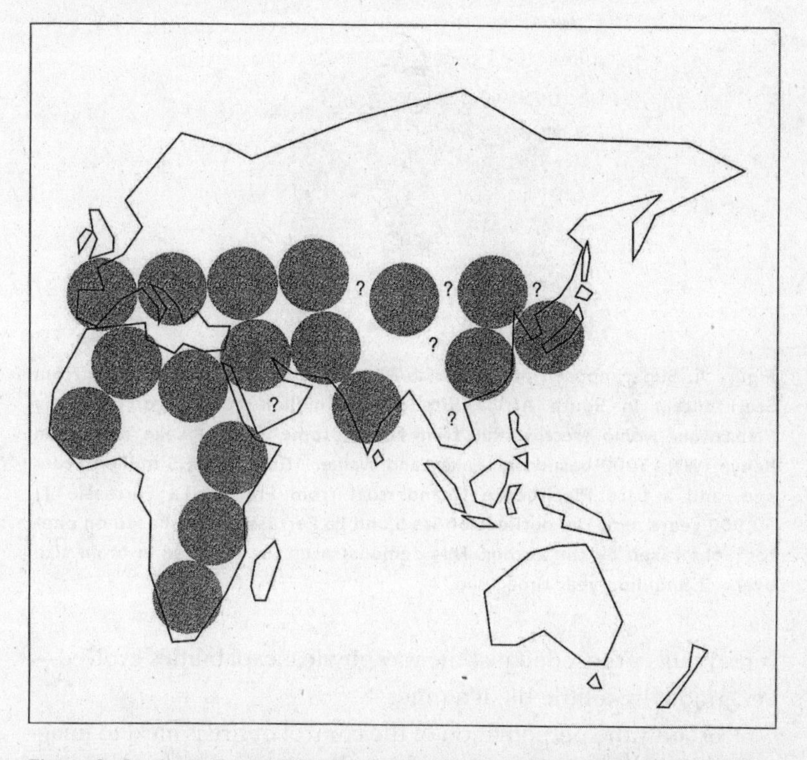

Figure 3. A schematic rendition of the distribution of shared stylistic similarities in the various inhabited portions of the Old World at the beginning of the Middle Stone Age in Africa and its contemporaries elsewhere (Brace 2000: 356). The question marks indicate areas for which little in the way of published research is available.

The evidence shows a slow, two-million-year increase in brain size (figure 4). At the same time, the gradual increase in technological sophistication shown by the tools in the archaeological record suggests that the growth of intellectual capability was itself the result of many tiny, favorable genetic alterations, and not something that happened suddenly. The verbal realm has its own selective forces, and those who cannot master it today are at a severe disadvantage. As a result, there are no intermediate representatives in the modern world. All peoples are equally adept at handling language, but there had to be a time when even a rudimentary capability was better than none at all, and accretion of

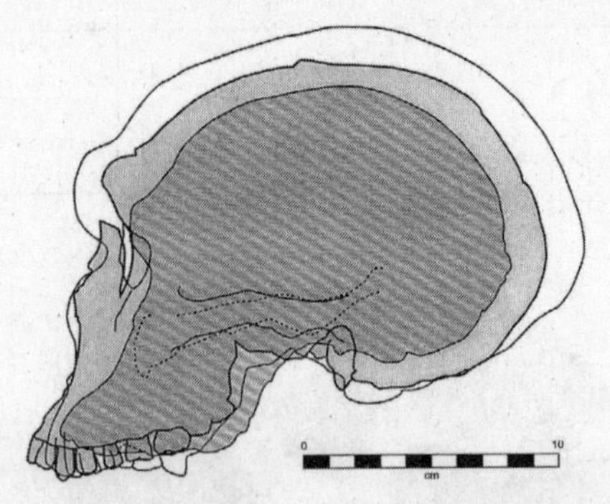

Figure 4. Superimposed outlines of a Pliocene australopithecine skull from Sterkfontein in South Africa (Sts 5) 2.5 million years ago; an Early Pleistocene *Homo erectus* skull from Nariokotome west of Lake Turkana in Kenya (WT 15000 based on Leakey and Walker 1997: 68) 1.5 million years ago, and a Late Pleistocene Neanderthal from France (La Ferrassie 1) 50,000 years ago. The outlines of Sts 5 and La Ferrassie were based on photographs taken by the author. This demonstrates the increase in brain size over a 2.5 million-year time span.

verbal skills proceeded just the way physical capabilities evolved—i.e., gradually, a little bit at a time.

Certainly the perpetuation of the control of fire is hard to imagine without a linguistic dimension. The nature of the appropriate fuels, tinder, the seasoning of the components, the requisite dryness, and the friction required for ignition constitute such a complex whole that it is unlikely that sheer imitation could ensure its successful transmission from generation to generation. Its transmission from one group to another was so rapid at the outset that it is impossible now to pinpoint just where it first occurred. Between 250,000 and 300,000 years ago, by the beginning of the Middle Stone Age in Africa (McBrearty and Brooks 2000, 453) and its northern contemporary, the Mousterian, in the temperate zone (Straus 1989, 489–90), virtually all the human groups in the world were using fire. Unlike the case for earlier *Homo erectus* form, there is nothing in the anatomy of the hominins of 250,000 to 300,000 years ago that would preclude their capacity for articulate speech. If, then, the hominins of the Late Middle Pleistocene had improved their chances for survival by the acquisition of language, they should be recognized as full human beings. This should qualify them for warranting the designation *Homo sapiens* no matter how robust and "primitive" they might appear. Interestingly enough, the two-million-year record of gradual brain-size increase stops at that point, and brain size in proportion to body size has remained essentially the same ever since.

The control of fire was one of the keys to permanent occupation of the temperate zone, which, because humans remain physiologically tropical mammals, had previously been unavailable for permanent habitation. When one compares the stone-tool assemblages of the African Middle Stone Age with the Mousterian in the north, the most striking contrast is in quantities of implements referred to as "scrapers" in the Mousterian and their near absence in Africa. The reasonable inference is that the northern people were preparing animal skins for use as clothing. That, plus some sort of shelter and the use of fire, were essentials for survival, particularly in regions where snow could be expected. One might question what incentive there would be for staying at latitudes that created survival problems for a physiologically tropical mammal. The reason clearly was the quantities of game animals available. Once they had figured out ways to endure the climate, there was an abundant supply of food.

Cooking and a Loss of Face

One of the environmentally related problems was the freezing of meat acquired by hunting activities. This was particularly a problem during periods of intensified glaciation. In such times, both people and animals were restricted to the southern portions of the temperate zone, such as Spain, Italy, and the Balkans. Even so, there was always the problem of food's freezing during the colder times of year. A hunting band could hardly eat a whole horse at a single sitting, and the uneaten part would have turned into a block of ice a few days later. One solution to this problem was the invention of cooking. One of the other major differences between the Middle Stone Age of Africa and the Mousterian of the north was the extensive evidence of hearths, earth ovens, and cooked animal bones in the sites associated with the latter. Archaeologists who work on African sites have never encountered such phenomena either in Middle Stone Age or Later Stone Age sites.

Today we tend to think of cuisine as the art of making food taste good, but initially it had to have developed simply to make it possible to eat what had previously frozen. This subsequently had surprising if unintended consequences. Cooked food requires less chewing than raw food, and the selective forces that had maintained jaw and tooth size for the previous two million years were relaxed, leading to a reduction in the intensity of the forces of selection. The consequences were a decrease in the size and robustness of the chewing machinery. Brain size in proportion to body size remained constant, but tooth size began a steady trajectory of reduction late in the Pleistocene, especially among those people who inhabited the northern reaches of the Old World. In fact, it was the consistent use of food-preparation techniques, including cooking, that led to the reductions that converted a Middle-Pleistocene–size face into what we regard as "modern" form (Brace 1979, 1995b, 2000).

Those people who invented cooking along the northwestern third of the area of human habitation as the Middle Pleistocene gave way to the Late Pleistocene around 120,000 years ago were the Neanderthals. It was the subsequent reduction in jaw and tooth—face size—that converted a Neanderthal face into that now visible in the people who range from the Middle East to the Atlantic coast of Europe, although the process took some 50,000 years to be accomplished (figure 5). Today, people with that facial configuration run north into Scandinavia and eastward through Russia and all the way to Siberia. Since they are the descendants of the first people who systematically cooked their food, it is no surprise to discover that they have the smallest jaws and teeth in proportion to body size of any of the world's people (Brace 1995a). Eventually, cooking was adopted for use farther south, in the warmer parts of the world. There it was not necessary to thaw what had become frozen, but people discovered that cooking could save food that had begun to spoil. The consequences were the same as

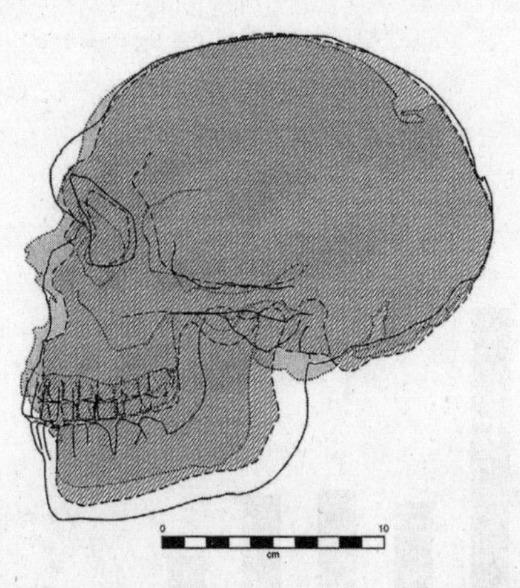

Figure 5. Superimposed outlines of the skulls and facial skeletons of a "classic" Neanderthal, La Ferrassie 1 from France at about 50,000 years ago; an early "modern," Predmostí 3 from the Czech Republic of just under 30,000 years ago, and a recent inhabitant of the Faeroe Islands, northwest of the northern tip of Scotland. The reduction in the jaw size at the same time that cranial size remains the same shows how a Neanderthal face was converted into a modern one. The outline of Predmostí was drawn from a cast; those of La Ferrassie and the Faeroe Islander were made from photographs taken by the author.

those in the north, and the jaw and tooth size began to reduce there as well. Everywhere in the world, now, there is a direct relation between the length of time that the ancestors of any given people have used cooking in food preparation and the degree to which the jaws and teeth of the people in question have decreased below the size levels that had remained so constant throughout the Lower and Middle Pleistocene (figure 6).

The Importance of String

The evolution of "modern" out of Neanderthal form involved more than just reducing the size of the face. Another critical component was the contemporaneous reduction of body robustness, which had remained constant throughout much of the Pleistocene, maintained by the selective pressures of hunting and the strength required for scavenging and bringing meat back to camp. As the last glaciation proceeded, however, technological

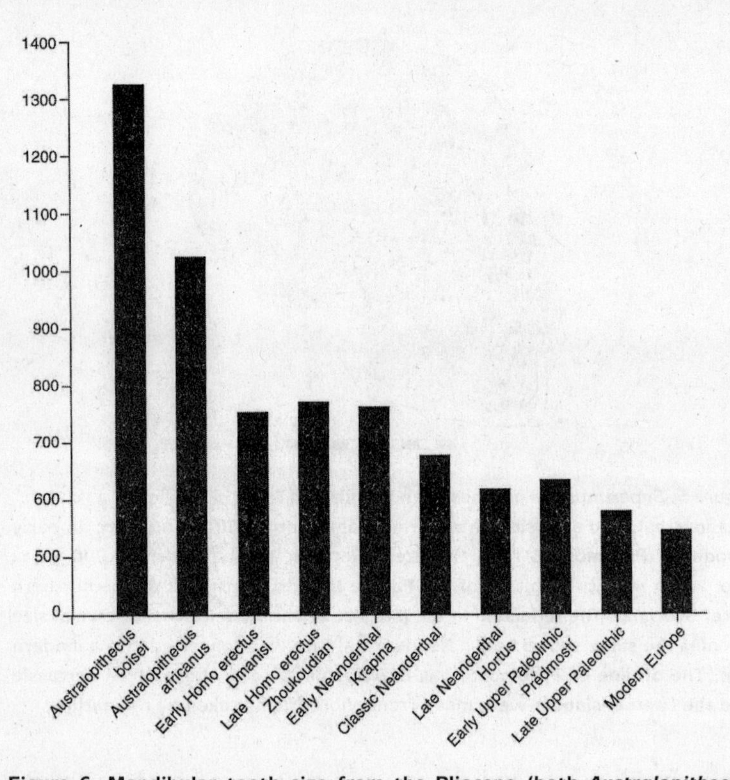

Figure 6. Mandibular tooth size from the Pliocene (both *Australopithecus boisei* and *africanus*); the Lower Pleistocene (Dmanisi, Republic of Georgia); the Middle Pleistocene (Zhoukoudian, just west of Beijing, China); the Late Middle Pleistocene early Neanderthals (Krapina, Croatia); "classic" Neanderthals of 50,000 years ago; late Neanderthals of 37,000 years ago (Hortus, France); the early Upper Paleolithic (Predmostí, Czech Republic); and recent Europeans. The figure on which the vertical bar is based is the sum of the average cross-sectional areas in square millimeters of all eight categories of tooth in the lower jaw for the group indicated. The measurements on which these are based are recorded in Matiegka 1934; Weidenreich 1937; Brace 1979; White et al. 1983; and Gabunia et al. 2000.

advances occurred that completely changed the nature of the game.

While at first it might not seem significant, one of the keys to human evolution was the creation of string (Brace 1995b, 272–73). Rolling plant fibers into twine made it possible to make snares and nets. With these, smaller game could be caught in sufficient quantity to feed the hunter's band with nowhere near the effort previously required. Birds, small mammals, and fish constitute a very large biomass, but catching them one at a time was more effort than it was worth. But netting a school of fish or a flock of quail would be well worthwhile, requiring nowhere near the amount of effort needed to impale a deer or a larger animal. We do not see nets and snares in the archaeological records because they are made of perishable material, but we do find beads, which can only be used when threaded in strings. From the archaeological record, we can see when cordage became known, and it corresponds with a huge increase in the presence of small mammal, bird, and fish bones at human living sites late in the Pleistocene. At the same time, the robustness of the human skeletons begins to show a marked decrease. The new technology reduced the selective pressures that had maintained great bodily strength ever since hunting had become one of the main hominin survival strategies at the end of the Pliocene, and the consequence was the appearance of modern human form. This technology spread more rapidly than had the earlier dissemination of cooking, with the result that there is much less disparity in bodily robustness than there is in jaw and tooth size among the living populations of the world.

Conclusion: The Cultural Ecological Niche and the Creation of the Human Condition

Human beings today would be absolutely unable to survive without possessing *culture*—in the anthropological sense of the term. That concept was first spelled out by the Victorian anthropologist Edward Burnett Tylor (1871, 1), and a slightly simplified version of his definition could be rendered in this fashion:

> Culture is "all those aspects of learning and experience accumulated by previous generations that people acquire by virtue of being members of a continuing society." (Brace 2000, 347)

We realize that our subsistence strategies and the principles regulating our behavior toward each other are not invented anew each generation. We learn these by means of the grammar and vocabulary that are transmitted to us, which constitutes one of the essential dimensions of culture. Language, of course, is a human creation, but the need to be able to learn a language has to constitute a powerful selective force in itself, and one that has had an equal effect on the ancestor of all living humans. To be sure, cultures vary enormously in many aspects of technological sophistication, but the requirements of learning a language and the rules of behavior have the same degree of difficulty in each, no matter how much their technologies may differ. Also, until about 10,000 years ago, with the beginnings of farming in circumscribed portions of the world, all humans had been living essentially the same kind of life for the previous two million years. Thus, we should expect that all the populations of the world have equal intellectual capabilities.

The shaping effects of a dependence on culture go right back to the australopithecines and their predecessors in the Late Miocene, nearly seven million years ago. The loss of a defensive role for canine teeth and the selection for a bipedal mode of locomotion were the likely consequence of a dependence on handheld defensive weapons. From that time on, the hominin line was committed to a dependence on increasing amounts of behavior learned from previous generations. The adoption of a significant component of hunting two million years ago depended on the employment of stone tools and the transmission of the traditions of their manufacture and use to subsequent generations. This added dimension to hominin survival strategies was accompanied by a slow increase in brain size. Body size increased as well, particularly in the females, who had to give birth to proportionally larger-headed infants.

At the point where we have reason to suspect that language as we know it had begun late in the Middle Pleistocene, brain size ceased expanding. Although people were decidedly archaic in appearance, a good argument can be made that at that point they warrant the designation *Homo sapiens.* The only thing necessary to transform them into recognizably modern humans was the reduction of Middle Pleistocene levels of tooth size and postcranial robustness and muscularity. The adoption of cooking as a standard method of food preparation took the pressures off maintaining those previous levels of jaw and tooth size—especially in the north, where the climate made it a necessary element in survival. Subsequently, the discovery and universal adoption of string-based techniques for game acquisition meant it was no longer essential to have the levels of skeletal and muscular robustness that had previously been necessary for survival. The reduction of Middle Pleistocene levels of sinew and fiber followed suit, and modern human form developed gradually out of their more robust predecessors over the last 50,000 years.

The advantages of adapting to the cultural ecological niche of course are obvious. The selection that produced the ability to learn a language had to be intense and equally applicable to all human populations. At the same time, learned accumulations in the cultural realm reduced some of the selective forces that had previously affected the human physique.

The emergence of what appears to us as quintessentially human form from something that looks far more like a chimpanzee takes place in unbroken incremental fashion over time—starting, as the evidence now available shows, nearly seven million years ago. The evidence for assessing the lifeways associated with that trajectory is preserved in an archaeological record starting 2.5 million years ago, and it shows how changes in the way of life generated the conditions that made the observed biological changes a predictable consequence.

The picture produced by the evidence is completely consistent with the operation of natural processes, with no need to posit any interventions of divine or miraculous origin—despite our wishes and desires to the contrary. The tangible record shows all the steps by which it has been achieved. The record even allows us to infer the circumstances that led to the development of our linguistic capabilities by natural means. We have every right to look at the whole thing as being quite marvelous, but there is no step that is not perfectly consistent with the cumulative results of natural processes. This is what led Charles Darwin to start the last sentence in his *On the Origin of Species* with the words, "There is a grandeur in this view of life" (Darwin 1859, 490). In that sense, those natural processes that shaped the unfolding of the world of living organism as well as the human component of that world can all properly be regarded as marvels, but not as the outcome of any divine or miraculous intervention.

References

Asfaw, B., Y. Beyene, G. Suwa, R. C. Walter, T. D. White, G. WoldeGabriel, and T. Yemane. 1992. The earliest Acheulean from Konso-Gardula. *Nature* 360: 732–35.

Brace, C. L. 1967. *The stages of human evolution.* Englewood Cliffs, NJ: Prentice Hall.

———. 1979. Krapina, "classic" Neanderthals, and the evolution of the European face. *Journal of Human Evolution* 8 (5): 527–50.

———. 1995a. Bio-cultural interaction and the mechanism of mosaic evolution in the emergence of "modern" morphology. *American Anthropologist* 97 (4): 711–21.

———. 1995b. *The stages of human evolution.* 5th edition. Englewood Cliffs, NJ: Prentice Hall.

———. 2000. *Evolution in an anthropological view.* Walnut Creek, CA: AltaMira Press.

Brain, C. K. 1981. *The hunters or the hunted? An introduction to cave taphonomy.* Chicago: University of Chicago Press.

Britten, R. J. 1986. Rates of DNA sequence evolution differ between taxonomic groups. *Science* 231: 1393–98.

Brunet, M., F. Guy, D. Pilbeam, H. T. Mackaye, L. Likius, D. Ahounta, A. Beauvilain, C. Blondel, H. Bocherens, J.-R. Boisserie, L. De Bonis, Y. Coppens, J. Dejax, C. Denys, P. Duringer, V. Eisenmann, G. Fanone, P. Front, D. Geraads, T. Lehmann, F. Lihoreau, A. Louchart, A. Mahamat, G. Merceron, G. Mouchelin, O. Otero, P. Pelaez Campomanes, M. Ponce De Leon, J.-C. Rage, M. Sapanet, M. Schuster, J. Sudre, P. Tassy, X. Valentin, P. Vignaud, L. Viriot, A. Zazzo, and C. Zollikofer. 2002. A new hominid from the Upper Miocene of Chad, Central Africa. *Nature* 418: 145–51.

Cachel C., and J. W. K. Harris. 1995. Ranging patterns, land-use and subsistence in *Homo erectus* from the perspective of evolutionary ecology. In *Evolution and ecology of* Homo erectus*: human evolution in its ecological context,* ed. J. R. F. Bower and S. Sartono. *Proceedings of the Pithecanthropus Centennial 1893–1993 Congress.* Leiden, Holland: Leiden University Press.

Cane, M. A., and P. Molnar. 2001. Closing the Indonesian seaway as a precursor to East African aridification around 3–4 million years ago. *Nature* 411: 157–62.

Carrier, D. R. 1984. The energetic paradox of human running and hominid evolution. *Current Anthropology* 25 (4): 483–95.

Cerling, T. E. 1992. Development of grasslands and savannas in East Africa during the Neogene. *Palaeogeography, Palaeoclimatology, Palaeoecology* 97 (3): 241–47.

Chomsky, N. 1972. *Language and mind* (enlarged ed). New York: Harcourt Brace.

Dart, R. A. 1925. *Australopithecus africanus*: The man-ape of South Africa. *Nature* 115: 195–99.

Darwin C. R. 1859. *On the origin of species by means of natural selection, or The preservation of favoured races in the struggle for life.* London: John Murray.

———. 1871. *The descent of man and selection in relation to sex.* 2 vols. London: John Murray.

Dubois, E. 1894. *Pithecanthropus erectus, eine menschenähnliche Übergangsform aus Java.* Batavia, Java, Netherlands Indies: Landesdruckerei.

Falk, D. 1990. Brain evolution in *Homo*: The "radiator" theory. *Behavioral and Brain Sciences* 13 (2): 333–81.

Gabunia, L., A. Vekua, D. Lordkipanidze, C. C. Swisher III, R. Ferring, A. Justus, M. Nioradze, M. Tvalchrelidze, S. C. Antón, G. Bosinski, O. Jöris, M. A. de Lumley, G. Majsuradze, and A. Mouskhelishvili. 2000. Earliest Pleistocene hominid cranial remains from Dmanisi, Republic of Georgia: Taxonomy, geological setting, and age. *Science* 288: 1019–25.

Gingerich, P. D., M. ul Haq, I. S. Zalmout, I. H. Khan, and M. S. Malkani. 2001. Origin of whales from early Artiodactyls: Hands and feet of Eocene Protocetidae from Pakistan. *Science* 293: 2239–42.

Goodall, J. 1967. *My friends the wild chimpanzees.* Washington, DC: National Geographic Society.

Huang, W., R. Ciochon, Y. Gu, R. Larick, Q. Fang, H. Schwarcz, C. Yongs, J. de Vos, and W. Rink. 1995. Early *Homo* and associated artifacts from Asia. *Nature* 378: 275–78.

Hunt, K. D. 1996. The postural feeding hypothesis: An ecological model for the evolution of bipedalism. *South African Journal of Science* 92 (2): 77–90.

Isaac, G. L. I. 1976. The activities of early African hominids: A review of archaeological evidence from the time span two and a half to one million years ago. In *Human origins: Louis Leakey and the East African evidence. Perspectives on human evolution, Vol. III,* ed. G. L. I. Issac and E. R. McCown. Menlo Park, CA: W. A. Benjamin.

Isaac, G. L. L, and B. Isaac, eds. 1997. *Plio-Pleistocene archaeology. Vol. 5: Koobi Fora research project.* Oxford: Clarendon Press.

Jarvik, E. 1980. *Basic structure and evolution of vertebrates.* 2 vols. London: Academic Press.

Jolly, C. J. 1993. Species, subspecies, and baboon systematics. In *Species, species concepts, and primate evolution,* ed. W. H. Kimbel and L. B. Martin, 67–107. New York: Plenum Press.

———. 2000. Baboon biogeography. *American Journal of Physical Anthropology,* Supplement 30: 190.

Keith, A. 1925. The Taungs skull. *Nature* 116: 11.

Kingston, J. D., B. D. Marino, and A. Hill. 1993. Isotopic evidence for Neogene hominid paleoenvironments in the Kenya rift valley. *Science* 264: 955–59.

Kortlandt, A. 1980. How might early hominids have defended themselves against large predators and food competitors? *Journal of Human Evolution* 9 (2): 79–112.

Larick, R., and R. L. Ciochon. 1996. The African emergence of early Asian dispersals of the genus *Homo. American Scientist* 84 (6): 538–51.

Lawler, A. 2001. Writing gets a rewrite. *Science* 292: 2418–20.

Leakey, L. S. B. 1959. A new fossil skull from Olduvai. *Nature* 184: 491–93.

Leakey, L. S. B., J. F. Evernden, and G. H. Curtis. 1961. Age of Bed I, Olduvai Gorge, Tanganyika. *Nature* 191: 478–79.

Leakey, M. D. 1966. A review of the Oldowan Culture from Olduvai Gorge, Tanzania. *Nature* 210: 462–66.

———. 1971. *Olduvai Gorge. Vol 3: Excavations in Beds I and II, 1960–1963.* Cambridge: Cambridge University Press.

Leakey, M., and A. C. Walker. 1997. Early hominid fossils from Africa. *Scientific American* 276 (6): 74–79.

Leonard, W. R., and M. L. Robertson. 1995. Energetic efficiency of human bipedality. *American Journal of Physical Anthropology* 97 (3): 335–38.

McBrearty, S., L. Bishop, and J. Kingston. 1996. Variability in traces of Middle Pleistocene hominid behavior in the Kapthurin Formation, Baringo, Kenya. *Journal of Human Evolution* 30 (6): 563–80.

McBrearty, S., and A. S. Brooks. 2000. The revolution that wasn't: A new interpretation of the origin of modern human behavior. *Journal of Human Evolution* 39 (5): 453–63.

McHenry, H. M., and R. S. Corruccini. 1978. The femur in early human evolution. *American Journal of Physical Anthropology* 49 (4): 473–88.

MacLarnon, A. 1993. The vertebral canal. In *The Nariokotome* Homo erectus *skeleton,* ed. A. C. Walker and R. Leakey, 359–90. Cambridge, MA: Harvard University Press.

MacLarnon, A., and G. P. Hewitt. 1999. The evolution of human speech: The role of enhanced breathing control. *American Journal of Physical Anthropology* 109 (3): 341–63.

Marean, C. W. 1989. Sabertooth cats and their relevance for early hominid diet. *Journal of Human Evolution* 18 (6): 559–82.

Matiegka, J. 1934. *Homo Předmostensis.* Fosilní Človek z Předmostí na Morave. L'homme fossile de Presmostí en Moravie (Tchécoslovaquie). I. Lebkky. Nákladem České Akademié Věd a Uměni v Praze.

Moreau, R. E. 1966. *The bird faunas of Africa and its islands.* New York: Academic Press.

Richmond, B. G., and D. S. Strait. 2000. Evidence that humans evolved from a knuckle-walking ancestor. *Nature* 404: 382–85.

Schick, K. D. 1998. A comparative perspective on paleolithic cultural patterns. In *Neanderthals and modern humans in Western Asia,* ed. T. Akazawa, K. Aoki, and O. Bar-Yosef, 449–60. New York: Plenum Press.

Semaw, S., P. Renne, J. W. K. Harris, C. S. Feibel, R. L. Bernor, N. Fesseha, and K. Mowbray. 1997. 2.5 million–year-old stone tools from Gona, Ethiopia. *Nature* 385 (6614): 331–36.

Shipman, P, and J. Rose. 1983. Early hominid hunting, butchering, and carcass-processing behaviors: Approaches to the fossil record. *Journal of Anthropological Archaeology* 2 (1): 57–98.

Smith, B. H. 1992. Life history and the evolution of human maturation. *Evolutionary Anthropology* 1 (4): 134–42.

———. 1993. Physiological age of KNM–WT 15000. In *The Nariokotome* Homo erectus *skeleton*, ed. A. C. Walker and R. Leakey, 195–220. Cambridge, MA: Harvard University Press.

———. 1994. Patterns of dental development in *Homo, Australopithecus, Pan,* and *Gorilla. American Journal of Physical Anthropology* 94 (3): 307–25.

Stanford, C. B. 1995. Chimpanzee hunting behavior and human evolution. *American Scientist* 83 (3): 256–61.

Stanford, D., R. Bonnichsen, and R. E. Morlan. 1981. The Ginsberg experiment: Modern and prehistoric evidence of a bone-flaking technology. *Science* 212: 438–40.

Straus, L. G. 1989. On early hominid use of fire. *Current Anthropology* 30 (4): 488–91.

Stringer, C., and C. Gamble. 1993. *In search of the Neanderthals.* London: Thames and Hudson.

Tattersall, I. 1993. Speciation and morphological differentiation in the genus *Lemur.* In *Species, species concepts, and primate evolution*, ed. W. H. Kimbel and L. B. Martin, 163–76. New York: Plenum Press.

Thieme, H. 1997. Lower Paleolithic hunting spears from Germany. *Nature* 385 (6619): 807–10.

Tobias, P. V. 1967. *The cranium and maxillary dentition of Australopithecus (Zinjanthropus) boisei.* In *Olduvai Gorge. Vol. 2*, ed. L. S. B. Leakey. Cambridge: Cambridge University Press.

Toth, N., and K. Schick. 1983. The cutting edge: An experimental elephant butchery with stone tools. *Interim Evidence* 5 (1): 8–10.

Tylor, E. B. 1871. *Primitive culture: Researches into the development of mythology, philosophy, language, art, and custom.* 2 vols. London: John Murray.

Vrba, E. S. 1979. A new study of the scapula of *Australopithecus africanus* from Sterkfontein. *American Journal of Physical Anthropology* 51 (1): 117–30.

Walker, A. C. 1984. Extinction in hominid evolution. In *Extinctions*, ed. M. H. Nitecki, 119–52. Chicago: University of Chicago Press.

Walker, A. C, and R. Leakey. 1993. The skull. In *The Nariokotome* Homo erectus *skeleton.* Cambridge, MA: Harvard University Press.

Watanabe, H. 1971. Running, creeping and climbing: A new ecological and evolutionary perspective on human locomotion. *Mankind* (8): 1–13.

Weidenreich, F. 1937. The dentition of *Sinanthropus pekinensis*: A comparative odontography of the hominids. *Palaeontologia sinica* Peking 101: 1–180.

Wheeler, P. E. 1984. The evolution of bipedality and the loss of functional body hair. *Journal of Human Evolution* 13 (1): 91–98.

———. 1993. Human ancestors walked tall, stayed cool. *Natural History* 102 (8): 65–67.

White, T. D., D. C. Johanson, and W. H. Kimbel. 1983. *Australopithecus africanus*: Its phyletic position reconsidered. In *New interpretations of ape and human ancestry*, ed. R. L. Ciochon and R. S. Corruccini. New York: Plenum Press.

White, T. D., G. Suwa, and B. Asfaw. 1994. *Australopithecus ramidus*, a new species of early hominid from Aramis, Ethiopia. *Nature* 371: 306–12.

White, T. D., B. Asfaw, and G. Suwa. 1996. *Ardipithecus ramidus*, a root species for *Australopithecus*. In *The first humans and their cultural manifestations*, ed. F. Facchini, 15–23. Forli, Italy: Colloquia of the XIII International Congress of Prehistoric and Protohistoric Sciences.

Whiten, A., and C. Boesch. 2001. The cultures of chimpanzees. *Scientific American* 284 (1): 60–67.

Whiten, A., J. Goodall, W. C. McGrew, T. Nishida, V. Reynolds, Y. Sugiyama, C. E. G. Tutin, R. W. Wrangham, and C. Boesch. 1999. Culture in chimpanzees. *Nature* 399: 682–85.

PART THREE

Understanding Science

God of the Gaps: The Argument from Ignorance and the Limits of Methodological Naturalism

Robert T. Pennock

Introduction: Creationism Old and New

WHY DO SCIENTISTS OFTEN APPEAR to lose when they debate creationists? They enter a debate confident that science will easily win over ignorance and are all too often surprised when at the end of the evening it isn't clear to everyone in the audience that they have prevailed. Scientists confront creationism by taking up its challenges one at a time; they marshal the data, work through the analysis, and overwhelm each objection. Familiar with the evidential basis of evolution, they easily recognize the creationist arguments as spurious. So why does the scientific evidence fail to carry the day? One reason is that it is impossible to explain more than a tiny fraction of the evidence during a debate. One cannot expect an audience to understand in an hour what may take a semester-long class to introduce properly. However, a more important reason is that the creation/evolution debate is only superficially about the science. At its base, it is about religion and it is about philosophy.

In this essay I will focus on two important aspects of the philosophical baggage that is carried along in any discussion of the creation/evolution issue. Both involve questions about the nature of scientific methodology. The first deals with creationists' contention that their creation hypothesis ought to be recognized not only as a legitimate scientific alternative to evolution but perhaps even a better alternative. The second deals with their contention that science, because of its methods, is itself an "established religion," and an atheist religion at that. In both cases, we will see that the contentions are unjustified.

Anti-evolutionism comes in many forms. Although the most common form is young-earth biblical literalism, it is by no means restricted to that specific view. Some creationists, for example, believe that Genesis can be interpreted to allow the standard scientific chronology. Nor is creationism limited to Bible-based views. Some fundamentalist Hindu sects reject evolution in favor of their own specific theistic account. Many Native American groups do as well, believing that it contradicts their own creation stories. Hindu creationists, Umatilla creationists, and other non-Christian creationists may appear less frequently in the public debate, but the logic of their argument is the same. Put simply, "creationism" refers to any view that rejects evolution in favor of the action of some personal, supernatural creator. (Note that modern theological views that accept scientific findings about evolution are not forms of creationism under this definition. Many mainline Christian denominations, for instance, hold that evolution was simply the process that God ordained for creating the biological world.) "Anti-evolutionism" includes creationism but is a broader category. The Raëlian Movement, for instance, rejects both evolution and creationism, holding that we were designed by extraterrestrial, but not supernatural, beings.

Despite the obvious and important differences among the views of these groups, they are remarkably similar in the way they attack evolution. In this paper, I will take most of my illustrations from the "intelligent design" (ID) movement, but most of the points apply with little modification to the other groups as well.

The "intelligent design" movement coalesced in the first few years of the 1990s, aiming to unite old-earth and young-earth creationists under a banner of "mere creation." In large part, it has been successful, with members of both camps in leadership positions in the movement, though one can detect some dissatisfaction among the more venerable creationist leaders who are leery of the strategy mapped out by the Discovery Institute, the ID think tank. Henry Morris, founder of the Institute for Creation Research (ICR), writes in "Design Is not Enough" that the idea behind the "intelligent design" movement is to begin with the design argument and postpone talking about the bedrock biblical doctrine.

> Any discussion of a young earth, 6-day creation, a world-wide flood and other biblical records of early history will turn off scientists and other professionals, they say, so we should simply use the evidence of "intelligent design" as a "wedge" to pry them loose from their naturalistic premises. Then, later, we can follow up this opening by presenting the gospel, they hope. (Morris 1999)

However, he is not completely happy with this wedging strategy. He writes:

> It is obvious that neither "intelligent design" nor "irreducible complexity" nor any other such euphemism for creation will suffice to separate a thorough-going Darwinian naturalist from his atheistic religion, in favor of God and special creation. (Morris 1999)

In these passages, Morris identifies many of the key terms in the ID vocabulary—the idea that evolution is an "atheistic religion" propped up by "naturalistic premises"; the strategy of "the Wedge"; the notion of "irreducible complexity" (IC); and the classic theist argument from "design" (which he notes is their "euphemism for creation").

The core contention of the ID movement is that evolution is not a good scientific theory at all, that it is really no more than dogmatic naturalist philosophy. This claim is the central, recurring complaint of retired University of California (Berkeley) law professor Phillip Johnson, who has been the most important leader of the ID movement since the publication of his book *Darwin on Trial* in 1991. In that book and in numerous subsequent books, articles, and speeches, Johnson tries to cast doubt upon the evidence for evolution and to suggest that the only reason scientists accept it is that they subscribe to a dogmatic ideology that says nature is all there is. They reduce everything to material forces, he says, and dismiss by fiat the possibility of designers who transcend nature. Because of their materialist blinders and desire to retain their cultural power, they promote evolution even though they know it is weak, rather than recognizing a transcendent designer as the better explanation.

One cannot overemphasize the importance of this philosophical component to the ID movement. Everyone in the movement repeats Johnson's complaint about the dogma of naturalism like a mantra. To give just one instance, ID leader William Dembski founded his own society early in 2002 to attract like-minded people to the ID cause. He promotes this new International Society for Complexity, Information, and Design as a "cross-disciplinary professional society that investigates complex systems apart from external programmatic constraints like materialism, naturalism, or reductionism."

These are tricky philosophical concepts, but "intelligent design" creationists (IDCs) treat them all as essentially synonymous with materialism. We will examine what these esoteric concepts mean when we return to the question of whether science's "naturalistic premises" make it equivalent to atheism. But first let us look at why Johnson, Dembski, and others in the ID movement think they are so important. This takes us to the first significant term in the ID vocabulary.

"The Wedge Document"

In "The Wedge document," an internal Discovery Institute manifesto written in 1999, we see what in the ID view is the connection among evolution, Christian theism, and materialism. The introduction states the fundamental position:

> The proposition that human beings are created in the image of God is one of the bedrock principles on which Western civiliza-

tion was built. Its influence can be detected in most, if not all, of the West's greatest achievements, including representative democracy, human rights, free enterprise, and progress in the arts and sciences.

Yet a little over a century ago, this cardinal idea came under wholesale attack by intellectuals drawing on the discoveries of modern science. Debunking the traditional conceptions of both God and man, thinkers such as Charles Darwin, Karl Marx, and Sigmund Freud portrayed humans not as moral and spiritual beings, but as animals or machines who inhabited a universe ruled by purely impersonal forces and whose behavior and very thoughts were dictated by the unbending forces of biology, chemistry, and environment. This materialistic conception of reality eventually infected virtually every area of our culture, from politics and economics to literature and art.

The document goes on to enumerate what they say are the "devastating" cultural consequences of materialism, including moral relativism and dismissal of personal responsibility. They say:

> The results can be seen in modern approaches to criminal justice, product liability, and welfare. In the materialist scheme of things, everyone is a victim and no one can be held accountable for his or her actions.

Citing product liability laws as one of the terrible problems of modern culture might seem a bit odd, but this follows from the promotion of laissez-faire capitalism, which is another of the Discovery Institute's main focuses. They also say that materialism

> . . . spawned a virulent strain of utopianism. Thinking they could engineer the perfect society through the application of scientific knowledge, materialist reformers advocated coercive government programs that falsely promised to create heaven on earth.

We then get a statement of the ultimate goal of the ID movement: "[It] seeks nothing less than the overthrow of materialism and its cultural legacies." It seeks to replace this view with a "theistic understanding of nature" (Discovery Institute 2002).

As they see it, evolution stands in the way of the hoped-for cultural renewal of the theistic worldview. How can we return to traditional conceptions of man and God if people accept the Darwinian view that life is the result of natural processes of chance variation and natural selection? How can the biblical belief that we are created in the image of God—*Imago Dei*—be returned to its foundational position if students learn that they are descended from apes? How can we get rid of moral relativism, the welfare system, and annoying product liability laws until people understand that God sets the rules and expects people to accept the blame if they disobey them?

In their view, Marx and Freud have been discredited, and only Darwin remains to be toppled. The world was designed not by natural evolutionary processes but by an agent who "transcends natural causes." The rest of "The Wedge Document" lays out a series of five-year plans to accomplish the overthrow of evolution and materialism and regain a place for the bedrock principle of design.

Although it is rare to see such a program set out in a manifesto, this sort of position is not unique to the ID movement, nor is it even new. Advocates of classic "creation science," like Henry Morris, have always seen things in exactly the same way, blaming evolutionism for moral relativism, loss of belief in God's authority, and a similar range of what they saw as social evils. Their terminology was slightly different (for instance, they used the terms *creation* and *creator* rather than *design* and *designer*), but their underlying strategy was identical. For example, when trying to introduce ID into the public-school science curriculum, they tried to disguise the religious basis of their position in the same way—by purporting to advocate only scientific arguments without any reference to the Bible and by not explicitly naming God as the agent. They attacked evolution by trying to cast doubt on its evidential basis and by claiming that it was propped up by dogmatic naturalism; Johnson's supposedly novel criticism had long ago been made by ICR debater Duane Gish, who wrote, "The reason most scientists accept evolution . . . is that most scientists are unbelievers, and unbelieving, materialistic men are forced to accept a materialistic, naturalistic explanation for the origin of all living things" (Gish 1978, 24). Moreover, creationists hoped to replace evolution in the same way—namely, by resurrecting the classic argument for design and presenting this theistic view as though it were an alternative scientific theory.

The Design Analogy

The design argument for the existence of God is a staple of philosophical theism. Some cite Romans 1:20 (which creationists regularly quote when opposing evolution) as a vaguely stated instance of the argument, but it was Thomas Aquinas who put forward one of the most important early versions of the argument in the thirteenth century. Aquinas argued that things in the world act to achieve the best result, and that they achieve their ends not by chance but by design. He invoked an analogy with an arrow, which cannot reach its target without an archer. Claiming that things without knowledge cannot move toward an end without the help of a being with knowledge, he concluded: "Therefore some intelligent being exists by whom all natural things are directed to their end; and this being we call God." The design argument is also known as the "teleological" argument, because it is based upon purported ends, goals, or useful functions that may supposedly be seen in the things of the world.

One finds variations of the design argument for the existence of God in the works of John Ray, Cotton Mather, and others who highlighted the intricate "perfections" to be found in the world, especially in the biological realm. Ray's *The Wisdom of God Manifested in the Works of Creation*, for instance, cited the human body that, Ray claimed, has no part without its "end and use." However, the most famous articulation of the argument comes in the work of William Paley, a driving force of the nineteenth-century movement of natural theology, which treated God via an analogy with a natural system. Like Aquinas, Paley thought that the arrangement of parts to an end was a sign of an intelligent designer.

> Wherever we see marks of contrivance, we are led for its cause to an *intelligent* author. And this transition of the understanding is founded upon uniform experience. We see intelligence constantly contriving; that is we see intelligence constantly producing effects, marked and distinguished by certain properties—not certain particular properties, but by a kind and class of properties, such as relation to an end, relation of parts to one another and to a common purpose. . . . [W]e conclude that the works of nature proceed from intelligence and design; because in the properties of relation to a purpose, subserviency to a use, they resemble what intelligence and design are constantly producing, and what nothing except intelligence and design ever produce at all. (Paley 1802)

In *Natural Theology*, Paley devoted chapter after chapter to examples of all manner of contrivances that he thought he saw in the biological and physical world—all of which purportedly revealed the perfections of their maker. God's benevolence, for instance, is evident in the sweet taste of peaches. It is also evident in the insipidity of water; had God given water a taste, said Paley, "It would have infected every thing we ate or drank, with an importunate repetition of the same flavor" (Paley 1802). Apparently God was contriving means to ends here in making water tasteless so that we would not get bored with it. Paley also pointed out how clever it was that God filled the seas with water rather than milk or wine, for otherwise, "Fish, constituted as they are, must have died" (Paley 1802).

The argument that drives all these examples is the famous watchmaker analogy. Were one to come upon a pocket watch and observe how the intricate complexities of its arrangement of gears served the purpose of turning its hands to tell the time, one would infer that there was some watchmaker who designed it. Similarly, in observing the complexities of the biological and physical world, such as those mentioned above, one ought to infer the existence of a worldmaker who designed them. This argument had been made earlier by Dutch theologian Bernard Nieuwentijt, but it was Paley who most thoroughly elaborated it.

The argument from design in its various forms has been criticized and rejected by philosophers over the centuries. Arguments from analogy are only as strong as the points of similarity, and there are far more points of disanalogy than analogy between natural objects and artifacts like a watch. One could as easily argue on analogy that the world was hatched from an egg as built by a designer. Even those who held to theistic beliefs on other grounds found flaws in the argument. Bishop George Berkeley, for instance, pointed out that if God willed a watch to run, it would need no more than an empty case—intricate material mechanisms are hardly a sign of divine design and power.

Of course, the design argument for the existence of God is not science. When a scientist infers that a pot found at an archaeological site is designed, the inference is based on precise arguments that rely upon specific knowledge about the properties and interests of human designers. However, as philosopher David Hume pointed out in the eighteenth century in his critique of natural theology, we have no knowledge of divine attributes and so have no basis for making a comparison. On the other hand, if one were to treat God anthropomorphically—that is, as having natural human characteristics—then, in looking critically at the waste and cruelty of the world, one would have to judge God as incompetent and uncaring, if not outright malicious. These and other problems with the teleological argument appear to be insurmountable. In general, Paley's argument is now only of historical interest.

As we will see, IDCs to date have offered nothing that goes the least beyond Paley's watchmaker analogy. Indeed, it seems that they offer a good deal less.

Design as the "Best Explanation"?

There are several ways that one could interpret the argument from design. The classic approach is to take it to be an analogical argument, as in the manner discussed above. This method has been shown to have so many flaws that it is more of a liability than a help. Even Dembski acknowledges this in his attempted defense of Paley's argument, writing, "If the design argument is nothing but an argument from analogy, then it is a very weak argument indeed" (Dembski 1999a, p. 273). For this reason, he and other ID creationists (e. g., Meyer, 1999) promote it under a second interpretation, discussed by philosopher of science Elliott Sober (1993) as what is known as an *inference to the best explanation.* This form of confirmation works by weighing the explanatory merits of competing hypotheses and concluding that the hypothesis that best explains the data is the true one (Harman 1965; Lipton 1991).

Ordinarily, when a scientist infers a certain "best explanation," the inference draws on contrasts among different causal hypotheses grounded within a strong body of background knowledge. Under certain conditions, such as in the case of archaeological inferences mentioned above, we can sometimes make a good case that a human being designed something. However, this interpretation of Paley's argument is even weaker than the analogical version when explanations are not constrained by natural causal processes. The moment one rejects the evidential requirement limiting appeal to lawful causal processes and opens the door to supernatural interventions—which is what creationists do when they reject methodological naturalism—explanatory chaos breaks loose. Since there are no known constraints upon processes that transcend natural laws, a supernatural agent or force could be called upon to "explain" any event in any circumstance; that is what miracles supposedly can do. However, the concept of a transcendent designer or other miraculous force that can explain any event under any set of conditions is no explanation at all (Pennock 1999, chap. 6). Moreover, because such a hypothesis neither makes any specific or general predictions nor rules out any possibility, no observation could count for or against it; it is in principle untestable. Thus, if the design inference is construed as the best explanation while rejecting methodological naturalism (as ID creationists do), it cannot possibly win in a comparative assessment of hypothesized explanations.

This is a fundamental and pervasive problem for all ID creationists. How do they attempt to avoid it? William Dembski has two replies, but each fails to address the central problem.

Design, Explanation, and Intentionality

When faced with the objection that transcendental design is a one-size-fits-all explanation, Dembski's first defense is to claim that it confuses design with intentionality. While everything that is designed must have been intended, Dembski argues that not everything that is intended is designed. To support this distinction, he gives an example of laying a mirror on a desk:

> The mirror's position is therefore intended. Nevertheless ordinary linguistic usage resists saying that the mirror's position is designed unless its position is also carefully calculated to accomplish some highly specific purpose. For instance, suppose someone I'm not too fond of is sitting at my desk and that I adjust the mirror so that it just manages to reflect the sun's rays into the person's eyes. In that case the mirror's position would not only be intended but also designed. (Dembski 1999a, 245)

Dembski claims that his notion of specified complexity is what distinguishes design from intentionality.[1] He concludes: "The objection that there is nothing that cannot be explained by invoking design is therefore mistaken. Whereas everything we observe might be intended, not everything we observe is designed" (Dembski 1999a, 245–46).

There are several significant flaws in this defense. First, it is not necessarily true that anything that is designed must have been intended. For instance, it is quite common for a designer to let chance or mechanized elements play a role in determining what specific, final design emerges. A design process, such as one done by a computer, could even be completely automatic, so that intentionality plays no role in the specific result. Indeed, the Darwinian

1. Immediately following this claim, Dembski writes "Designed things are both complex and specified" (1999a, 245). What he meant to say was that designed things *must be* both complex and specified, because his claim holds only if he is making it true by definition.

mechanism may be thought of as such a design process. I have elsewhere shown how various examples that Dembski gives of specified complexity (which he claims can only result from intelligent design) can be produced by an evolutionary mechanism (Pennock 1999, 2001a), and I will not repeat those arguments here. It is only if one *defines* design as being a kind of intentionality that Dembski's claim above holds. However, even Dembski's own examples do not qualify under his definition.

To make matters worse, Dembski does introduce a nonstandard, technical definition of design, which figures in a second problem with his defense. Central to his argument is that something must be "carefully calculated to accomplish some highly specific function" for it to count as designed. However, a person may properly be said to have designed something even if very little thought ("calculation") were required and even if the result fulfilled only a very general function (or sometimes no function at all). Dembski is simply *assuming* his idiosyncratic definition of design, which requires his "specified complexity" as an integral component (see Elsberry, in this volume). However, we commonly may say that a thing was designed irrespective of whether it is likely or unlikely, specified or not. Moreover, in his formal treatment, Dembski defines design by negation, as the "set-theoretic complement of necessity and chance" (Dembski 1998). This is far removed from ordinary linguistic usage, but Dembski needs this odd, negative definition to legitimize the negative form of his "design inference," which we will examine shortly.

Leaving aside these and other problems now, I will grant for the sake of argument that we are limiting discussion to the kind of intentional design (such as in the mirror example, above) that Dembski gives in his rebuttal. Has he successfully shown that "there is nothing that cannot be explained by invoking design"? He has not, as can be seen when one notes the illegitimate modal slide he makes (in the major conclusion of his argument, quoted above) from the statement that everything we observe "might be intended" to the statement that not everything we observe "is designed." In fact, the distinction Dembski tries to draw does nothing at all to avoid the problem ID theorists face: *Anything might be designed in any circumstances in just the same way that anything might be intended in any circumstances.* For instance, no matter in what specific, unlikely spot the mirror reflected the sun's light, one could always suppose that someone had some specific purpose for shining it there rather than somewhere else, just as in Dembski's example. However, that something *might* have had a purpose does not make it so. One may not just read purpose or function off the phenomena of the world (Pennock 1999, 255–56). Effects do not come with function labels. Functions must always be judged relative to a specific theory.

It is of course correct that in a particular case (such as his mirror example) a true, nonvacuous explanation may be that Dembski intended and designed the position of the mirror to reflect the sun into the eyes of the person he disliked. He could report to us that this was his intended design. But we have no such information for ordinary objects, unless God (or some other transcendent designer) has been revealing His plans to Dembski. With regard to human beings, we do not always require a report from the "designer" to draw an inference to intentional design because we have considerable background information about human desires and purposes. It is that kind of specific knowledge about the nature of the purported designer that allows such inferences to work in the social sciences. But, again, we have no such knowledge of God or any other transcendent agent and so have no grounds for an inference. As is said, God's ways are mysterious.

The "mere design" hypothesis (or "mere creation" hypothesis, as Dembski puts it elsewhere) has no explanatory traction—not because there is something intrinsically wrong with every sort of design inference but because Dembski and other IDCs cannot (for conceptual or strategic reasons) give their hypothesis any content. Again, this is a problem not only for them but also for any other observer who dispenses with methodological naturalism. The moment one opens the door to gods, vital forces, or any other transcendent designer or power, one gives up the possibility of a scientific test (Pennock 1999, 289–94).

Design, Explanation, and the Supernatural

William Dembski's second defense against the objection that transcendental design is a one-size-fits-all explanation follows his discussion of inference to the best explanation and constitutes the closing section of *Intelligent Design: The Bridge Between Science and Theology* (Dembski 1999a). This is one of only a very few places in which he directly addresses the problem of the explanatory emptiness of appeal to the supernatural.[2] However, his defense here is significant in that it reveals the real nature of the ID argument. Two-thirds of Dembski's three-page discussion is nothing more than an extended quotation from Thomas Reid's eighteenth-century defense of the inference to God's "intelligent design" against the cogent arguments of David Hume. Dembski admits that Reid had no "full-fledged criterion of design" but claims that he and Michael Behe now have such a criterion in their notions of *specified complexity* and *irreducible complexity*. We will examine these shortly, but here the key point to note is that Dembski thinks he does not have to address what such a transcendent intelligent agent would or would not do, or anything at all about it. He writes:

> This is not an argument from ignorance. Behe and I offer in-principle arguments for why undirected natural causes (i.e., chance, necessity or some combination of the two) cannot produce irreducible and specified complexity. Moreover we offer sound arguments for why intelligent causation best explains irreducible and specified complexity. The ontological status of that intelligent cause simply does not arise in the analysis. (1999, 276–77)

But Dembski cannot sidestep this issue. As we have noted, the nature of the purported cause is crucial for making an inference to the best explanation. However, one will rarely find ID theorists

2. One other place is his article "Who's Got the Magic?" (Dembski 2001), in which he responds to an objection I made in *Tower of Babel*. In Pennock 2001A, I show why his reply fails.

making any specific claim about the hypothesized designer—especially not anything that could form the basis for any empirical test. The single exception that I know of in all the ID literature is Phillip Johnson's suggestion that God's "whimsy" is a better explanation for the peacock's tail than any natural alternative. Theirs is an inference to the best-explanation argument only in the most attenuated sense; ID explanations begin and end with an appeal to the featureless hypothesis, "It was designed."

The reason ID theorists think they can get away without specifying the nature of the designer is that they hope to win by default. Other than Johnson's tacit argument involving divine whimsy, they actually offer little in the way of positive content or arguments (sound or otherwise) for their alternative to evolution. Instead, they rely on purely negative arguments—variations of their claimed "in-principle" refutation of the power of *any* undirected natural cause to produce biological complexity. That is to say, what ID theorists rely upon is actually an eliminative god-of-the-gaps argument. It is to this (now third) interpretation of the design argument that we now turn.

Design in the Gaps

The form of the design argument that concerns us here is a syllogism that usually looks something like the following:

1. X is some (typically complex, functional) feature of the world.
2. There are only two possible explanations of *X*: natural (e.g. evolution) or transcendent (intelligent design).
3. Science has (in principle) no natural explanation of *X*.
4. Thus, a transcendent intelligent designer (God) designed/created *X*.

We should note in passing that, under some circumstances, it would be useful to draw a distinction between designing *X* (in the sense of conceiving the plan for *X*) and creating *X* (in the sense of building or manufacturing *X*). However, creationists typically do not draw that distinction in their arguments—or, if they do, they hold that both are required. One representative example of the dual-model premise (2) upon which the argument turns is the following from ID theorist Walter Bradley. With regard to the Second Law of Thermodynamics and possible explanations for how biological order can arise, he writes: "Either there is some as-yet undiscovered energy coupling mechanism or self-ordering mechanism, or else, God accomplished this part of His creation in a supernatural way" (Bradley and Olsen 1984). The presumption is that no mechanism will be discovered, leaving the second as the only alternative. By using some variation of this eliminative argument, anti-evolutionists avoid the hopeless task of giving positive evidence for their preferred view and expect to win by default simply by trying to poke holes in the scientific account.

Once one learns to recognize this negative pattern of argument, one finds it behind every one of the standard creationist arguments. One of the most common challenges involves the origin of life; *Of Pandas and People,* the "intelligent design" junior-high and high-school textbook (soon to be retitled *The Design of Life*), devotes a section to this. Another of the most common examples is the so-called Cambrian explosion. Phillip Johnson, Stephen Meyer, and other ID creationists regularly cite the contrast in the richness of the fossil record between the Cambrian and earlier periods as something evolution cannot explain. The Institute for Creation Research challenges evolution to explain the origin of different languages, while others, such as ID theorist John Omdahl, challenge it to explain the origin of language itself.

Another common class of challenges regarding what evolution supposedly cannot explain involves complex organs. Walter Brown of the Center for Scientific Creation includes some of these in his list of *Twenty Questions for Evolutionists.*

> How could organs as complicated as the eye or the ear or the brain of even a tiny bird ever come about by chance or natural processes? How could a bacterial motor evolve? (Brown n.d.)

It is surprising that creationists continue to bring up these challenges, many of which go back to Paley. Darwin had himself considered the difficulties that "organs of extreme perfection" like the eye presented for his view. By now, scientists have a good picture of how Darwinian processes can account for the evolution of the eye (Salvini-Plawen and Mayr 1977; Goldsmith 1990; Land and Fernald 1992; Nilsson and Pelger 1994). Research into the evolution of the eye is just one example of how science continually progresses in figuring things out, and how gaps in scientific knowledge are filled. This progress exposes these arguments from ignorance as intellectually weak and dooms them to certain failure. But when one puzzle is explained, creationists quietly move on to a different one. For years, creationists brought up the bombardier beetle as purportedly beyond explanation by Darwinian mechanisms. Now, as we see in Brown's challenge, they have turned to the inner workings of the cell itself—"molecular machines" of such intricate complexity that they claim it is inconceivable for them to have been produced by evolution.

Behe's Gap Argument

On the ID side, it is Michael Behe who has made the most of the molecular challenge. Behe also cites the bacterial motor as a key example. His novel contribution is to argue that such systems had to have been designed because they purportedly are "irreducibly complex" (IC). Behe defines irreducible complexity as

> . . . a single system composed of several well-matched, interacting parts that contribute to the basic function, wherein the removal of any one of the parts causes the system to effectively cease functioning. (Behe 1996, 39)

He claims that such systems cannot be produced by gradual modification "because any precursor to an irreducibly complex system that is missing a part is by definition nonfunctional" (Behe 1996, 39). Behe claims that the bacterial flagellum is irreducibly complex and that it, as such, could not have been produced by a Darwinian mechanism. He concludes that it therefore must be the result of "intelligent design." Behe's argument is covered in another article in this volume (see Dorit, in this volume), so here I will add only a few points.

For Behe's definition of irreducible complexity even to make sense, it must be interpreted generously, and even on the most generous reading, it fails outright as a conceptual refutation of Darwinian mechanism (Pennock 1999, 2001b). In a 2001 article, Behe conceded that a counterexample I gave did undermine his notion of irreducible complexity as he had defined it (Behe 2001). He pledged that a revised definition would repair the problem, but he did not provide one at the time, nor has he done so in the years since. Moreover, I showed conceptually how a gradual, stepwise process, using a simple natural scaffolding, could produce an IC system (Pennock 2000).

If taken as an empirical challenge that hangs on the specific examples he gives (e.g., the bacterial flagellum), "irreducible complexity" reduces to what Richard Dawkins has called "the argument from personal incredulity." Because Behe *himself* cannot imagine a way to explain how the flagellum arose, he concludes that God did it. Miller, Orr, Doolittle, Shanks and Joplin, Kitcher, and many others have published other criticisms of Behe's concept as well as of the specific examples he gave. Moreover, my colleagues and I have experimentally demonstrated the evolution of an IC system; we can now observe in fine detail what Behe claims is impossible in principle (Lenski et al. 2003). As a supposed "in principle" argument against Darwinian evolution, the "irreducible complexity" gap has already closed.

Dembski's Gap Argument

The other version of the design discussion that ID creationists put forward is William Dembski's argument that what he calls "complex specified information" (CSI), or "specified complexity," is a reliable sign of intelligent design. He claims to find CSI in DNA and in complex, functional organs such as the bacterial flagellum. Dembski says that inferring that these were intelligently designed is perfectly scientific—no different from what SETI (search for extraterrestrial intelligence) scientists attempt to do in searching for a radio signal that would indicate intelligent beings on another planet.

Again, this is not a new argument. Some version or other of it has long been a staple of creationism.[3] For instance, we find a variation of this very question on Walter Brown's classic list for evolutionists:

> What evidence is there that information, such as that in DNA, could ever assemble itself? What about the 4000 books of coded information that are in a tiny part of each of your 100 trillion cells? If astronomers received an intelligent radio signal from some distant galaxy, most people would conclude that it came from an intelligent source. Why then doesn't the vast information sequence in the DNA molecule of just a bacteria [sic] also imply an intelligent source? (Brown, n.d.)

The premise underlying this and similar arguments is that "information" is a sign of design because no natural mechanism is capable of explaining its existence. We get a more explicit articulation of this reasoning in the writings of Walter Bradley and Roger Olsen, two of the early pioneers of the "intelligent design" movement.

> How much danger is there in attributing to miraculous activity on God's part that which science cannot explain today? The specific observations demanding either supernatural activity by God or a new mechanism all have to do with increasing information content and ordering, both in the origin of life and the development from simple to complex life forms. While it is possible that God used some natural but subtle mechanism to organize living systems, it seems reasonable at present to assume no such mechanism exists and attribute as we have the organizing of the major types of living systems to a supernatural activity on God's part. (Bradley and Olsen 1984)

The "danger" that Bradley and Olsen weigh is the problem (mentioned previously) that any argument from ignorance faces—namely, that things science cannot explain today may well

3. In Pennock 1999, 251–52, I trace the constellation of elements of Dembski's argument back to Norman L. Geisler. Many of the specific points go back even further.

be explainable tomorrow. This ought to be a serious worry for any theist who relies on this kind of argument. The history of scientific progress is such that, if one insists upon basing belief in God in a current explanatory lacuna, it is rather likely that one's belief will be crushed as scientific investigation progressively closes the gap. Creationists have suffered this fate multiple times, but they are undeterred and are now placing their bets on the origin of information as a problem that will never be explained by science.

The one apparently novel feature of Dembski's presentation is that he tries to justify the inference to the activity of a designer by what he calls his "explanatory filter," but this turns out not to be viable for its intended purpose. (Indeed, since Behe's irreducible complexity is just a special case of Dembski's specified complexity, the latter fails immediately together with the former, so no additional arguments are required.) In *Tower of Babel* (Pennock 1999), I showed some of the problems with Dembski's filter and with his notion of information. My book was already in press before his book *The Design Inference* was published, so my criticisms had been based upon the way he had explained his proposal in talks and articles. His newer book gave a more formal presentation, but the basic flaws I cited remained. Dembski has since written another book, *No Free Lunch*, in which he claims to improve and supplement his earlier arguments, but nothing there corrects the fatal flaws. In *Intelligent Design Creationism and Its Critics*, I include several additional articles (Fitelson et al. 2001; Godfrey-Smith 2001; Pennock 2001a) that criticize other aspects of Dembski's argument and show how the Darwinian mechanism can produce biological information. I will not repeat these or other criticisms here but will highlight one point that is relevant to our current discussion.

Dembski's various presentations of his explanatory filter are inconsistent in some of the details, but he presents his basic argument in the form of a flow chart. Here are the steps: Consider some event *X*. Try to explain *X* in terms of lawful regularities, the first stage of the filter. If that fails, go to the second stage in the filter and consider whether *X* might have occurred by chance. If *X* cannot be explained by chance either, then conclude that *X* was the result of "intelligent design." That is, the filter is just a multistage process of elimination. The problem, however, is that their eliminative arguments only have a chance of working if the options are mutually exclusive and jointly exhaustive.[4] To try to avoid this problem, Dembski does not give the "design" basket in his filter a positive definition but defines it negatively as the set-theoretic complement of necessity and chance—that is, it really just means "anything else." By rigging the definition in this way, Dembski excuses himself from having to provide any positive evidence for intentional design.

In other words, Dembski's argument works like this: If you cannot think of a way for natural regularities and/or chance to explain something, then say that a "designer" did it. Dembski's "design inference" is nothing more than a formalization of a simple god-of-the-gaps argument. It is the standard argument from ignorance put in the form of a flow chart.

In the past, Dembski has tried to deny the charge that his is an argument from ignorance (e.g., Dembski 1999a, 276), but in a recent statement in the *Chronicle of Higher Education*, he admits but tries to excuse this approach, saying, "An argument from ignorance is still better than a pipe dream in which you're deluding yourself. I'm at least admitting to ignorance as opposed to pretending that you've solved the problem when you haven't" (Monastersky 2001).

However, design theorists do not just "admit to ignorance" but rather claim to find a transcendent designer in the purported gaps in our knowledge. When arguing against evolution, Dembski and other anti-evolutionists are fond of quoting the old saw that nothing can come from nothing. However, in this basic,

4. In science, for instance, to test causal hypotheses one begins with the hypothesis that the variables of interest are related by chance alone and then checks to see whether the data allow one to reject this null hypothesis. This eliminative argument is valid because the hypotheses of mere chance versus causal regularity (where the latter may include indeterministic causal processes) are indeed mutually exclusive and jointly exhaustive.

recurring argument, they ignore their own rule and make an exception for design, which they leave unexplained. Given the religious assumptions that underlie the ID movement's hope for a "theistic science," it is not surprising that we find at its core this epistemic counterpart of *creation ex nihilo.*

"Naturalistic Premises"

Dembski does briefly consider the objection that ID commits a god-of-the-gaps fallacy, but his discussion of this is notable in that he does not deny that his "design inference" is a gap argument. Rather, his defense is that it is not a fallacy. Sometimes, he says, we are right to give up searching for a natural way to fill a gap. He writes:

> A full and efficient use of our empirical and theoretical resources for discovery should be made before we accept a proscriptive generalization. But once this has been done, to suppose that all the gaps in extraordinary explanations must be fillable by natural causes cannot be justified. Nor is it the case that one is necessarily blocking the path of inquiry by putting forward a proscriptive generalization which asserts that natural causes are incapable of filling a certain gap. Not all gaps are created equal. To assume that they are is to presuppose the very thing that is in question, namely, naturalism. (Dembski 1999a, 245)

This last statement about "the very thing that is in question" takes us back to the core philosophical and theological concern of the ID movement. A scientist would likely find the statement incomprehensible in the present discussion. Wasn't it *evolution* that the ID movement was calling into question? In fact, as we noted at the beginning of this article, the Wedge strategists' fundamental objection to and desired defeat of naturalism and materialism goes hand in glove with their opposition to evolution.

In its concern that contemporary society is materialistic, the Wedge does not take on overconsumption or the accumulation of possessions. Rather, its worry is materialism in the metaphysical sense, which is the view that matter is the sole underlying substance of the world, of which all other things are composed, and that everything works by merely mechanical principles. In rejecting a metaphysical materialism, those who take God to be the ultimate creator and sustainer of the world would opt for either an idealist metaphysics (which holds that spirit or "mind," not matter, is the ultimate substance) or a dualist metaphysics (which accepts that spirit and matter are both basic substances).

ID creationists typically use the term *naturalism* interchangeably with *materialism*, even though metaphysical naturalism is a richer concept that says that nature and its laws are all that exist, but it allows that nature may not be limited to matter *per se*. More important, they regularly conflate these *metaphysical* concepts with the related *methodological* norms that are actually employed by science. (More on this shortly.)

ID creationists also typically equate reductionism with both materialism and metaphysical naturalism, though that, too, misconstrues the relationships among the philosophical concepts. Reduction, as it is understood in philosophy of science, is an explanatory relation: Some phenomenon is reduced to something else if the latter explains the former. It is indeed common in science to reduce, for example, macroscopic properties of objects to the properties of their microscopic components, but nothing in the concept of reductionism *per se* limits what the *relata* are. What is really going on when creationists complain about reductionism is that they are objecting to ultimate reduction to mere matter, or to material mechanisms where they believe supernatural intervention took place.

How is this related to evolution? Whether young-earthers or old-earthers, biblical creationists all reject Darwinian evolution because they see it as reducing human beings to purposeless products of material processes instead of individuals who were specially created, as Scripture says, in the image of God.[5] As Dembski has put it:

5. See Pennock 2004 for a recent discussion of how this is exemplified in the ID movement.

> The world is a mirror representing the divine life. The mechanical philosophy was ever blind to this fact. Intelligent design, on the other hand, readily embraces the sacramental nature of physical reality. Indeed, intelligent design is just the Logos theology of John's Gospel restated in the idiom of information theory. (1999b, 84)

In other words, ID theory rejects natural explanatory premises in favor of supernatural theistic ones.

Once one understands the difference between naturalism as a metaphysical view and naturalism as a methodological view, it becomes clear that it is ID theory, not science, that is in the thrall of a dogmatic ideology. Unlike metaphysical naturalism, methodological naturalism does not assume, as Dembski claims, that all explanatory gaps will be discovered to be fillable by natural processes. Science is agnostic with regard to whether there will be gaps in nature that are filled by some supernatural power. It is the ID theorists who make an absolute claim about gaps in our knowledge that science will never be able to explain. Moreover, they have given no evidence that a designer, supernatural or not, could actually fill the gap. Given that they have no method to test their preferred theistic alternative, their argument remains as no more than an assertion that, in Dembski's words, "natural causes are incapable of filling a certain gap"—the old argument from ignorance. But to claim that "full and efficient use of our empirical and theoretical resources" has already been entirely exhausted in some particular case (such as the bacterial flagellum), and that a transcendent designer must have done the trick, is simply to say that one has given up the search. Certainly it is to give up on science rather too quickly.

Science, in its modern sense, is actually a relatively new term. What we now think of as the natural sciences used to be called "natural philosophy." That earlier term was clearer, in that it explicitly highlighted the distinctive naturalistic methodology of the approach that we have been talking about. What is significant is that this specific approach to discovering empirical knowledge contrasted with and improved upon the older approach, which was known as the "occult philosophy" in which natural phenomena were to be explained in terms of the actions of spirits and other supernatural powers. The nonnaturalistic alternative that ID wants to "renew" is nothing but an ill-considered return to that older philosophy.

Is Science an "Atheistic Religion"?

On the basis of a consistent conflation of quite different forms of naturalism, Phillip Johnson and his Wedge movement claim that naturalism is a scientific "dogma" that has become an "established religion." Their charge is of a piece with creation scientists' claim that Darwinian evolution is equivalent to atheism. However, as we have seen, by hewing to methodological naturalism and eschewing metaphysical conclusions, science avoids that charge.

In a commentary on *Tower of Babel* (where I had laid out this position in more detail), philosopher Sandra Mitchell nicely summarized the conclusions of my arguments about the difference between metaphysical and methodological naturalism in science:

> So I agree with Pennock's argument that modern science does not rule out the existence of God, and hence polemical ascriptions of atheism to science are indeed unwarranted. Methodological naturalism more modestly claims that empirical test, repeatability, etc., are reliable ways to get at the nature of nature only if there are laws to be discovered, or at least stable regularities, and we seem to be able to generate explanations that hold up under the application of the method, so there must be natural processes that explain what happens that permit us to do so. Minimally, then, what methodological naturalism can claim is that there is an account of the world that does not require supernatural intervention that can be accessed by scientific means. (Mitchell 2001)

She went on, however to make one criticism:

> However, what I think he has not completely secured is the view that the existence and operation of supernatural forces in the world is something that can be settled by means of scientific method. (Mitchell 2001)

I agree that I have not secured this, but it was never my intention to say that I had done so. Quite the opposite. My claim was that scientific methods, by themselves, do indeed remain neutral about such metaphysical possibilities. The success or failure of science might ultimately play a role in an argument that draws some conclusion about the metaphysical nature of the world, but that argument would necessarily be philosophical, not scientific.

So where did the confusion arise? I think that Mitchell took me to be saying something stronger in my definition of methodological naturalism that she quoted above. I had written that, unlike the metaphysical naturalist, who commits to an ontology (matter in motion, for example) and adds a closure clause—"That is all there is"—the methodological naturalist commits only to a method of inquiry and indirectly to what that method discovers.

However, that is not the same as saying that the method can discover all things. There are limits to scientific methods. To illustrate this, take the example from the Roman Catholic view of the Eucharist. The faithful believe that during service the communion wine they drink turns into Jesus' blood. Can a scientist test the truth of that belief? If a chemist analyzes samples to try to detect a change of alcohol to blood plasma, and grape tannins to hemoglobin molecules, her instruments will detect no such transformation. She will find no religious truth in communion wine. It will appear as it always did. Does that mean that science has disconfirmed the religious belief? Not necessarily. Although it may have sounded like an empirical claim, Church doctrine does not regard it as such. Rather, the Church holds that God miraculously intervenes to turn the wine into the blood of Christ in such a way that its physical properties (or "accidents") remain the same but that its metaphysical essence is transubstantiated. Indeed, the term *transubstantiation* refers to a change in *substance*—that which metaphysically *stands under* physical properties (which themselves can remain unchanged). When the Eucharist is conceptualized in this way, no scientific observation or test could apply. Similarly, many young-earth creationists argue that the 6,000-year-old world was simply created by God so that it appears to be old (or "mature"). The radically ambiguous ID hypothesis of "design" or "mere creation" is of a similar kind. Each of these cases evades the possibility of empirical test by rejecting the methodological constraints of naturalism. There is no way that science could ever confirm or refute such possibilities. Such ideas are not science, but religion.

Thus, science may not draw the *metaphysical* conclusion of no intervention ever—perhaps there are supernatural designers who do surreptitiously manipulate DNA in ways we cannot determine empirically. Instead, *qua* scientists, we must stick with the methodological constraint that we may not rationally appeal to that bare possibility in practice.

In my experience, most scientists are usually scrupulous about not overstepping the limits of scientific method. Consciousness of the strictures of evidence makes the scientist cautious about advancing even relatively simple empirical claims, let alone grand metaphysical ones. However, in every generation there are some scientists who push beyond the boundary and use their scientific authority to promote broad religious or philosophical conclusions. Most of these try to argue that science has confirmed some particular theistic or religious view, but there are always a few who draw the opposite conclusion. Not surprisingly, it is members of the latter group whom creationists cite in their complaints that scientists are improperly promoting atheism.

Wedge proponents often name scientists whom they claim promote the idea of purposeless, Godless nature. In several cases, I have shown that such charges misrepresented the scientist's actual position, but this is not the place to pursue such details. For those cases in which scientists have indeed promoted metaphysical conclusions, theistic or atheistic, in the name of science, I would second the ID complaint. (However, I would also note a significant inconsistency, in that ID theorists complain about scientists who draw an atheistic conclusion while at the same time they promote their own theistic conclusion.) My advice to scientists is simple: Scientists need to recognize and respect, as most do, the limits of methodological naturalism. If individual scientists wish to dive into deeper metaphysical waters, then they should be clear when they are doing so—a simple, "Now for a philosophical aside . . . ," will often do—and not suggest that their conclusions are drawn strictly from within science. If they want to keep out of trouble, however, it is probably best for scientists to leave the metaphysics to metaphysicians and theologians.

Conclusion: "Euphemisms for Creation"

Although the terminology has shifted slightly, the Wedge strategy of the ID theorists is exactly the same as that of the creation scientists—to argue simultaneously that the creation hypothesis is scientific and that Darwinian evolution is a religious dogma. In rebuttal, I defended two broad conclusions:

1. Creationists' negative attacks upon evolution miss their mark—they have not advanced anything close to a positive scientific alternative to evolution, but have simply given an argument from ignorance. The design inference fails to confirm a transcendent designer—whether it is interpreted as an argument by analogy, an inference to the best explanation, or an eliminative argument. In the end, their version is no more than a spurious god-of-the-gaps argument.
2. That their attack upon scientific method as being dogmatically atheistic is also misplaced. Science, properly understood, is not, as Phillip Johnson puts it, an "established" religion; it is indeed religiously neutral.

"Intelligent design theory" fails as science, but it may yet succeed in its political goal of wedging its brand of anti-evolutionism into the public schools. It remains to be seen whether school boards and the courts will allow them to wedge open the door to their euphemisms for creation.

References

Behe, M. J. 1996. *Darwin's black box: The biochemical challenge to evolution.* New York: Free Press.

———. 2001. Reply to my critics: A response to reviews of Darwin's black box: The biochemical challenge to evolution. *Biology & Philosophy* 16: 685–709.

Bradley, W. L., and R. Olsen. 1984. The trustworthiness of scripture in areas relating to natural science. In *Hermeneutics, Inerrancy, and the Bible,* ed. E. D.

Radmacher and R. D. Preus, 283–348. Grand Rapids, MI: Zondervan. Reprinted at www.origins.org/articles/bradley_trustworthiness.html.

Brown, W. n.d. *Twenty questions for evolutionists.* Phoenix: Center for Scientific Creation. Available at www.creationscience.com/onlinebook/main.html (last accessed July 27, 2005).

Dembski, W. A. 1998. *The design inference: Eliminating chance through small probabilities.* New York: Cambridge University Press.

———. 1999a. *Intelligent design: The bridge between science and theology.* Downers Grove, IL: InterVarsity Press.

———. 1999b. Signs of intelligence: A primer on the discernment of intelligent design. *Touchstone* 12 (4): 76–84.

Discovery Institute. 2002. *The Wedge Strategy* 1999.

Fitelson, B., C. Stephens, and E. Sober. 2001. How not to detect design—Critical notice: William A. Dembski, *The design inference.* In *Intelligent design creationism and its critics: Philosophical, theological and scientific perspectives,* ed. R. T. Pennock. Cambridge, MA: MIT Press. [Original edition, *Philosophy of Science* 66 (September 1999): 472–88.]

Gish, D. 1978. *Evolution? The fossils say NO!* 3rd ed. San Diego: Creation-Life Publishers.

Godfrey-Smith, P. 2001. Information and the argument from design. In *Intelligent design creationism and its critics: Philosophical, theological and scientific perspectives,* ed. R. T. Pennock. Cambridge, MA: MIT Press.

Goldsmith, T. H. 1990. Optimization, constraint, and history in the evolution of eyes. *Quarterly Review of Biology* 65: 281–322.

Harman, G. 1965. The inference to the best explanation. *Philosophical Review* 74 (1): 88–95.

Johnson, P. E. 1991. *Darwin on trial.* Washington, DC: Regnery Gateway.

Land, M. F., and R. D. Fernald. 1992. The evolution of eyes. *Annual Review of Neuroscience* 15: 1–29.

Lenski, R., C. Ofria, R. T. Pennock, and C. Adami. 2003. The evolutionary origin of complex features. *Nature* 423: 139–44.

Lipton, P. 1991. *Inference to the best explanation.* New York and London: Routledge.

Meyer, S. C. 1999. The return of the God hypothesis. *Journal of Interdisciplinary Studies* XI (1/2): 1–38.

Mitchell, S. D. 2001. Comment on Pennock's *Tower of Babel.* Paper read at Association for Informal Logic and Critical Thinking (AILACT). American Philosophical Association Conference, Atlanta, GA.

Monastersky, R. 2001. Seeking deity in the details. *Chronicle of Higher Education* 48 (17).

Morris, H. 1999. Design is not enough. *Back to Genesis* 127: a–c.

Nilsson, D. E, and S. Pelger. 1994. A pessimistic estimate of the time required for an eye to evolve. *Proceedings of the Royal Society of London* 256: 53–58.

Paley, W. 1802. *Natural theology.* London: Faulder.

Pennock, R. T. 1999. *Tower of Babel: The evidence against the new creationism.* Cambridge, MA: MIT Press.

———. 2000. Lions and tigers and APES, oh my! Creationism vs. evolution in Kansas. *Science, teaching and the search for origins.* American Association for the Advancement of Science Dialogue on Science, Ethics and Religion. www.aaas.org/spp/dser/evolution/perspectives/pennock .shtml

———. 2001a. The wizards of ID: Reply to Dembski. In *Intelligent design creationism and its critics: Philosophical, theological and scientific perspectives,* ed. R. T. Pennock. Cambridge, MA: MIT Press. [Original published in *META* (now *Metanexus*) Metaviews 089. 2000.]

———. 2001b. Whose God? What science?: Reply to Behe. *Reports of the National Center for Science Education* 21 (3–4): 16–19.

———. 2004. DNA by design? Stephen Meyer and the return of the God hypothesis. In *Debating Design,* ed. M. Ruse and W. Dembski. New York: Cambridge University Press.

Salvini-Plawen, L. V. , and E. Mayr. 1977. On the evolution of photoreceptors and eyes. *Evolutionary Biology* 10: 207–263.

Sober, E. 1993. *Philosophy of biology,* ed. N. Daniels and K. Lehrer. Dimensions of Philosophy Series. Boulder, CO: Westview Press.

Is Evolution "Only a Theory"? Scientific Methodologies and Evolutionary Biology

Norman A. Johnson

DURING THE 1980 PRESIDENTIAL CAMPAIGN, Republican candidate Ronald Reagan was asked about evolution and whether it should be taught in public schools. He responded, "Well, it's a theory, it is a scientific theory only, and it has in recent years been challenged in the world of science and is not yet believed in the scientific community to be as infallible as it once was believed" (*Science* 1980, 1214).

Anti-evolutionists often have used statements similar to former President Reagan's in their attempts to banish or marginalize evolution in public-school science curricula. After all, they will argue, if Darwinian evolution and creation science are "just theories," then each should be given equal consideration in the classroom. In the first edition of *Scientists Confront Creationism*, Robert Schadewald (1983) provides statements by "creation scientists" and "flat earthers" proclaiming that they want nothing but facts to be taught in schools. According to them, science is just fact, and theory is not science. Some supporters of evolution have resorted to responding that evolution is fact, not theory. These and similar uses of the words fact and theory by both sides of the debate are not new and actually trace back at least as far as the Scopes trial (Numbers 1992; Larson 1997).

In this chapter, I explore the meaning of theory in scientific methodology and discuss the roles of observation and inference in science in general. I use the case example of genetic mapping to demonstrate how observations and inferences have been used in science. I also highlight the evidence supporting evolutionary biology, focusing on new evidence gathered since President Reagan's statement. In summary, evolutionary biology is both fact and fact-based theory.

Theory and Scientific Methodologies

From a scientific definition, just what is a *theory*? The anti-evolutionists' claim that evolution is "just a theory" can be persuasive due to the general public's lack of understanding about the methodologies and practices of science and the different connotations words have when used in science in contrast to their use in everyday life. Unlike the everyday use of *theory*, the formal scientific definition of *theory* is not mere speculation, or "an educated guess." An educated guess that can be tested is what a scientist would formally call a *hypothesis*. A scientific theory is something more: It is a coherent set of hypotheses, tested by evidence and reasoning, that possesses explanatory power.

In one of the premier scholarly textbooks on evolutionary biology, Douglas Futuyma provides an elegant definition of a theory. He defines a theory as "a mature coherent *body of interconnected statements*, based on reasoning and evidence, that explains a variety of observations" (Futuyma 1998, 11; emphasis in the original). A good working definition of a *fact* would be a statement that is well supported by evidence and reasoning.

So, there are at least three aspects to the contents of scientific theory. Theories are interconnected statements that are: based on evidence or observations; based on reasoning or logic; and coherent or internally consistent.

More than just searching for patterns, doing science involves attempting to explain why these patterns exist. Scientists search for processes that underlie these patterns. In doing so, they construct and test hypotheses. Groups of related hypotheses that withstand the scrutiny of these tests become theory. Some scientists, and some branches of science, are more interested in uncovering patterns. Some are more interested in process. Scientists use many different approaches to uncover patterns and processes. These include, but are not limited to, observations, analysis, manipulative experiments, demonstration experiments, computer simulations, and mathematical theory (see Pickstone 2001).

Inference

Science begins with observations, but it is more than that. It also involves making sense of these observations, for which scientists rely heavily on inference. In our daily lives, we also make inferences (logical deductions) as we try to make sense of our world. We base these inferences on what we observe with our senses, our prior experiences, and logic. Suppose you were in a windowless office and could see people walking in and out of the office but had no direct access to observing the outside conditions. Even without communicating with anyone else, you could still make inferences about the weather based upon your observations of the people walking in and out, your prior knowledge, and logic. If most people coming into the office were dressed in heavy coats, you could infer that it was cold outside. If they were wet or were carrying umbrellas, you could make the inference that it was raining. You could even make inferences about the intensity of the rain. If you saw that a single person was wet but others were dry, you might be in doubt. If uncertain, you could test the inference by collecting more data (watching more people). You could confirm these inferences by collecting other types of data, such as by talking to people who were outside or by going outside yourself. Note that these inferences can be used as hypotheses, subject to more tests (confirmation).

Inferences are also made when we try to predict future events. Returning to the example of the weather, if you noticed that the skies were darkening, the wind was picking up, and thunder was booming in the distance, a logical inference would be that it would rain within the hour. The inference may not be perfect. It is possible that the storm may pass over you. One could also formulate a probabilistic inference—rain is more likely under the present conditions than under others. Such an inference could be tested by observing what happens over multiple times when similar conditions are present. Weather forecasters often will report such probabilistic inferences when they tell us, "There's a 70 percent chance of showers tonight." All that means is that 70 percent of the time when conditions were similar, it rained. Meteorologists will also update forecasts, refining their predictions based upon additional and more recent data.

We also use inferences to explain the past. If you observed water on the floor of your apartment right below your open windows, you could infer that it probably had rained earlier that day even if you did not witness (by direct observation) the rain.

Police work and our legal systems also commonly use inference. In their search to find "whodunit," detectives sift through the "evidence" (or observations) and try to make sense of it. Based on the evidence, they make inferences. These inferences often can be tested by being based on other evidence. Evidence that does not come from direct eyewitness testimony is called "circumstantial evidence." Circumstantial evidence can be very powerful, and people are often convicted of crimes based solely on it. A striking example of the power of circumstantial evidence is the case of Timothy McVeigh. Nobody saw McVeigh bomb the Murrah Building in Oklahoma City, yet he was convicted. While the prosecutor's case was purely circumstantial, it was resoundingly compelling.

Inferences in Science

"*All* science, when you boil it down, is inference, but it works nonetheless" (Tudge 2001, 152; his emphasis).

All fields of science rely heavily on inference. It is a common failing, sometimes perpetuated by scientists, not to recognize the extent to which their work is based upon inference. In the discussion that follows, I will point out examples of the inferences made in genetic mapping.

One of the central tenets of Mendelian genetics is particulate inheritance. Traits are determined by entities that are transmitted from parent to offspring; the expression of those entities (now called genes) may be masked, but the entities themselves are not lost in transmission. Soon after the rediscovery of Mendel's work at the turn of the twentieth century, Walter Sutton and Theodor Boveri proposed that genes were carried on chromosomes—bodies in the cell that were visible when viewed under a light microscope. Thomas Hunt Morgan, then a young physiology professor, was initially skeptical of Mendelian genetics, and in particular Sutton and Boveri's chromosome hypothesis. So Morgan set out to test and challenge these theories using what was then called the vinegar fly and is now referred to as the fruit fly, *Drosophila melanogaster.*

Around 1910, Morgan found a male mutant with white eyes—in contrast to the normal red color of eyes in both male and female flies. Morgan crossed this male to a red-eyed female and found that all of the first-generation (F1) offspring had red eyes, consistent with this trait's being recessive—not expressed in individuals with both types of genetic variants. (These genetic variants are also called *alleles.*) When the F1 flies were crossed with each other, the resulting second-generation (F2) offspring were an interesting mix: All of the females had red eyes but some of the males had white eyes and some had red eyes. Morgan inferred that there was a connection between sex and white eyes. However, it was not that only males could have white eyes, because in crosses of the F1 females with red eyes to their white-eyed fathers, white- and red-eyed individuals were found in roughly equal numbers in both sexes in the resulting backcross progeny. This result and others led Morgan to the inference that sex and a genetic factor determining eye color are inherited as a unit (Kohler 1994). We now would call the genetic factor (or locus) that determines whether eyes are white or red *sex-linked* (or X-linked). This inference was strengthened with the observation that male flies had one X chromosome and female flies had two X chromosomes (Kohler 1994; Snustad et al. 1997).

In 1916, Calvin Bridges, one of Morgan's students, presented even stronger proof of the chromosomal theory of inheritance in the first paper published in the journal *Genetics*. Bridges's proof is an example of how observation and inferential reasoning are used in science. Recall that in the standard cross of white-eyed females to red-eyed males, one expects white-eyed males (1X, with a "white" allele) and red-eyed females (2Xs, one with a "white" allele and one with a normal or "red" allele, with red dominant to white). Bridges, however, observed that this pattern did not always fit. Approximately one in 2,000 offspring was an exception—either a male with red eyes or females with white eyes. Bridges went further and undertook microscopic examinations of the chromosomes of these exceptional flies. He observed that the exceptional white-eyed females had two X chromosomes plus a Y chromosome. The exceptional red-eyed males had just a single X chromosome and were missing the Y chromosome. This occurred because X chromosomes would, on rare occasion, fail to separate during meiosis (a phenomenon called nondisjunction). From this and other observations, Bridges showed that not only did the expected genetic results fit with the expected cytological (pertaining to the cell) observations, but the exceptional genetic results also corresponded with exceptional cytological results (Bridges 1916; Kohler 1994). The inference that can be drawn from these correspondences of extraordinary results was stronger than that based solely on correspondences of ordinary results.

The Bridges and later studies also showed that sex in *Drosophila* is determined solely by the number of X chromosomes, and not the presence or absence of the Y chromosome. In contrast, in humans and other mammals, sex is determined by whether or not an individual possesses a Y chromosome—or, more specifically, a particular portion of the Y chromosome. In mammals, XO individuals are female and XXY individuals are male (just the reverse of what is found in flies). Despite these differences, the inferences used in mapping the region of the Y chromosome responsible for sex determination are similar to those Bridges used: looking for correspondence among exceptional phenomena. An occasional human male is XX and an occasional human female is XY. Some XX males have a region of the Y chromosome exchanged with a part of one of their X chromosomes. Some XY females are missing part of the Y chromosome. The inserted part of the Y chromosome in XX males corresponds to the missing piece in XY females. (More detail can be found in Snustad et al. 1997.)

Another one of Morgan's students, Alfred Sturtevant, constructed the first genetic map soon after Morgan's discovery of sex-linkage genetic loci in flies. Sturtevant produced this map from inferences he made based on data generated by crosses in Morgan's lab. For instance, females that carried two different alleles at both the X-linked white-eye mutation and the yellow-body mutation would produce about 50 percent white-eyed sons and 50 percent red-eyed sons, and 50 percent dark-bodied sons and 50 percent yellow-bodied sons. That was not surprising and is indeed expected under Mendelian genetics. The sons that had yellow bodies tended to have white eyes and those that had dark bodies tended to have red eyes. That was also not surprising, because the genetic loci controlling body color and eye color are on the same chromosome. What *was* surprising was that some (about 3 percent) of the sons had yellow bodies and red eyes or dark bodies and white eyes. The traits had recombined. Sturtevant inferred that the genes themselves could recombine and that the frequency of the genes' recombination was a function of the distance between them on the chromosome. Recombinants between white and yellow did not occur very frequently, and thus these loci were close together. Sturtevant then took the data from several different crosses and found that he could indeed construct a rough map based upon the recombination of traits. Results of further crosses confirmed the inferences made from the map. Until recently, virtually all genes were mapped in this way. Less than twenty years after Sturtevant's first map, cytological proof of genetic recombination was published by Harriet Creighton and Barbara McClintock working with maize and Curt Stern working with *Drosophila* (Kohler 1994; Snustad et al. 1997).

Current genetic work is often based upon similar inferences. Consider this passage from Davies (2001) on the mapping of genetic loci associated with disease:

> By identifying random segments of DNA that vary slightly in sequence between different people, the inheritance of these markers could be followed in families with a genetic disease. If the inheritance pattern of the marker corresponded to that of the disease (in other words, the mutant gene) and the chromosomal location of the marker was known, then *by inference,* the map position of the disease gene was also revealed. (p. 50; emphasis added)

Genetic mapping is not usually presented as theory, but it is theory nonetheless. A genetic map is a collection of interconnected statements based on observations and inference. These statements are coherent and help us make sense of the data presented. They also allow us to make predictions. If one is told the map positions of genetic loci, one can, in principle, predict the frequencies of the different types of offspring between crosses that involve markers at those loci. These predictions depend upon various assumptions, including those made about the viability of different genotypes and the extent to which recombination at one part of the chromosome affects recombination elsewhere. These considerations and others all go into the theory of genetic mapping.

The types of inferences used in genetic mapping are also used in other fields of science. A striking example was the electronegativity scale, which was formulated and used by the Nobel laureate chemist Linus Pauling in his studies of the chemical bond. Pauling was attempting to categorize and understand the different types of chemical bonds. One type of bond, the ionic bond, arises from the electrostatic attraction between positively charged and negatively charged ions. In such bonds, electrons are much closer to the positively charged ion. An example is the *ionic bond* formed in table salt by the positively charged sodium ion and the negatively charged chloride ion. A very different type of bond is the *covalent bond*, in which electrons are shared more or less equally among noncharged atoms. For example, the bonds formed between different carbon atoms in carbohydrates are covalent. In the 1930s, Morgan and his group moved to the California Institute of Technology, where Pauling was a new assistant professor. Influenced by Morgan's idea of mapping by recombination, Pauling

> . . . now borrowed from the mapping idea [of the geneticists] to create his own scale of the relationship between pairs of elements. The more ionic character he calculated in the bond between two atoms, the greater the difference in their ability to attract electrons, and the further apart they were on his scale. (Hager 1995, 166–67)

Ionic and covalent bonds could now be viewed as two extremes of a continuum of bond character.

Pauling then used this electronegativity scale to make novel predictions about bond strength and character in other compounds. Bonds between elements that were far apart on the scale would be more ionic in nature, while bonds between those that were close in electronegativity would tend to be covalent. This scale made the mass of information about chemical bonds sensible and understandable. In other words, it helped formulate theory.

Models and Mathematical Theory

All science, including evolutionary biology, uses models. Why? Think of the mind-boggling complexity of our bodies, filled with trillions of cells that can "communicate" to varying levels with each other. Think of the complexities of trying to predict prices in the stock market, where many thousands of people trade billions of shares each day and the behavior of each player can influence the behaviors of others. We do not have the capability to fathom these systems in all of this complexity. Science's reaction to this overwhelming complexity is not to give up in despair. Instead, researchers work in a stepwise progression to try to understand the essence of various simplified parts of the world. Scientists recognize that modeling is necessary to reduce this complexity to manageable levels and get at the essence of the system. What is a model? Futuyma (1998, 228) provides a good definition: a "deliberately oversimplified representation of the real world."

Because, by definition, some detail is omitted, no model is perfect—that is, no model explains everything about the things it sets out to explain. However, some models are better than other ones. By what criteria do scientists judge models? The best models are those that are realistic, general, and precise (Levins 1968). In practice, however, a scientist might be forced to sacrifice one criterion for another. For instance, we can set up a model with many parameters precisely defined that explains the dynamics of populations of yeast. Such a model may be very precise; it might predict within a small margin of error the sizes of yeast populations given particular starting conditions. On the other hand, the model would probably be applicable to only a small range of yeast types and to a particular laboratory setting. It would be lacking in generality. Scientists also seek simplicity; they look for the simplest possible model that can explain the widest range of phenomena of interest. This does not mean that they will select the simplest conceivable model but rather the simplest one that successfully explains the essence of the system or set of phenomena.

Sometimes when scientists speak of "theory," they are referring to mathematical or analytical formulations of a theory. Theories can certainly be verbal, but there are advantages to mathematical theory. As Futuyma (1998, 229) notes, the major virtue of mathematical models is that "the conclusions are true within the framework of the model's assumptions. . . . [Thus] if we can be confident that a real system obeys, or nearly obeys, the assumptions, then we may be sure that the system will follow the predictions of the model." Given knowledge of various aspects about two strains of yeast—including their respective population-growth trajectories and initial population sizes, and the extent to which they interfere with each other's growth—we can predict whether one strain will outcompete the other or whether they can maintain coexistence. The quality of these predictions depends upon the extent to which our assumptions about the properties of yeast strains fit reality. When formulating mathematical models, scientists often discover that they have made certain assumptions that were not well formulated in previous verbal models of the system. The rigors of mathematics can make assumptions more apparent and more precise (Haldane 1964; Lewontin 1974; Futuyma 1998).

Coherence, Facts, and Theories

Consider the following statements:

> "The earth is about 7,900 miles (12,740 kilometers) in diameter."
>
> "The earth is approximately 93 million miles (150 million kilometers) from the sun."
>
> "The speed of light is approximately 186,000 miles (300,000 kilometers) per second."

Most people would call these "facts." They are statements that have been verified by observation and reasoning. None of these "facts," however, could be determined *solely* by observation. Inference was involved.

Knowing the facts of science without an understanding of the theories that unify them, and without an understanding of how we came to know the facts, is as sterile as being able to recite Shakespeare without an understanding of what the lines mean and how they relate to one another. It is more important to understand the processes behind how the speed of light was calculated and the implications the speed of light has for physics than it is to know what the speed of light is. The theories provide the structure needed to organize the facts so that their relationship to other facts can be better understood, explained, and predicted. Without theories, it would be impossible to calculate the earth's diameter, the earth's distance from the sun, or the speed of light.

The late astronomer Carl Sagan has often been quoted as saying, "Extraordinary claims require extraordinary evidence" (e.g., Goldsmith 1997; Schopf 1999). What did he mean? Because scientists are fallible humans, errors in observation or logical inference can be made. Science, however, is a self-correcting process. Scientists can go back and try to replicate observations or experiments made by themselves or others. They can try to confirm those results using complementary methods. When the different methodologies or approaches lead to the same or similar results, the confidence in those results is strengthened. However, scientists do not have the time or other resources to repeat all or even most studies. Those claims that are most important, as well as those that fly in the face of conventional wisdom, will receive the most scrutiny. The recent controversy over whether AH84001 (the Martian rock found in Antarctica) revealed the existence of life on Mars highlights this process. Goldsmith (1997) does an excellent job describing the types of evidence used in the debate (see also Schopf 1999). This claim was held to a higher standard than many claims because of its immense implications. While science is organized skepticism, scientists save their skepticism for those things that matter (Goldsmith 1997).

Although there is some degree of unity among the sciences—what E. O. Wilson calls "consilience" (Wilson 1998)[1]—important changes in one field of science do not necessarily affect the theories of other fields. For example, physicists have recently discovered that the fine-structure constant, one of the major cosmological constants, may have been minutely smaller (ten parts in a million) eight billion years ago (Webb et al. 2001). If confirmed, this result would have monumental implications for cosmology. As the old cliché goes, cosmology textbooks would have to be rewritten. On the other hand, despite the immense impact on cosmology, this slight malleability of the fine-structure constant in the long-distant past does not affect genetics, evolutionary biology, and the rest of the biological sciences. It also does not affect the principles of some nonbiological sciences, such as chemistry and geology.

1. Wilson took the term *consilience* from the nineteenth-century polymath William Whewell.

Evidence, Theory, and Evolution

Evolution is undoubtedly the central organizing principle of biology, and Darwin—more so than any other person—deserves the credit for making it so. In a few short years after the publication of *On the Origin of Species*, evolution *per se* was widely accepted among scientists (Numbers 1998). Why does Darwin deserve the credit? He was by no means the first to discover evolution. In fact, the general idea of evolution permeated debate in Victorian society. Even the idea of natural selection, Darwin's preferred mechanism for evolutionary change, was not unique to Darwin (Mayr 1991; Dennett 1995). Natural selection also was not as easily accepted as evolution *per se* by scientists of the nineteenth century. Darwin is correctly considered the father of evolutionary biology because he was the first to make a compelling case for evolution. He did so by marshaling evidence from diverse sources (geology, the emerging fossil record, biogeography, embryology, animal behavior, plant and especially animal breeding). Darwin also drew upon logical inference, most famously the Malthusian arguments about the "struggle for existence" (Mayr 1982).

In the nineteenth century, Darwin's theory faced a serious challenge—a lack of understanding of how traits and genes are transmitted. While he knew that individuals resembled their parents, he did not know *why* they did. Most people of Darwin's day thought inheritance was like the blending of paints: individuals would have traits that were roughly halfway between those of their parents. The loss of variation through successive generations of blending inheritance was a difficulty with which Darwin struggled in the later editions of the *Origin*. The rediscovery of Mendel's laws at the turn of the twentieth century showed that individuals possessed particles (genes) that they inherited from their parents. The expression of these particles may be masked, but the particles themselves persisted. After some initial rancor, Mendelian inheritance and Darwinian natural selection were shown to be very much compatible. Darwin's variation problem was solved (Provine 1971; Mayr 1982; Smocovitis 1996).

In the 1920s and 1930s, a trio of mathematically inclined biologists—Ronald Fisher, John Burdon Sanderson Haldane, and Sewall Wright—established the field of population genetics by providing the theoretical foundation for the way evolutionary forces act upon populations to cause changes in the frequencies of genetic variants (Provine 1971; Smocovitis 1996). These three men and subsequent population geneticists showed that mutation was the raw material upon which natural selection (among other forces) could and does act to cause evolutionary change. Smocovitis, a historian of biology, underscores the importance of this development with her declaration: "The title of Fisher's 1930 book, *The Genetical Theory of Natural Selection*, raised the status of natural selection from fact to theory, as Fisher articulated his 'fundamental theorem of natural selection' " (Smocovitis 1996, 123–24). In other words, natural selection became not just a set of well-tested inferences based upon many observations. Instead, with Fisher's "fundamental theorem" and other theoretical studies of population genetics, biologists had acquired a systematic and quantitative understanding of how natural selection can affect the frequencies of genetic variants over the course of time. The facts were organized into a coherent structure—theory.

The other difficulty that Darwin faced was that nineteenth-century physicists—most notably Lord Kelvin—thought the solar system had not been in existence long enough to allow for evolution by natural selection to produce the known diversity of life. The sun, they thought, could only burn for a few million years before exhausting its fuel. Kelvin and the other physicists thought the sun would burn like coal or some other fossil fuel. Now we know that the source of the sun's heat is nuclear fusion and that it can persist for billions of years. Radioactivity from other nuclear decay also allows scientists to determine the age of rocks on earth (see Dalrymple, in this volume). The oldest rocks are more than four billion years old. This fact comes from the independent confirmation of multiple lines of evidence based on the patterns of isotopes left in the rocks and confirms the age of the earth (Miller 1999, chap. 3).

A creationist may argue that "nobody has ever found an organism changing into another." Strictly speaking, that statement is largely accurate.[2] But it is also true that nobody has ever taken a photograph that will prove someone is walking; the person is standing still in the photo. Yet, suppose a series of still photographs, tightly placed in time, is taken of that person walking. If those pictures are laid out in chronological order, they can provide evidence that the person is walking. The more pictures and the less change in the person's position in each successive picture, the better the quality of the evidence that the person is indeed walking. Indeed, cinematography is this process taken to the limit—simply a succession of still pictures where the interval between them is so small that it appears continuous to the human eye and brain. Paleontologists can use fossils (the equivalent of those snapshots) in a similar way to provide evidence for large-scale morphological change over long periods of time.

Evolution 1980–2001

The evidence for evolution by Darwinian processes is even stronger today than it was when Ronald Reagan made his statement more than twenty-five years ago. For instance, recent discoveries have made the fossil record much richer. Paleontologists have found fossils as old as 3.4 *billion* years and have added substantially to our knowledge of the Precambrian fossil record (Schopf 1999; see Lazcano, in this volume). Paleontologists have also found many more fossils that show transition features from one form to the another (see Padian and Angielczyk, in this volume). The current hominin fossil record, often attacked by creationists, provides very strong evidence that humans indeed evolved from an apelike ancestor (see Brace, this volume). We know much more about the "transitional" features between our

2. Speciation—the formation of two species from one—can arise in a single generation from the doubling of chromosome numbers (polyploidy). Polyploid speciation has been observed numerous times and appears to be very important in plant evolution. However, for the most part, speciation requires a time scale that is longer than that amenable to direct human observation.

ape-like ancestor and ourselves (Futuyma 1998). While there are competing explanations for why we evolved big brains, the fossil record leaves no doubt that brain size in the lineage leading to us has increased in fits and starts over the past five million years. Moreover, we know the transition to bipedalism (upright walking) occurred when our ancestors had brains not much larger than those of chimps. The first clearly bipedal hominins are *Australopithecus anamensis* and *A. afarensis*, at about 4.1 million years ago (Richmond and Strait 2000). The record continues to improve. In fact, Haile-Selassie (2001) very recently has found fossils of a hominin that dates back to more than five million years ago, near the time when DNA evidence dates the split between humans and chimps (Hacia 2001).

Much of the new evidence in support of evolution—in particular, common descent—comes from molecular data and systematics. Ironically, in the very same issue of *Science* in which President Reagan's criticism of evolutionary theory appeared was a letter from several molecular biologists. This letter promoted a demonstration of their database that contained DNA sequences (Dayhoff et al. 1980). Sequence databases, such as the one compiled by Dayhoff et al., have changed the way biology is done. Many biologists spend their days and nights using these databases to compare the DNA sequences of different organisms to infer their evolutionary relationships. Systematics, the discipline in which scientists compare and classify organisms, has been around for a long time, but the molecular data have helped to spur a recent boom in systematics. Evolutionary biologists are continuing to develop new methodologies for classifying organisms and for giving quantitative measures of how well the data support their inferences about these evolutionary relationships. In general, the methodologies are based upon the principle of parsimony—the idea that the most likely evolutionary relationship among organisms is usually the one that requires the fewest evolutionary changes (Hillis et al. 1996; Futuyma 1998). These studies using molecular data and the methodologies of systematics have forcefully demonstrated common ancestry of life and confirmed much about what we knew regarding the relationships of organisms.

Comparative developmental genetics has also provided strong support for evolution. Like many of the nineteenth-century zoologists, Darwin had a strong interest in embryology. Indeed, Darwin used the similarity in features of embryology across widely different animals as evidence for common descent and evolution. Biologists after Darwin, and especially in the last twenty years, have learned a great deal about how animals develop and the genes that are involved. Results from these studies confirm and reinforce the support for the deep similarity of many features of development, even across very diverse animals. A major process in the development of all multicellular organisms is the modulation of expression of genes. Different genes have their expression (or activity) turned off and on in different cell types at different times through the course of development. This modulation is due in part to the protein products of other genes binding to DNA. Some of these genes that produce DNA-binding proteins are very similar to each other. Called homeobox genes, they are found in virtually all multicellular animals (reviewed in Raff 1996). The mid-1980s discovery of these homeobox genes has led to an acceleration of studies of the evolution of development. These studies have shown that certain aspects of developmental genetics systems are remarkably similar—or what evolutionary biologists call *conserved*—across very diverse animals (Raff 1996). More recently, biologists have been starting to examine how these developmental systems can evolve via natural selection and other well-established evolutionary genetic principles (Johnson and Porter 2001).

In the past twenty years, field-oriented evolutionary biologists have produced more case studies of evolution in action. The quality and scope of these studies have greatly improved. For instance, Peter and Rosemary Grant and their colleagues have been studying Darwin's finches in the Galapagos Islands for the past three decades (Grant 1986; Weiner 1994). By observing and measuring these birds, the Grants have demonstrated measurable and heritable morphological and behavioral change over time. They have been able to tie these changes to natural selection caused by changes in the seeds available to the finches owing to climatic fluctuations. Bruce Grant and others have improved our understanding of how natural selection has led to changes in the frequencies of different color morphs of the peppered moth (Grant et al. 1996; Grant 1999). Other classic studies include research on guppies (e.g., Reznick et al. 1997), salmon (e.g., Hendry et al. 2000), and stickleback fish (e.g., Schluter 2000). Science writer Jonathan Weiner, discussing the work examining natural selection in nature, wrote:

> Taken together, these new studies suggest that Darwin did not know the strength of his own theory. He vastly underestimated the power of natural selection. Its action is neither rare nor slow. It leads to evolution daily and hourly, all around us, and we can watch. (1994, 9)

As a testimony to the extent of activity measuring evolution in the wild, Kingsolver et al. (2001) analyzed 2,500 estimates of selection gradients published in the technical literature between 1984 and 1997. Selection gradients are statistical estimators of the intensity of selection acting on a particular trait. They state that these 2,500 estimates are but a subset of the known studies chosen that satisfied rigorous criteria for inclusion in the study. In addition to the sheer volume of studies, the measurement and analysis of evolutionary change has increased steadily in rigor and sophistication. Analytical (mathematical) theory, simulations, and statistical analysis are heavily used in these studies. It is also quite evident from these studies on rates of evolution that the problem is not explaining the observed change in the fossil record but explaining why that change can be so slow for long periods of time (Hendry and Kinnison 1999; Kingsolver et al. 2001; Merilla et al. 2001).

Since Darwin, and particularly in the past twenty years, evolutionary biologists have also made great strides at better understanding speciation (see Howard and Berlocher 1998). The most widely accepted model of speciation is that natural selection and other evolutionary forces operate on geographically isolated populations, causing them to diverge. As these populations diverge, their genes become progressively more incompatible. These incompatibilities can manifest themselves when the protospecies is unable to produce viable and fertile offspring with each other (Johnson 2000; Turelli et al. 2001). Our knowledge about the genes involved in this process, and the interactions among these genes, is rapidly increasing, particularly in model systems, such as species related to *Drosophila melanogaster* (Johnson 2000). Orr (1995) and others have developed a rather sophisticated analytical theory of this process. Such a theory aids our understanding of this process and generates testable predictions (see also Turelli et al. 2001). The genetic divergence among geographically isolated populations can also contribute to speciation by changing aspects of the behavior of each protospecies so that were they to come back into contact, they would be unlikely to mate with each other. Substantial progress has also been made studying this phenomenon called "premating reproductive isolation" and teasing out the genes involved in isolation (e.g., Noor 1997; Wells and Henry 1998). Nearly all biologists studying speciation think that the standard model above can and does account for species formation at least some of the time. Debate instead is focused on the validity and generality of alternative models. For instance, biologists are asking such questions as, To what extent, and under what circumstances, can speciation occur without geographic isolation? They ask about the role of different classes of genes and genetic elements in speciation (see Howard and Berlocher 1998; Turelli et al. 2001). They are exploring these questions on many fronts: laboratory experiments, field studies, computer simulations, and mathematical modeling.

Coda

Ronald Reagan's aforementioned statement is seriously incorrect for several reasons. First, evolution is not a "scientific theory only." It is based on facts and inference. Second, recent advances, far from invalidating Darwinian evolution, demonstrate why Darwinian evolution continues to be the central organizing principle of biology. Finally, this statement illustrates the common misunderstanding that a "theory" is not as reliable or as worthy to be "taught" as "facts" are. By criticizing evolution as "only a theory" and stating that only fact-based science should be included in the curriculum, anti-evolutionists pervert the nature of science.

Evolutionary biology is good science, replete with facts and made coherent by a strong theoretical framework. The areas of progress listed above are but a small sample of what evolutionary biologists have done and continue to do.

Acknowledgments

I thank Laurie Godfrey and Andrew Petto for inviting me to participate in this forum. I thank Julie Froehlig, Linden Higgins, Chad Hoefler, Michael Wade, and anonymous reviewers for their comments and/or discussions of this chapter.

References

Bridges, C. 1916. Non-disjunction as proof of the chromosome theory of heredity. *Genetics* 1: 1–52.

Davies, K. 2001. *Cracking the genome: Inside the race to unlock human DNA.* New York: Free Press.

Dayhoff, M. O., R. M. Schwartz, H. R. Chen, L. T. Hunt, W. C. Barker, and B. C. Orcutt. 1980. Nucleic acid sequence bank. *Science* 209: 1182.

Dennett, D. C. 1995. *Darwin's dangerous idea.* New York: Simon and Schuster.

Futuyma, D. J. 1998. *Evolutionary biology,* 3rd ed., Sunderland, MA: Sinauer Associates.

Goldsmith, D. 1997. *The hunt for life on Mars.* New York: E. P. Dutton.

Grant, B. S. 1999. Fine tuning the peppered moth paradigm. *Evolution* 53: 980–84.

Grant, B. S., D. F. Owen, and C. A. Clarke. 1996. Parallel rise and fall of melanic peppered moths in America and Britain. *Journal of Heredity.* 87: 351–57.

Grant, P. R. 1986. *Ecology and evolution of Darwin's finches.* Princeton, NJ: Princeton University Press.

Hacia, J. 2001. Genome of the apes. *Trends in Genetics* 17: 637–45.

Hager, T. 1995. *Force of nature: The life and times of Linus Pauling.* New York: Simon and Schuster.

Haile-Selassie, Y. 2001. Late Miocene hominids from the Middle Awash, Ethiopia. *Nature* 412: 178–81.

Haldane, J. B. S. 1964. A defense of beanbag genetics. *Perspectives in Biology and Medicine* 7: 343–59.

Hendry, A. P., J. K. Wenburg, P. Bentzen, E. C. Volk, and T. P. Quinn. 2000. Rapid evolution of reproductive isolation in the wild: Evidence from introduced salmon. *Science* 290: 516–18.

Hendry A. P., and M. T. Kinnison. 1999. Perspective: The pace of modern life: Measuring rates of contemporary microevolution. *Evolution* 53: 1637–53.

Hillis, D. M., C. Moritz, and B. K. Mable, eds. 1996. *Molecular systematics,* 2nd ed. Sunderland, MA: Sinauer Associates.

Howard, D. J., and S. H. Berlocher, eds. 1998. *Endless forms: Species and speciation.* New York: Oxford University Press.

Johnson, N. A. 2000. Gene interaction and the origin of species. In *Epistasis and the evolutionary process,* ed. J. B. Wolf, E. D. Brodie III, and M. J. Wade, 197–212. New York: Oxford University Press.

Johnson, N. A., and A. H. Porter. 2001. Toward a new synthesis: Population genetics and evolutionary developmental biology. *Genetica* 112/113: 45–58.

Kingsolver, J. G., H. E. Hoekstra, J. M. Hoekstra, D. Berrigan, S. N. Vignieri, C. E. Hill, A. Hoang, P. Gilbert, and P. Beerli. 2001. The strength of phenotypic selection in natural populations. *American Naturalist* 157: 245–61.

Kohler, R. E. 1994. *Lords of the fly: Drosophila genetics and experimental life.* Chicago: University of Chicago Press.

Larson, E. J. 1997. *Summer for the gods: The Scopes Trial and America's continuing debate over science and religion.* Cambridge, MA: Harvard University Press.

Levins, R. 1968. *Evolution in changing environments.* Princeton University Press, Princeton, NJ.

Lewontin, R. C. 1974. *The genetic basis of evolutionary change.* New York: Columbia University Press.

Mayr, E. 1982. *The growth of biological thought: Diversity, evolution, and inheritance.* Cambridge, MA: Belknap Press.

———. 1991. *One long argument. Charles Darwin and the genesis of modern evolutionary thought.* Cambridge, MA: Harvard University Press.

Merilla, J., B. C. Sheldon, and L. E. B. Kruuk. 2001. Explaining stasis: Microevolutionary studies in natural populations. *Genetica* 112/113: 199–202.

Miller, K. 1999. *Finding Darwin's God: A scientist's search for common ground between God and evolution.* New York: HarperCollins.

Noor, M. A. F. 1997. Genetics of sexual isolation and courtship dysfunction in male hybrids of *Drosophila pseudoobscura* and *D. persimilis. Evolution* 51: 809–15.

Numbers, R. L. 1992. *The creationists: The evolution of scientific creationism.* Berkeley: University of California Press.

———. 1998. *Darwinism comes to America.* Cambridge, MA: Harvard University Press.

Orr, H. A. 1995. The population genetics of speciation: The evolution of hybrid incompatibilities. *Genetics* 139: 1805–13.

Pickstone, J. V. 2001. *Ways of knowing: A new history of science, technology, and medicine.* Chicago: University of Chicago Press.

Provine, W. B. 1971. *The origins of theoretical population genetics.* Chicago: University of Chicago Press.

Raff, R. 1996. *The shape of life.* Chicago: University of Chicago Press.

Reznick D. N., F. H. Shaw, F. H. Rodd, and R. G. Shaw. 1997. Evaluation of the rate of evolution in natural populations of guppies (*Poecilia reticulata*). *Science* 275: 1934–37.

Richmond, B.G., and D. S. Strait. 2000. Evidence that humans evolved from a knuckle-walking ancestor. *Nature* 404: 382–85.

Schadewald, R. J. 1983. The evolution of Bible-science. In *Scientists confront creationism,* ed. L. R. Godfrey, 283–99. New York: W. W. Norton.

Schluter, D. 2000. *The ecology of adaptive radiation.* New York: Oxford University Press.

Schopf, J. W. 1999. *Cradle of life: The discovery of earth's earliest fossils.* Princeton, NJ: Princeton University Press.

Science. 1980. Republican candidate picks a fight with Darwin. *Science* 209: 1214.

Smocovitis, V. B. 1996. *Unifying biology: The evolutionary synthesis and evolutionary biology.* Princeton, NJ: Princeton University Press.

Snustad, D. P., M. J. Simmons, and J. B. Jenkins. 1997. *Principles of genetics.* New York: John Wiley.

Tudge, C. 2001. *The impact of the gene: From Mendel's peas to designer babies.* New York: Hill and Wang.

Turelli, M., N. H. Barton, and J. A. Coyne. 2001. Theory and speciation. *Trends in Ecology & Evolution* 16: 330–43.

Webb, J. K., M. T. Murphy, V. V. Flambaum, V. A. Dzuba, J. D. Barrow, C. W. Churchill, J. X. Prochaska, and A. M. Wolfe. 2001. Further evidence for cosmological evolution of the fine structure constant. *Physical Review Letters* 87: 091301.

Weiner, J. 1994. *The beak of the finch: A story of evolution in our time.* New York: Alfred A. Knopf.

Wells, M. M., and C. S. Henry. 1998. Songs, reproductive isolation, and speciation in cryptic species of insects. In *Endless forms: Species and speciation.* ed. D. J. Howard and S. H. Berlocher, 217–33. New York: Oxford University Press..

Wilson, E. O. 1998. *Consilience: The unity of knowledge.* New York: Alfred A. Knopf.

The Invisible Bible: The Logic of Creation Science

J. Michael Plavcan

Introduction

CREATION SCIENTISTS ARE A SMALL MINORITY among the many creationists in the world. Most creationists are simply people who choose to believe that God created the world—either as described in Scripture or through evolution. Creation scientists, by contrast, strive to use legitimate scientific means both to disprove evolutionary theory and to prove the creation account as described in Scripture. Their influence on society is far greater than their numbers would warrant. To many people, they provide a scientific legitimacy to a belief in Divine creation. There are many who proselytize Christianity using the work of creation scientists, without really understanding the evidence for evolution, the way that creation scientists distort such evidence, and the process of science itself. Organizations such as the Institute for Creation Research, Answers in Genesis, and the Discovery Institute's Center for Science and Culture are very active in proselytizing their beliefs and lobbying either directly or indirectly for political measures to promote their beliefs in public schools.

Recently, Niles Eldredge lamented that nothing in the anti-evolution rhetoric of creation scientists has changed over the years (Eve and Harrold 1990; Eldredge 2000). He points out that arguments invoking anatomical complexity as proof of a creator have been proffered since St. George Mivart asserted in 1871 that the complexity of the eye can only be explained as the work of a creator. Such arguments are well-published, widely discussed, and believed by a substantial number of people. The advent of the Internet offers free and easy access not only to an enormous body of anti-evolutionary writing but also to a wide selection of detailed rebuttals, and even the archives of debate between Internet correspondents. (www.talkorigins.org is probably the most comprehensive source on the Internet for both creationist and anti-creationist material.)

The academic community has vigorously and intensely scrutinized creation-science claims and arguments, forcefully and repeatedly rejecting all (see Dalrymple, Elsberry, Padian and Angielczyk, and Stenger, in this volume). It sees the arguments of creation science not merely as blatantly false but also as reflecting gross ignorance and misunderstanding of scientific principles. Even so, creationists have been successful again and again in convincing policymakers and the public of two apparently contradictory conclusions: (1) that creation science stands as an equal to evolutionary biology as a true and valid scientific endeavor, and (2) that evolutionary theory stands as an equal to creationism as a mere historic dogma that represents religious belief rather than science.

Because anti-evolutionary arguments have been refuted so completely, one might ask why people vehemently persist in their claims that scientific evidence both supports creation and refutes evolution? Is it blind faith alone? Or is there something about the creation scientists' logical approach that allows people to consider their arguments as valid scientific refutation of evolutionary theory? An exploration of these questions is important for two reasons. First, understanding the creation scientists' point of view—the way they see the world—is essential if scientists are to confront them effectively in the public forum. Second, education policymakers must understand that allowing the teaching of creation science is not a matter of giving equal treatment to conflicting theories; rather, it involves the dissemination of a worldview that, while internally consistent, is antithetical to the current practice of modern science.

A number of works have explored the mindset and philosophy of creationists and creation scientists (Eve and Harrold 1990; Toumey 1994; Locke 1998; see Kehoe, in this volume). In this essay, I briefly review and extend this work by exploring the logic of creation scientists. I argue that the logic and approach to science that are typical of creation scientists are characterized by two important phenomena. The first is "cognitive dissonance." In order to resolve the dissonance created by the conflict between their conviction that Scripture is literally true and the overwhelming scientific evidence to the contrary, creation scientists accept at best marginal, and often bizarre, "scientific" evidence against evolution and for creation. The second is that creation scientists approach science in the same way that they approach biblical scholarship. Literal interpretation of the Bible leads to an absolutist view of evidence. If something is true, then all evidence must be consistent with it, therefore creation scientists believe that any uncertainty in research disproves the theory.

Creation Scientists in Contrast to Other Creationists

Before evaluating the views of creation scientists, it is essential to define those of whom we speak. In the United States, *creationism*—the belief that the world was created in its current form by the biblical Judeo-Christian God as described in the first book of Genesis—is longstanding and still common. It is important to distinguish *creationism* from *creation science.* Creation science is but one manifestation of creationism. Creationism is often portrayed as a Protestant American phenomenon (Eve and Harrold 1990; Toumey 1994; Eldredge 2000), but those who believe in special creation may be of any faith or denomination. In my own teaching experiences, I have seen a belief in creation expounded by Protestants, Catholics, Orthodox Jews, and Muslims alike. Creationists are people from all walks of life. Apart from evolutionary biology itself, creationists may be found in every scientific, technical, and engineering profession.

Creationism may be a strict, pure, and literal belief in the Genesis account, or it may be a more general belief that God created the universe, without any adherence to the specifics of the Genesis account. For example, many people believe that evolution is the mechanism through which God created the world and humanity. These are often referred to as "theistic evolutionists," and they are held in contempt as "sellouts" by the more vocal creationists who hold to a literal interpretation of Genesis. A visit to the Answers in Genesis Web site (www.answersingenesis.org) will reveal almost as much verbal abuse hurled at "theistic evolutionists" as against "evolutionists."

For most people, creationism is a belief system—there is no attempt or need to evaluate carefully the scientific evidence for or against creation. They simply believe. Arguments supporting evolution are easily dismissed as rhetorical devices favoring a morally dangerous substitute belief system, while arguments against evolution are accepted as supportive of their personal beliefs. Most creationists resemble the American public as a whole, in that they are not well educated in the sciences and so are unable to judge between creation-science arguments and the rebuttals offered by scientists (Scott 1987; Eve and Harrold 1990; Toumey 1994). I would argue that this also extends to professionals who have earned graduate degrees. In eight years of teaching in medical school, I was surprised by the number of medical students who were not trained in the philosophical or even practical form of science, but rather in the assimilation and management of facts. Though many physicians consider themselves scientists (and indeed many are), there is broad confusion between training that involves learning and skillfully using large amounts of scientifically derived information in contrast to that which involves the actual process of generating such information.

Among creationists, a minority of people hold a view of literal creation very deeply, and they aggressively proselytize others to believe in such a view (Duncan and Geist 2004). However, not all of these are creation scientists. Many biblical literalists proselytize strict creationism and simply shrug off scientific arguments as irrelevant because the Word of God is inerrant (Toumey 1994). Others proselytize their belief in biblical creation, aggressively using creation-science arguments as a tool but with no real understanding of the scientific validity of the arguments. For example, Toumey (1994) notes that the Bible-Science Association adheres to a philosophy of defining science as that which agrees with a literal interpretation of the Bible. Like creation scientists, they accept creation as fact *a priori*, but science itself is defined on the basis of its relationship to that authority.

Creation scientists are a distinct subset among creationists in that they proselytize their religious convictions by aggressively attempting to prove the validity of the creation model and refute evolutionary theory using scientific methods and arguments. Most are conservative Protestant Christians. Participant listings from the Web sites for Answers in Genesis (www.answersingenesis.org), the Institute for Creation Research (www.icr.org), the Discovery Institute (www.discovery.org), and DI's Center for Science and Culture (www.discovery.org/csc/) demonstrate that most people in these groups have at least some postgraduate training in the sciences, or engineering, including some with earned doctoral degrees from reputable programs and institutions (though, as often noted, there are few who were trained in evolutionary biology).

Creation Science as Science

Attempts to reconcile natural observations with biblical creation are by no means novel, and they certainly were not inspired by the advent of modern evolutionary theory. In the nineteenth century, such attempts were a key part of the development of intellectual support for evolutionary theory. As noted by Niles Eldredge (2000), the debate about creation versus evolution was fought intensely in the nineteenth century, with evolutionary theory clearly emerging victorious. The modern movement called "scientific creationism" began in 1961 (Eve and Harrold 1990; Toumey 1994; Eldredge 2000) with the publication of *The Genesis Flood: The Biblical Record and Scientific Implications* by Henry Morris and John Whitcomb, which insisted that *scientific* observation clearly supports special creation as well as a worldwide flood that wiped out all life except that which was protected on Noah's Ark. Contemporary anti-evolutionists further argue that scientists are blinded by their belief in evolution as a metaphysical doctrine, so they cannot objectively evaluate the scientific evidence against evolution. (Some go so far as to accuse scientists of engaging in an active conspiracy against God, implying or openly stating that scientists are liars [Kofahl 1977].)

There are several common threads to creation scientists' arguments that are critical to the thesis expounded here. It is well known that creationists generate very little original research, relying instead on critiques supporting particular evolutionary interpretations and the age of the universe. The types of critiques vary somewhat. One current and important theme is that the evidence for evolution is not "scientific" but rather is an interpretive construct that arises from a faith in evolution. Creationists then assert that because historical processes and events are not repeatable, and because evolutionary and creationist interpretations of such evidence necessarily depend on historical inference, the evidence brought to bear on the history of life cannot be said to support one processual interpretation over another. Creation scientists are generally inconsistent in their application of this type of argument. For example, radiometric dating is commonly brushed aside as invalid because it relies on "assumptions" (see Dalrymple, in this volume), while polonium halos are hailed as evidence of a young earth. Likewise, creation scientists commonly argue that scientists refuse to see that the laws of physics prove that evolution cannot happen (see Stenger, in this volume). Related to this are claims that there are no transitional fossils in the fossil record, and that evolutionary theory cannot explain the "sudden" appearance of bony vertebrates in Cambrian formations (see Padian and Angielczyk, in this volume). Creationists insist that examples of transitional fossils are nothing more than fictional constructs arising from a belief in evolution. They argue that, instead, such fossils are fully formed and distinctive entities that do not fit a prediction of gradual intermediates between species.

Another important theme is the attack on evolution as unscientific because, in their view, it can be molded and changed ad infinitum to accommodate any observation from nature. Creationists often use the fact that there are technical debates among evolutionists as an indication that evolution is not well supported, and therefore that scientists adhere to a belief in evolution as a matter of faith (see Petto and Godfrey, in this volume). They make this claim without irony, despite the fact that creation science is also open to the same charge because it begins with a belief in biblical creation and then tries to force evidence to fit this belief. In contrast to the active debate among scientists about the details of evolutionary change, creation scientists do not normally engage in internal critique to evaluate "highly divergent arguments of colleagues who happen to share the same conclusions" in order to develop a consensus (Smith 2000). (The recent disagreement between Kent Hovind and Carl Wieland and others at Answers in Genesis is a rare exception: www.answersingenesis.org/docs2002/1011hovind.asp.) Instead, creation scientists see disagreements within evolutionary biology as evidence that scientists are desperately struggling in a losing battle to find internally consistent evidence to support their preconceived view that life evolved without the agency of God as creator.

These common threads among the various proponents of creation science often manifest themselves as a double standard in attempting to garner "equal time" for teaching creation and evolution in schools. Creation scientists insist that their views are equally as or more scientifically valid than any supporting evolution, while at the same time claiming, ". . . believers in evolution or creation must accept either view by faith" (Kofahl 1977, 14). Creation scientists are insistent that their "research" is systematically censored from scientific journals because of an open hostility to religious beliefs. This ignores the fact that scientists have repeatedly evaluated creationist claims, finding most of them erroneous, disingenuous, misrepresentative, and even ludicrous (as demonstrated in this book). At the same time, scientists are puzzled that scientifically compelling rebuttals of naïve creation-science claims consistently fall on deaf ears. Why do creation scientists continue to repeat positions that have been independently discredited so many times?

Several explanations are common—in both private circles and public settings—and it is enlightening to examine a number of them. Several works have elegantly described creation-scientist logic and philosophy (Kitcher 1983; Marsden 1984; Toumey 1994; Locke 1998). Two important points are consistently reiterated. First, creation science has grown out of a perception that evolutionary theory threatens the conservative Protestant moral and ethical belief system (see Cole, Kehoe, Numbers, in this volume). Second, creation scientists in practice follow a model of science that differs dramatically from that embraced by modern science (see Elsberry, Johnson, in this volume).

The first point is important because it underscores that creation scientists are not interested in science *per se* but rather in defending their religious views and the moral and ethical belief system that arises from them. Thus, even though creation scientists present the rather thin façade that they are crusading for the teaching of good science (this is especially true of the current push for "intelligent design"), in fact they are continually seeking any means to convince people that evolution is not true in order to support their faith and proselytize it.

The second point—that creation scientists promulgate a model of science different from the mainstream—is critical, because this is effectively what they want taught in public schools as "true" science. What constitutes "true" science is a matter of considerable philosophical debate; however, there are a few simple points that help to illustrate the differences between mainstream science and creation science. Though "science" has been practiced for thousands of years, the modern "scientific method" developed out of the natural philosophy of Francis Bacon

(1620). Baconian science relies on inducing natural laws from observations of the material world. This is a very practical, empirical view that shuns theory and deduction. It comes as no surprise that Protestant American ministers in the pre-Darwinian era were often naturalists, gathering evidence from nature of the glory of God's creation (see Toumey 1994) in a manner that fit well with the Baconian scientific framework.

The contemporary practice of the "scientific method" can be traced to the work of Karl Popper (1959). Popperian science emphasizes a deductive framework in which theoretical models based on observation are used to generate falsifiable hypotheses. Pending falsification or corroboration of hypotheses, models are refined and further tested. Above all else, the concept of falsifiability pervades Popperian science. If a model or theory cannot be falsified by some observation, it is often not considered scientific. However, while Popperian science is often heralded as a gold standard of whether one is doing "true science," modern science in fact represents a mix of inductive and deductive processes, coupled with both the corroboration of falsifiable hypotheses and the ability of models to explain a wider range of natural phenomena.

Creation scientists exploit simplistic notions of science in a simultaneous attempt to portray evolution as unscientific and creation science as the most reasonable interpretation of the natural world. For example, arguments about design implying a designer are consistent with Baconian philosophy (Eve and Harrold 1990; Toumey 1994; Eldredge 2000). Creation scientists point to the complexity of the eye, the "perfection" of anatomical and physiological design, and the codependence of many animals and plants as evidence for a creator. They induce a creator through the claim that complex design could only arise through deliberate creation—hence the oft-quoted statement that a watch found on the moor implies a watchmaker. Such arguments are easy for almost anyone to understand, regardless of his or her educational background.

More recently, proponents of "intelligent design" have attempted to recast this type of argument in Popperian fashion.

William Dembski (see Elsberry, Pennock, in this volume) and Michael Behe (see Dorit, in this volume) both argue that "irreducible complexity" is a set of falsifiable hypotheses (trying to *deduce* God instead of *induce* Him), corroborating the suggestion that complex biological systems must have been deliberately designed (Behe 1996; Dembski 1999). While the scientific community is virtually unanimous in its opinion that the model is purely inductive and relies on false assumptions about biological function, nevertheless creation scientists have put forth the model as scientifically superior to evolutionary studies because it is supposedly falsifiable.

Creation scientists commonly try to coopt Popperian science to prove creation and disprove evolution. For example, they argue that the hypothesis that geological stratification was caused by a global flood is corroborated by the observation that sediments from the eruption of Mount Saint Helens were rapidly stratified (Austin 1986). They also claim that the ancient age of the earth, and hence evolutionary theory, is falsified by the fact that in some areas older rocks are found above younger rocks (Kofahl 1977). They claim frequently that the absence of transitional forms (as they define them) in the fossil record falsifies evolutionary theory (Kofahl 1977).

In contrast, creation scientists maintain that evolutionary theory is unscientific because it cannot be falsified. The basis of this claim is that evolution is a product of historical events that cannot be observed directly or influenced experimentally. However, this is a subtle misrepresentation of the concept of falsification, preying on the public's misunderstanding that falsifiable hypotheses include only those that can be replicated in an experiment in a laboratory. In fact, the concept of experimental replication focuses on the validity of observations. Observations need not be generated from experiment—they need to be verifiable by independent observers. Thus, the concept that historical events are not subject to falsifiability is highly misleading. For example, the hypothesis that life evolved generates numerous predictions about the fossil record that can be falsified (Eldredge 2000).

Each time fossils are found in sequence in the geologic record, the hypothesis of evolution has been corroborated.

Left out of creation-science arguments is the ability of evolution (and an ancient universe) to explain so much about the physical and biological worlds. While claiming that explanatory models are just unscientific stories, creationists conveniently ignore or deny the fact that *their* explanatory models fail miserably to explain observations as simple as the fact that light from distant stars records events that must be billions of years old; that fossils in the geological column appear in nonrandom patterns; and that organisms contain a host of bizarre and seemingly nonsensical design features. Each of these observations reliably and repeatedly presents an opportunity for the hypotheses derived from evolutionary theory to be falsified. And evolutionary theory has withstood the exposure to masses of new observations.

Given the success and resilience of evolution as a scientific theory, one can logically ponder how creation scientists can honestly maintain their arguments in the face of repeated and forceful refutation. To many scientists, creation scientists seem to be liars, hypocrites, or people too ignorant to understand simple science. Opinions such as these are expressed, often in private, and sometimes in thinly veiled language in public. However, these are largely untrue and counterproductive to understanding the logic of creation scientists.

For example, Duane Gish has had enough of a following over the years for people to carefully compare notes about what he has claimed and what he has acknowledged as erroneous (Arthur 1996). Apparently Gish persists in repeating the same errors over and over, even when shown explicitly that he is misrepresenting facts. It is difficult to conclude from this that Gish is not a liar. However, the majority of creation scientists are fundamentalist Christians who hold strong values of honesty and integrity, so we might conclude that Gish is convinced that his arguments are true—not because they conform to science but because they conform to the Bible.

We can be certain that creation scientists are not stupid

people—their tactical effectiveness in working against evolution in the public arena testifies to this. While it is true that many creationist arguments reflect a clear ignorance of the details of particular scientific disciplines, careful perusal of literature published by the Institute for Creation Research alone indicates that many of these people keep abreast of current events in science—even if they do not have deep knowledge and mastery of the specific topics. It is true that creation-science publications often are sorely lacking in citations from the primary literature. But many creation scientists seem sufficiently well read to avoid this excuse. The nature of many of the citations suggests that the general absence of citations reflects selective omission rather than ignorance.

Eliminating the more derogatory explanations, we are left with the hypothesis that creation scientists indeed interpret the world, and scientific evidence, in a different framework from that used by mainstream scientists. Indeed, creation scientists maintain that the only difference between creation science and evolutionary biology is a matter of unbiased interpretation of evidence. Yet scientists overwhelmingly and forcefully reject the claim that these differences stem from "reasonable" interpretation of evidence.

Creation Science and Cognitive Dissonance

One explanation of how creation scientists can maintain their arguments in the face of repeated forceful rebuttals is that they are so deeply committed to their literal belief in Scripture that their interpretation of the world—their acceptance of what is real and what is unreal —is driven by a need to maintain this belief system. An understanding of this can be gained from exploring the concept of "cognitive dissonance"—a term that refers to the phenomenon whereby disconfirmation of a strongly held conviction actually reinforces belief and leads to increased proselytizing activity.

The theory was put forth and tested by Leon Festinger and others (1964) in an elegant study of a modern space-alien cult that predicted the end of the world at a specific time and date.

Briefly, the theory holds that when people hold a belief in which they have made a substantial public and personal commitment, and when that belief is disconfirmed, their adherence to the belief will actually be strengthened, and they will vigorously proselytize it in order to reduce the "dissonance" between the belief and the disconfirmation. The disconfirming evidence is explained away with some sort of internal rationalization, and group reinforcement is used to uphold the behavior.

The cognitive-dissonance model does not apply fully to creation scientists, because the pattern of belief and disconfirmation does not fulfill all of the requirements of the model in strictest terms. The literal interpretation of the Bible does not provide a clear, absolutely falsifiable prediction, as would a prediction that the world will end on a specific date. The evidence for evolution—while interpreted as falsifying the Genesis story if true—is complex and not directly observable (most creation scientists have never done field or laboratory work in geology, paleontology, or evolutionary biology), and therefore is easily denied as proof. Furthermore, the commitment that most creation scientists have made is one of public and personal belief, rather than one where subjects give away all their possessions, quit their jobs, and sever ties to family and friends (though some creation scientists' beliefs may have led to some career difficulties). That said, we should not underestimate the power of personal conviction as a commitment. Belief systems are obviously very important to everyone, and it can be very traumatic to have an entire belief system overturned. Furthermore, creationists are clear and unambiguous in arguing that if evolution is true, their belief in original sin and the atonement offered by Jesus Christ must be untrue (though this view is not accepted by all Christian denominations). To most scientists, the scientific evidence against a literal interpretation of the Genesis story is so overwhelming as to provide a clear disconfirmation of the belief.

In this light, it is an interesting exercise to evaluate creation scientists' behavior in the cognitive-dissonance model. In the first place, creation scientists are strident not only in their belief in a literal interpretation of biblical Creation but also in the acceptance of "science" as a valid and important phenomenon. Hence, there seems to be a dissonance forced by their simultaneous acceptance of literal biblical authority and the overwhelming opinion held by the scientific community of a disconfirming theory. Second, many creation-science arguments are post-hoc constructions, appealing only insofar as they can be twisted into a "logical" rebuttal of evolutionary theory, pending a laserlike focus on the single argument outside of a larger context. In the cognitive-dissonance model, what is important is not the relative plausibility of the argument to others, but only the importance of the evidence in reconciling the dissonance between the belief and the disconfirmation. In other words, as long as the creation scientists' argument is plausible in their own belief system, it validates the belief and reconciles the dissonance. To the creation scientist, evolution is proved untrue by the existence of an explanation that is "scientific." All other matters become irrelevant. At the same time, this "disproof" allows the creation scientist to accept other "science" as a real and valid enterprise.

The dissonance model is appealing to me because it suggests an explanation for the zeal with which creation scientists pursue "scientific" explanations for why evolution must be untrue—their vigorous political activity in attempting to convert others to their views (by both argument and legal force), their insistence that they are doing "true" science while evolutionists are not, and their seemingly desperate acceptance of what is at best trivial, marginal evidence against evolutionary theory. What is striking about creation scientists is not their rejection of evolutionary theory; for example, millions of people are perfectly comfortable accepting the truth of the biblical Genesis story on faith alone, regarding scientific arguments for evolution as irrelevant. Rather, creation scientists are unique among creationists in their emphatic, aggressive attempts to reconcile the literal interpretation of Genesis with scientific evidence from the natural world and to convince others that they are right.

The flip side of creation scientists' pursuit of scientific evidence against evolutionary theory is their attempt to explain contradictory evidence from the Bible itself. Scientific evidence for evolutionary theory can only lead to cognitive dissonance if the creation scientist adheres to an absolutely literal interpretation of the Bible. Obviously, any contradictions among biblical passages could themselves be viewed as disconfirmation. It is no surprise, then, that creation scientists go to great lengths to explain away contradictions in the biblical text that seem self-evidently irreconcilable even to most Christians and theologians. To nonliteralists, their explanations can seem ludicrous and even humorous (Kitcher 1983). But to biblical literalists, it is sufficient only that the explanation reconcile the accounts and support their belief.

The cognitive-dissonance model is insufficient to explain completely the beliefs and behavior of creation scientists. Nevertheless, it offers potential insight into the selective reasoning of a very small but very aggressive and active group of anti-evolutionists whose behavior is difficult to explain, given their intelligence and education.

Creation Pseudoscientists Operate in a Different Logical Framework

All of us operate in a system of beliefs. We create a worldview and interpret evidence in the framework of that worldview. In the sciences, this is familiar from the work of Thomas Kuhn (1970), who postulated that science operates within large paradigms that are only overthrown through scientific revolutions. Without referring to Kuhn, the creation scientists argue fervently that evolution is a paradigm that is equivalent to creation. In a way, they are right. Both evolution and biblical creation are paradigms within which evidence is interpreted. However, a fundamental difference between science and creation science is that science is based on the premise of self-correcting critique (Smith 2000), whereas creation science is based on the premise of the absolute truth of biblical authority.

Creation scientists have a fundamentally different way of approaching "science" that transcends simple distinctions between

Baconian and Popperian science. This is the most insidious problem that they present to science education. It is not just that they are attempting to control the factual material that is taught in high schools and colleges. Rather, if creation science is successfully introduced into public education, then students will be mistaught about the fundamental process of science that has made it so successful for the past 150 years. This applies not just to evolutionary biology but in fact to all science—including biology, physics, chemistry, geology, and so on. Given that our entire economy is fundamentally based on advances in science, creation science should be viewed not only as a threat to academic freedom and truth but also as a dangerous philosophy with enormous destructive potential.

To understand creation-science logic, one must first describe its tenets. It is obvious that creation scientists are deeply religious people who believe wholeheartedly that God created the world exactly as described in Genesis. However, the approach to science espoused by creation scientists appears derived from their absolutist approach to the Bible (Eve and Harrold 1990). Perusal of creation-scientist literature quickly reveals the common argument that the Bible encapsulates moral authority, that it is the literal Word of God, and, most important, that faith in redemption through Jesus is contingent on the literal truth of the biblical account. If the authority of the Bible were not absolute, biblical text would be open to interpretation, and the absolute authority of the Bible would be in danger. Once this happens, then people can begin to pick and choose which biblical accounts they believe are literal truth and which are not. In other words, once one accepts a single contradiction in the Bible, the authority of the entire document is undermined, and faith itself is destroyed.

In a way, the faith of the creation pseudoscientist might be viewed as weaker than that of the nonliteralist. It is possible, and common, to accept that the Bible teaches truth through story, parable, and historical account. This makes sense, because stories and parables are extremely efficient and effective tools for teaching. For example, in the Gospels, Jesus taught truth through parables, not through long, philosophical discourses. A parable can teach in a few sentences what whole chapters could fail to convey. To the nonliteralist, faith is conveyed through a belief that truth is contained within the Scripture, regardless of the literal truth of the stories. The literal existence of Job, for example, is immaterial to the messages that are clearly conveyed in the text. The creation story set out in Genesis is viewed by most theologians as an antipolytheistic tract loaded with subtle allegorical nuances of enormous meaning to the ancient Hebrews and to modern Christians (Hyers 1983). The nonliteralist has already demonstrated that faith can be upheld in spite of challenges. The creation scientist, however, holds that the text must be literally true, or faith collapses. It is almost as if the creation scientists place God within a box of the literal truth of the Bible. They have in a sense erected "Schrödinger's God." Open the box (by questioning the absolute literal truth of the Bible) and one might find that God is dead!

The creation scientists apply the same logic to the natural sciences. They first set up a fundamental tenet—the earth was created exactly following the account in Genesis. All evidence from the natural world is then interpreted in a way that supports this tenet, regardless of how bizarre the explanation is. But the logic extends further. If a single textual contradiction can undermine the authority of the entire Bible, and faith in God and Jesus, then a single contradiction in evolutionary models undermines the whole theory itself. To creation scientists, evolutionary theory is disproved by almost any apparently contradictory evidence. For example, if a research produces an erroneous radiocarbon date, then radiocarbon dating is wrong. If radiocarbon dating is wrong, then all dating methods are wrong. In fact, because all dating methods rely on "assumptions," and assumptions can be wrong, then all dating methods that indicate that the earth is more than a few thousand years old are wrong. No amount of evidence for the validity of dating is even relevant, because the lack of perfect consistency itself is the proof against the model. The same logic applies to evolutionary biology, physics, astronomy, and geology.

Creation scientists treat evolution as a belief system identical to literalist Christianity. This fallacy has been central to the argument that creation science should be taught alongside evolution in public schools. There is a commonality between the two, in that both require a degree of faith that a model, or paradigm, is true, but the similarity stops there. Science constructs models that explain evidence, but it also accepts that particular models may be restricted to particular conditions. The failure of Newtonian mechanics to explain numerous observations indicated that the model was inadequate, but it did not undermine the truth of basic Newtonian principles under particular conditions. Science progresses by the search for contradictions and the modification of models to explain the totality of evidence better. But this implies by its very nature that scientific models are imperfect.

Creation science, by contrast, argues that a model is wrong if there are any contradictions at all. This same logical approach explains why creation scientists comfortably cite outdated scientific works (Zuckerman 1970; Oxnard 1975) as evidence against evolutionary theory. Truth, to the creation scientist, is an absolute. As such, it is self-evident and unchanging. Once something has been demonstrated, it should remain externally valid and true. By the logic of creation science, once a scientific work has been published, subsequent findings that conflict with the work do not invalidate it, but rather stand as a testimony to the uncertainty of the model. It is little wonder that creation scientists continually press the issue that true science must be repeatable. Replicability of scientific experiments is entirely consistent with an absolutist view of scientific truth. But creation scientists refuse to accept any premise that evolutionary theory is any way testable. To do this, they must distort the concept of testability to a limited notion of the prediction of future events and deny that observation of the natural world (such as findings from the fossil record, genetics, comparative anatomy, population biology, geology) could in any way fulfill the requirements of a testable model. To do this, they equate the concepts of replicability and testability, coopting the rhetoric of

Popperian science in their critique of evolutionary theory while distorting the normal practice of science.

Scientific models like evolutionary theory are constantly debated and refined. It is the uncertainty of evolutionary theory that, to the creation scientist, proves its lack of truth. Yet creation scientists also fail to acknowledge that all modern scientific enterprises involve the same type of model building, debate, disagreement, and refinement as evolutionary theory. Thus, in their zeal to attack evolutionary theory, they inadvertently attack modern science itself. Those defending the teaching of evolutionary theory need to demonstrate to the public and to politicians that "creation science" represents an attack on all science. If politicians advocate teaching creation science in schools, they indirectly undermine science education by endorsing a warped and twisted view of science. Like the Soviets in the era of T. D. Lysenko, the United States could face damage not from simple ignorance but from the miseducation of students into a counterproductive and damaging philosophy of science. It took the Russians decades to recover from the misguided attempts to dictate science as a sociopolitical state tool. It is disturbing to contemplate that the creation scientists wish to impose a similar model.

References

Arthur, J. 1996. Creationism: Bad science or immoral pseudoscience? *The Skeptic* 4 (4): 88–93. Available at http://mypage.direct.ca/w/writer/gish.html (last accessed October 22, 2001).

Austin, S. A. 1986. Mount St. Helens and catastrophism. *Institute for Creation Research Impact No. 157.* Available at www.icr.org/pubs/imp/imp-157.htm (last accessed October 22, 2001).

Bacon, F. 1620. *Novum organum* [translation]. In *The Works* (Vol. VIII), ed. J. Spedding, R. L. Ellis, and D. D. Heath. Boston: Taggard and Thompson [1863].

Behe, M. J. 1996. *Darwin's black box: The biochemical challenge to evolution.* New York: Free Press.

Dembski, W. A. 1999. *Intelligent design: The bridge between science and theology.* Downers Grove, IL: InterVarsity Press.

Duncan, O. D, and C. Geist. 2004. The creationists: Who, where, how many? *Reports of the National Center for Science Education* 24 (5): 26–33.

Eldredge, N. 2000. *The triumph of evolution and the failure of creationism.* New York: W. H. Freeman.

Eve, R. A., and F. B. Harrold. 1990. *The creationist movement in modern America.* Boston: Twayne Publications.

Festinger, L., H. W. Riecken, and S. Schachter. 1964. *When prophecy fails.* New York: HarperCollins.

Hyers, C. 1983. Biblical literalism: Constricting the cosmic dance. In *Is God a creationist?* ed. R. M. Frye. New York: Scribner.

Kitcher P. 1983. *Abusing science: The case against creationism.* Cambridge, MA: MIT Press.

Kofahl, R. E. 1977. *Handy dandy evolution refuter.* San Diego: Beta Books.

Kuhn, T. S. 1970. *The structure of scientific revolutions,* 2nd ed. Chicago: University of Chicago Press.

Locke, S. 1998. *Constructing "The Beginning": Discourses of creation science.* London: Laurence Erlbaum.

Marsden, G. M. 1984. Understanding fundamentalist views of science. In *Science and creationism,* ed. A. Montagu. New York: Oxford University Press.

Morris, H. M., and J. C. Whitcomb. 1961. *The Genesis flood: The biblical record and scientific implications.* Philadelphia: Presbyterian and Reformed Publishing Co.

Oxnard, C. 1975. The place of the australopithecines in human evolution: Grounds for doubt? *Nature* 258: 389–95.

Popper, K. R. 1959. *The logic of scientific discovery.* New York: Harper and Row.

Scott, E. C. 1987. Anti-evolutionism, scientific creationism, and physical anthropology. *Yearbook of Physical Anthropology* 30: 21–39.

Smith, K. C. 2000. Can intelligent design become respectable? *Reports of the National Center for Science Education* 20: 40–43.

Toumey, C. P. 1994. *God's own scientists: Creationists in a secular world.* New Brunswick, NJ: Rutgers University Press.

Zuckerman, S. 1970. *Beyond the ivory tower.* New York: Weidenfeld & Nicolson.

Why Target Evolution? The Problem of Authority

Alice Beck Kehoe

"SATAN HIMSELF IS THE ORIGINATOR of the concept of evolution," stated Dr. Henry Morris, founder of the Institute for Creation Research. "[T]he entire monstrous complex . . . pantheism, polytheism, astrology, idolatry, mysteries, spiritism, materialism . . . was revealed to Nimrod . . . by demonic influences" at Babel (Morris 1974, 72, 74–75). How did "evolution" become Satan's gospel?

Evolution, as a term, is wholly compatible with orthodox Bible-based doctrines. Its Latin source, *evolvere,* simply means "unrolling," as one does with a scroll. It implies that what is perceived was already present but not yet visible. St. Augustine accepted evolution in the sense that everything that is, was, and ever will be on earth was created by God, as described succinctly in Genesis, but it takes the unrolling, the evolution, of history to reveal all creation (Gilson 1984, 50). Charles Darwin, understanding the word in this classic sense, did not use "evolution" in his 1859 *On the Origin of Species,* since he did not accept the Augustinian notion that all species were created during the fabled six days of Genesis (Gilson 1984, 52).

Darwin argued for transmutation, the development of new species from existing populations through descent with modification. Presenting natural selection as the principal factor in speciation, his work supported "the development hypothesis" popularized fifteen years before *The Origin of Species* by the Scottish essayist Robert Chambers. Published anonymously in London, in 1844, Chambers's *Vestiges of the Natural History of Creation* became the talk of the country, went through eleven editions, and outsold Darwin's *Origins* until the end of the nineteenth century (Secord 2001, 526). Chambers, in turn, took "The Law of Variety-Production" from an 1834 *Bridgewater Treatise* by Peter Mark Roget; far from being atheistic, the *Bridgewater Treatises*, supervised by the Archbishop of Canterbury and the president of the Royal Society of London, discoursed "on the power, wisdom, and goodness of God as manifested in the creation" (Kehoe 1998a, 49). Chambers fulsomely praised the Deity for initiating natural laws; Darwin carefully distanced his mundane observations from theological controversies.

How, then, did "evolution" and "Darwinism" become appellations for ungodly damnation? The answer lies in the construction of science as a secular discipline during the second half of the nineteenth century. Its acolytes denounced their elders' premise that the pursuit of knowledge should be a godly enterprise. Scientific data do not carry moral lessons, and the scientist *qua* scientist (a phrase coined by William Whewell in 1833), the public person, should not be affected by whatever spiritual beliefs he holds as a private person. ("He," because women were held to be incapable of sustained rational scientific work [Richards 1989].) This is the position exciting the horror of conservative Christians. It came to be tagged "evolution" because it was invoked by proponents of a doctrine of innate evolutionary progress, and "Darwinism" because secularists held up Charles Darwin as hero and model. Darwin's upper-middle-class position in London society gave him leadership in a research coterie that included several ambitious men of lower social standing.

"The Development Hypothesis" Controversy

The 1840s was a decade of change. It saw the emergence of Manifest Destiny ideology, erupting to whitewash the United

States's aggression against Mexico (the Mexican–American War and colonization of California). In Britain, magnates of industrial capitalism wanted trade to dominate the entire society—in Charles Dickens's words, "What you couldn't state in figures, or show to be purchaseable in the cheapest market and saleable in the dearest, was not and never should be" acknowledged by government (quoted in Hobsbawm 1962, 225). Merchant lobbying persuaded Parliament to repeal the protectionist Corn Laws in 1846, improving Britain's export position and consolidating industry's dominance over agriculture in Britain's economy. Aristocrats' land-based income diminished. Factories and their owners' ostentatiously ornamented mansions were the cathedrals of the age.

Robert Chambers was a man of those times. Born in 1802 in a Scottish Borders village to a cotton jobber, he moved with his family to Edinburgh when mechanized weaving mills doomed his father's business. Love of books led him to sell them from a stall, then assist his older brother William to start a printing shop. W. & R. Chambers pioneered mass printing, binding, and distribution, using the latest mid-nineteenth-century technology. Cheap basic-education books were their bread and butter, the weekly periodical *Chambers's Edinburgh Journal* their ticket to bourgeois households. Robert wrote most of the material for the *Journal*, remarking, "It was my design from the first to be the essayist of the middle class—that in which I was born, and to which I continued to belong" (quoted in Kehoe 1998a, 10).

Fretting under his brother's hard business drive, Robert claimed ill health in 1842 and retired, with his family of eleven children, to a suburban villa on the golf links near St. Andrews University. Announcing recovery in 1844, the family returned to Edinburgh and Robert resumed writing—usually unsigned news, essays, and stories—for the *Journal*. Never did he acknowledge authorship of the book everyone discussed, *Vestiges of the Natural History of Creation*. "Anonymous" on the title page, "Mr. Vestiges" excited speculation: Was he Sir Richard Vyvyan, gentleman scientist–philosopher? Ada, Lady Lovelace, daughter of Lord Byron?

Her children's tutor, the biologist Carpenter who wrote for W. & R. Chambers's educational series? Prince Albert, once an earnest student of the great scientist Alexander von Humboldt? By 1847, a Scottish periodical printed the glaringly obvious parallels between passages in *Vestiges* and in early 1840s issues of *Chambers's Edinburgh Journal.* Robert Chambers ignored the charges; he could not jeopardize W. & R. Chambers's reputation for sound, solid, useful publications.

The scandal in *Vestiges* was its uncompromising repudiation of the doctrine of repeated, miraculous special creations. Its Deity (Chambers's preferred term) was not a Father continuously guiding, and punishing, His erring children; He was not a Patriarch. The Deity in *Vestiges* created a marvelous machine that never broke down, even if minor parts jammed sometimes. It operated on two mechanical principles—the Law of Gravitation for inorganic matter and the Law of Development for organic matter. The Deity was no more involved in day-to-day operations than would be the owner of a huge factory. Men, like the employees of a factory, have free will; they can work or try to get on some other way. The Deity would not be intervening.

Vestiges challenged the very foundation of the Established Churches of England and Scotland, their patriarchal authority. The British state had, since the seventeenth century, been conceptualized as analogous to a landed family, its head a benevolent master guiding his womenfolk, children, servants, and tenants (Fletcher 1995, 293; Murray 1995, 77, 97). The king was father to his country. God was father to the world. Each intervened to correct the wayward, otherwise leaving ordinary business to the stewards. Another image of the state considered the body politic to be like a human body, with a controlling head; a torso with its various processes such as digestion, arms (police, the armed forces, the judiciary); legs (carters, porters); and feet (peasants and laborers). These two images were compatible, as can be seen in the phrase *head of state.* In the mid-nineteenth century, instead of an anthropomorphized body politic, the human body and the body politic came to be imaged in mechanical metaphors, with the flow of commodities and the flow of blood in the circulatory system both likened, by Herbert Spencer, to systems of drains, such that obstructions—idle poor or clots—would cause damaging toxic reactions (Shuttleworth 1990, 56, 58).

Capitalism insidiously undermined British sociopolitical structure, loosening earlier licensed ties between class status and consumption. Any Briton who had the money could display dress, housing, food, and recreation once reserved, by sumptuary laws, for the dominant class. Men—and a few bold women, such as Mary Ann Evans (George Eliot)—could defy their fathers to strike out independently; in a rapidly expanding capitalist economy, many succeeded. Samuel Smiles (so aptly named!) glorified impecunious youths whose determination and hard work brought them fame and a good bourgeois living (Smiles 1859; Shuttleworth 1990, 51). Property, labor, products, and the means of production had become quantifiable "things" no longer irrevocably bound to kin and patron relations. The dialectic between burgeoning markets and political acts such as the Corn Laws and the 1832 Reform Bill increasingly demonstrated how obsolescent was the image of the state as a landed proprietor's household (Roberts 1973; Appleby 1992, 171). If the "invisible hand" of the market rules, how does God rule? As a father in His mansion? As an invisible hand? Or as the creator of the market wherein the invisible hand adjusts according to lawlike principles?

"The development hypothesis" proposed a universe built on a wonderfully capitalist model. Beginning with a nebulous fire–mist, Chambers said, matter gravitated and condensed into stars and planets, and life forms were sparked by electrical discharges in chemical solutions. The Law of Development operated as continuously and uniformly on organic life as the complementary Law of Gravitation operated on inert matter. Thus, new species of animals and plants appear; after millions of years, humans develop out of higher primate races; and over thousands of years, the elements of civilization are invented by occasional geniuses arising in accordance with the Law of Development (Chambers 1994 [1844]). One can envision young Robert with his few books on the stall on Leith Walk, developing that meager capital into a larger and more varied stock, then William gets a printing press and they produce a book, and W. & R. Chambers develops a steam-press printing plant and a range of books and periodicals. . . . After Ralph Waldo Emerson read *Vestiges*, he characterized it as "that book we have wanted so long," its narrative of innate power congruent with his own famous call for "self-reliance" (Secord 2001, 425). *Vestiges* describes a fascinating and beautiful world untainted by guilt, optimistically looking forward—like Britain's rapidly developing capitalist economy. It was anathema to the landed classes, to hierarchical authoritarian churches, and to the hidebound structure of inherited privilege.

Conservatives and "Baconian Science"

Francis Bacon, Lord Chancellor of Britain under King James I, published treatises on natural philosophy—what we now call *science*—advocating a strongly empirical approach. Arguing forcefully like the lawyer he was, employing sexually charged language, Bacon persuaded readers that men can penetrate Nature's secrets by systematically collecting observations and then inductively deriving generalizations from the facts. Two centuries later, in a series of studies culminating in the *History of the Inductive Sciences* (1837) and *The Philosophy of the Inductive Sciences* (1840), Cambridge natural philosopher William Whewell pointed out the necessity of some concept guiding the collection of observations—not merely put together but "colligated," or bound by a premise or hypothesis to be tested. Collecting and classifying observations, postulating and testing generalizations, came to be termed "Baconian science" and "the scientific method."

Nature, for the Baconian, sits out there. All that is, is; natural philosophers can bit by bit elucidate it. Astronomer John Herschel, a colleague of Whewell, stated, "To ascend to the origin of things, and to speculate on the creation, is not the business of the natural philosopher . . . [it is to know] what *are* these primary qualities originally and unalterably impressed on matter" (quoted in Fisch 1991, 73; Herschel's emphasis).

Robert Chambers took issue with this position. Whewell had insisted, in his 1840 *Philosophy*:

> Philosophers have never demonstrated, and, so far as we can judge, probably never will be able to demonstrate, what was the primitive state of things from which the progressive course of the world took its first departure. . . . [I]t becomes not only invisible, but unimaginable. (Chambers 1994 [1844], 127, quoting Whewell)

Chambers acerbically reminds us that Whewell was "the superior of one of our greatest academical institutions" (master of Trinity College, Cambridge). "The professional position of Dr. Whewell may be held . . . accommodated as far as possible to the prepossessions expected in a large class of persons" (Chambers 1994 [1844], 128). In other words, it was the business of the master of Trinity to be conservative.

Chambers does not mince words: Whewell's denial curries favor with the class that sends its sons to Cambridge—a class at odds with Chambers's Scottish reformist group. "The development hypothesis" in *Vestiges* carried unmistakable political overtones. Historian Adrian Desmond agrees with Chambers. Darwin's caution—his long delay in publishing his theory already committed to paper when *Vestiges* appeared—can be attributed to his being

> too worldly-wise not to realize the wider social implications. . . . [I]n a materialist evolutionary sweep Darwin invited being identified with Dissenting or atheistic lowlife, with activists campaigning against the "fornicating" Church [of England], with teachers in court for their politics, with men who despised the "political archbishops" and their corporation "toads."

Ultimately, Darwin was frightened for his respectability (Desmond 1989, 413).

Over the course of the next century and a half, the conservatives who opposed "palætiological science" shifted from upper-class Britons to middle-class Americans. The vanguard of the shift comprised Scottish immigrants to nineteenth-century America. They carried Presbyterian Protestantism and Thomas Reid's "common sense" philosophy, especially as interpreted by his disciple Dugald Stewart at the turn of the nineteenth century. Reid argued that men generally share both mental processes and common experience of the physical world. Men hold in common a sense of the real world. "Baconian" scientific method verifies this common sense.

Scotsmen, enjoying a relatively more democratic society than the contemporary government of England, found this feature of Reid's philosophy particularly attractive: Observation of the natural world is readily accessible and free; anyone could use "the scientific method" to gain secure knowledge. Scottish emigrants observed American society and its markets, colligating (as Whewell would have said) their facts under an understanding of economics promulgated by Reid's colleague Adam Smith. They, along with "Yankee" manufacturers descended from earlier immigrations, aimed for a mass market by divesting products of nonfunctional ornamentation. Mass markets and mass production culminated in the "American system" of standardized parts and the assembly-line process. An American middle class supported by industrial capitalism grew during the nineteenth century and gained ascendancy in the bloody Civil War. Engineering and science worked in conjunction with industrial and consumer capitalist expansion, achieving high status (Noble 1979).

Although the development hypothesis and The Law of Variety-Production might seem congenial to these Americans, many felt threatened by the instability inherent in the hypothesis. Roman Catholics could be confident in the stability of their Church and its apostolic succession. Protestants might struggle in a morass of contesting doctrines—among them, the proposition that material success and social standing are worldly signs of moral worth and salvation (Appleby 1992, 116). Seeking to relieve doubts and anxieties, the Presbyterian General Assembly declared in 1910 that there are five fundamental tenets of Christian faith: the miracles performed by Christ; the virgin birth; Christ's bodily resurrection;

his sacrifice on the cross, atoning for humanity's sins; and the inerrancy of the Bible. The Bible's indubitable truth is the authority for the need and the promise of salvation. As the revealed Word of God, it had to be a rock of stability.

Inerrancy and Authority

The Presbyterians' Five Fundamentals (from which we get the label "Fundamentalist") were a direct response to the "higher criticism" of the Bible coming from sophisticated theologians and historians of religion in the universities, especially in Germany (Torrey 2003, a recent reissue of the original volumes). Their challenges to the miracle of Revelation came to be coupled with contemporary science's challenges to the Bible's account of the miraculous genesis of species. True to the legacy of Scottish "common sense," Fundamentalists state: "When the plain sense makes common sense, we seek no other sense" (Kehoe 1985, 165). William Jennings Bryan urged, in 1924:

> Commit your case to the people. Forget, if need be, the high-brows both in the political and college world, and carry this cause to the people. They are the final and efficiently corrective power." (quoted in Numbers 1992, 44)

How would "the people" of America exercise their corrective power? Through their representatives in legislative bodies. Personal commitment to Jesus Christ may be the axiom of Protestant Christianity, but mainstream churches operate a structure of authority, mediating the Word of God. The 1910 decision by the Presbyterian General Assembly to promulgate the Five Fundamentals is exactly such an authoritative interposition between the individual and God. The context of the declaration—scholastics' "higher criticism"—made it clear that laypeople should not presume confidence in their private judgment, but rely instead on the wisdom of their representatives in legislative church assemblies.

Henry Morris, founder of the Institute for Creation Research, sanctifies authority:

> It is precisely because Biblical revelation is absolutely authoritative and perspicuous that the scientific facts, rightly interpreted, will give the same testimony as that of Scripture. There is not the slightest possibility that the *facts* of science can contradict the Bible. (Morris 1974, 15; his italics)

"Rightly interpreted" is of course the critical issue: Whose interpretation is right? Morris *implies* that "Baconian science" will produce scientific facts, as indubitable as Scripture. In actuality, a body of men decree the right interpretation.

Authority, and obedience to authority, is the crux of organized scientific creationism. The Creation Research Society—which requires voting members to hold at least a master's degree "or equivalent in experience" in a science—enjoins members to sign assent to the statement,

> . . . the Bible is the written Word of God, and . . . all of its assertions are historically and scientifically true in all of the original autographs. . . . This means that the account of origins in Genesis is a factual presentation of simple historical truths. All basic types of living things, including man, were made by direct creative acts of God during Creation Week as described in Genesis. . . . The great Flood described in Genesis . . . the Noachian Deluge, was an historical event, worldwide. . . . Finally, we are an organization of Christian men of science, who accept . . . salvation can come only thru accepting Jesus Christ as our Savior. (quoted in Kehoe 1985, 170; reprinted in full in Numbers 1992, 230–31)

"The original autographs" is meant to distinguish acceptable versions of Scripture—primarily the King James and American Standard translations—from those corrupted by scholarly "higher criticism" or liberal usage, such as inclusive language.

A favorite theologian for scientific creationists was Francis A. Schaeffer (1912–1984), who insisted: "The Christian world view . . . [offers] the certainty of something 'there'—an objective reality—for science to examine. . . . [I]n doing so one was investigating God's creation" (1976, 140). Schaeffer emphasized:

> The basic [premise] was that there really are such things as absolutes . . . in the area of Being (or knowledge), and in the area of morals. . . . Absolutes imply antithesis . . . right and wrong, . . . true and false. (1968, 14–15)

To oppositional dualism and certainty, Schaeffer added a third tenet of evangelical Christianity: "God himself had told mankind to have dominion over nature" (1976, 140). Authority inheres in the relations between men and the world they investigate (and in the relations between men and women, but that theme would take another essay).

Up until *Vestiges* and the bourgeois revolution it accompanied, "natural theology" had comfortably taught that God had set out the Book of Nature to be read alongside Scripture. Paired with Scripture, the Book of Nature was "not about what God can do, but about what can be done by natural agents, not elevated above the sphere of nature," explained devout seventeenth-century natural philosopher Robert Boyle (quoted in Shapin 1996, 105). Europeans' Age of Exploration made the Book of Nature more and more recondite and confusing. *Vestiges* articulated a new reading: Nature was not a prepared document to be read but a pageant transpiring—unrolling, evolving—around us. The God of *Vestiges* is an "absentee landlord" (Shapin 1996, 149) in whose venue the pageant takes place. Only in the remotest sense does he have authority; he has no business interfering in, stage-managing, the improvisational pageant. The pageant of *Vestiges* has a single, unifying theme. Antithesis is not the story. Chambers's rejection of theologians' Manichaean antithesis of Good versus Evil, God versus Satan—absent also in both Spencerian and Darwinian evolution—goes against the grain of oppositional dualism hallowed in the Old Testament.

A century after Chambers's death, American "scientific creationists" pushed what they termed the "two models" of biology: "evolution" and "creation science." They claimed that creationists' model of the Noachian Flood producing geological and fossil remains is an alternative scientific interpretation conforming to observational data. In their view, suppression of this model in the science classroom amounts to censorship by adherents of the "religion of secular humanism." Their theater presents a drama of competition between a plain, honest chronicler and a conniving, amoral seducer. Although they publicly eschew any biblical ties to their creation model (the Institute for Creation Research published two editions of Henry Morris's 1974 textbook—one with and one without a preface explicitly connecting the text to Scripture), scientific creationists' insistence that there are two and only two models for scientific interpretation fits their campaign into Schaeffer's theology: Antithesis implies absolutes, right and wrong, true and false.

Scientific creationists (or creation scientists) want to read from the published script of an exceptional Author, an Authority. Their world of antitheses cannot accommodate indeterminancy, the hallmark of twentieth-century science. One must accept the King James Bible as God's Word, a divinely guided translation, or one will flounder in darkness. A scientist should colligate observations to demonstrate Scripture. Nature yields to man. In their pairing of antitheses, they omit discussing the role of pastors, men who teach right interpretations. Signing assent to the inerrancy of the Bible is the litmus test of obedience to authority—authority vested in the Author of the Word but manifested through ministers of faith. These radical Christian Right pastors battle to conserve the patriarchal tradition that gives them status and dominion over women, the unchurched, and the realm of Nature.

The War Between Science and Religion

The British Association for the Advancement of Science met at Oxford University in 1860, five months after publication of Darwin's *On the Origin of Species.* The assembled gentlemen of science were enlivened by a spirited debate between Oxford's Bishop Samuel Wilberforce and Darwin's friend and champion, Thomas Huxley. Many in attendance would have remembered Wilberforce's well-received sermon at the association's 1847

Cambridge meeting. Preaching on "Pride a Hindrance to True Knowledge," he warned against those who dare "to deal boldly by means of nature, instead of reverently following her guidance" (quoted in Secord 2001, 436). With the highly respectable and respected Charles Darwin now openly arguing for the validity of transmutation, the 1860 audience found Wilberforce reactionary. What was an ordinary learned contest was turned into a mythical battle by John William Draper, an American chemist and author of *The History of the Intellectual Development of Europe* (1862), who, earlier the same day, had presented a lengthy paper entitled "On the intellectual development of Europe considered with reference to the views of Mr Darwin and others that the progression of organisms is determined by law." Another American chemist, Edward Livingston Youmans, invited Draper to contribute a volume to an "International Scientific Series" he was publishing. Herbert Spencer already was represented in the series by his *Study of Sociology*. Draper accepted, and in 1874, Youmans published Draper's *History of the Conflict Between Religion and Science.* Selling fifty printings in the United States, twenty-one in Britain, translated into ten languages, the book was nearly as widely read as *Vestiges* (Moore 1979, 20–28).

Two years later, Andrew Dickson White, president of Cornell University, published *The Warfare of Science,* elaborated from his popular 1869 New York lecture, "The Battle-fields of Science." Twenty years later, White published the two-volume *History of the Warfare of Science with Theology in Christendom.* Neither Draper nor White was much concerned with evolution—Draper was attacking the Roman Catholic Church's temerity in proscribing scientific interpretations such as Galileo's, and White was advocating divorce of higher education from sectarian controls—but their military metaphor resonated with deep-seated Western cultural themes (Moore 1979, 31–48). Their books chimed with citizens' discomfit over the revolution in political economy—in Britain at midcentury and in America after the Civil War. Those who longed for the olden days of a cultivated gentry and deferential workers felt religious authority had been betrayed, overthrown by Satan's rebellious spawn. Those who embraced the new order of entrepreneurs and technocrats fulminated against woolly-headed divines preaching against progress.

That professed Christians should favor metaphors of war ought to be astounding, yet Protestantism no less than the Roman Catholic Church comes out of fierce combativeness. Martin Luther said,

> The Scriptures . . . see the temporal sword aright. . . . It is God's servant for vengeance, wrath, and punishment. . . . I am called a clergyman and am a minister of the word, but even if I served a Turk and saw my lord in danger, I would forget my spiritual office and stab and hew as long as my heart beat. If I were slain in so doing, I should go straight to heaven. (quoted in Marrin 1971, 102–7)

Oppositional dualism permeating Semitic tradition merges with Germanic ideology glorifying death in battle (Kehoe 1986, 161). War is premised to be the natural state, and peace only its interruptions (Kehoe 1992, 55). Considering the depth of these cultural traditions, it is not surprising that a Protestant testifying to include creationism in California science curricula should explain that he wants to "see the forces of evil being met in these last days [before Christ's Second Coming] with an aggressive, explosive reaction of men who are led and filled by the Spirit of God . . . not simply to withstand their attack, but to attack them" (quoted in Kehoe 1983, 10).

The notion of unavoidable warfare between science and religion owes more to the embedment of the premise of "natural" conflict in our culture than to historical events. James Moore minutely examined the evidence in Darwin's Britain, concluding,

> the military metaphor has taken a dreadful toll in historical interpretation. . . . [E]ach of its three major "associated implications" is entirely misleading if not utterly false. There was not a polarization of "science" and "religion" . . . but a large number of learned men, some scientists, some theologians, . . . who experienced various differences among themselves. There [were] . . .

> deep divisions among men of science . . . and a corresponding and derivative division among Christians who were scientifically untrained, with a large proportion of leading theologians quite prepared to come to terms peacefully with Darwin. (Moore 1979, 99)

Moore's peroration is couched in strong words:

> [T]he military metaphor perverts historical understanding with violence and inhumanity, by teaching one to think of polarity where there was confusing plurality, to see monolithic solidarity where there was division and uncertainty, to expect hostility where there was conciliation and concord. . . . What are the attitudes and assumptions which "conflict" has expressed for historians of the post-Darwinian controversies? Most obviously, perhaps, it reveals the absence of any deep moral aversion from war. (Moore 1979, 99–101)

Scientific Creationism in Its Social Context

In the Bible, God seems to have no aversion to war. Most of the psalms call for vengeance, even conjuring slasher phraseology ("The Lord hath them in derision. . . . Thou shalt break them with a rod of iron; Thou shalt dash them in pieces like a potter's vessel" [Psalm 1:2]). American theologian Robert Jewett discerned an archetypal American hero he dubs "Captain America." This "good guy" is portrayed as "a perfectly clean and basically passive hero, committed to lawful obedience," nevertheless

> carrying out his highest form of faithfulness by violating cleanliness, law, and passivity. . . . [He is] curiously immune to criticism if he happens to break one of the ideals or laws in battle. . . . His initial desire is to be passive, but when he receives the clear call to battle he must faithfully but regretfully obey. (Jewett 1973, 153)

Think Superman or Popeye; think Duane Gish, fearless debater for the Institute for Creation Research.

Christian Heritage College, affiliated with the Institute for

Creation Research, advertises that it emphasizes "the foundational importance of special creationism in every subject." Its founding president, Tim LaHaye, teaches:

> The theory of evolution is the philosophical foundation for all secular thought today, from education to biology and from psychology through the social sciences. It is the platform from which socialism, communism, humanism, determinism, and one-worldism have been launched. . . . Accepting man as animal, its advocates endorse animalistic behavior such as free love, situation ethics, drugs, divorce, abortion, and a host of other ideas that contribute to man's present futility and despair. . . . It has wrought havoc in the home, devastated morals, destroyed man's hope for a better world, and contributed to the political enslavement of a billion or more people. (LaHaye 1975, 5)

Battlelines are unequivocally drawn in this simple antithesis between "evolution"—encompassing all forms of evil—and Christian Heritage College's faithful. It's only a short step to the *Left Behind* action novels that LaHaye writes with Jerry Jenkins—books that sell three to seven million copies per title, with movie adaptations following (*Time*, July 1, 2002, 44–45).

Scientific creationists focus on God's workmanlike constructions of the present world (Toumey 1994, 68–74) rather than the *Left Behind* books' melodrama of the unsaved facing damnation at Christ's Second Coming. Scientific creationists are mostly white, middle-class members of established, conservative, evangelical denominations (Toumey 1994, 245). They are sufficiently mainstream that science matters to them. Many earn their livings as engineers or working in research-and-development laboratories—settings where "Baconian science" works well. They know, directly, how science has contributed to Americans' security, ease, and longevity; God and science cannot be antitheses. Not science, then, but evolution is the evil doctrine. Science, in Christopher Toumey's perspicacious phrase, "sanctifies" these Americans' Christian faith by testifying to God's work (Toumey 1994, 74). The Book of Nature describes "intelligent design,"—obvious to

Thomas Reid, to Robert Boyle in 1688 (Young 1989, 390), and notably to Cambridge philosopher William Paley, famous for his 1802 *Natural Theology* (Voltaire had already used Paley's line, "a clock must have a clockmaker" [Gilson 1984, 106]).

The vast majority of Christians are classed as theistic evolutionists: that is, like Robert Chambers, they assume the Deity of Judeo–Christian–Islamic tradition created the universe. That deity set in motion evolution, a principle of change impelled by external and internal factors. Adaptation through natural selection can be one of these factors. (Inorganic matter changes, too, as Chambers discussed with the nebular hypothesis and geological processes.) Evangelical Christian scientists organized the American Scientific Affiliation (ASA) in 1941 to affirm the compatibility of commitments to the Bible and to contemporary science. Stricter brethren in the Creation Research Society, spun off from the ASA in 1963, and the Institute for Creation Research perceive too much latitude in ASA's interpretations of Scripture. Theistic evolution reached orthodoxy for Roman Catholics when the Vatican embraced the thesis of German Jesuit theologian Karl Rahner that organic evolution is the Incarnation, the coming into being of Christ in human form (Kehoe 1985, 174). That position, and a doctrine of progress and perfectibility, had been taken already in the mid-nineteenth century by Scottish physicists, including Lord Kelvin and James Clerk Maxwell (Smith 1998, 313).

Scientific creationists are distinguished by two characteristics: (1) an insistence on narrow literal reading of the King James and American Standard versions of Genesis (e.g., "day" must refer to a twenty-four-hour day), and (2) a set of empirical observations that neatly fit into the Genesis narrative. Under Henry Morris's dedicated leadership, scientific creationists see rocks and fossils illustrating the worldwide Noachian Flood: Geologic strata are made up of the sediments laid down by the waters, and fossils are the remains of organisms trapped by the deluge—marine species swimming down in the Flood and mammals and birds located higher in the sediments where they raced to escape their doom.

Morris earned a doctorate in hydraulic engineering in order to better understand the Flood.

Taking Genesis literally, scientific creationists abjure radiometric dating. They reject probability theory (as did Whewell [Richards 1997, 57]), pragmatically equating low probability with impossibility. They believe that Rudolf Clausius's Second Law of Thermodynamics, the principle of entropy, invalidates evolution: This is the "law" (regularity) that heat energy transfers from hotter to colder bodies, dissipating and becoming unavailable for conversion into mechanical work. Adam and Eve's disobedience in the Garden of Eden (the "original sin") threw "a curse upon all things, by which the entire cosmos was brought into a state of gradual deterioration"—i.e., entropy (Morris 1974, 109). Robert Chambers didn't deal with Clausius's laws because they were published years after *Vestiges.* Ironically, in the mid-twentieth century, the Second Law of Thermodynamics was used by anthropologist Leslie White (1959) to argue *for* evolution—the tendency toward entropy is reversed in living organisms. James Clerk Maxwell had stated in the 1878 *Encyclopædia Britannica,* "[The] notion of dissipated energy could not occur to a being [God] . . . who could trace the motion of every molecule and seize it at the right moment" (quoted in Smith 1998, 240).

Why do these American heirs of common-sense philosophy (Bozeman 1977) reject today's common sense? The perspective of anthropology suggests that Charles Darwin was quite right in perceiving

> . . . the probability of the constant inculcation in a belief in God on the minds of children producing so strong and perhaps an inherited effect on their brains not yet fully developed, that it would be as difficult for them to throw off their belief in God, as for a monkey to throw off its instinctive fear and hatred of a snake. (quoted in Moore 1989, 2003)

Fundamentalist churches reiterate unceasingly the threat of eternal damnation, instilling and reinforcing anxiety. A "Bible-based, Christ-centered Sunday school curriculum" published for conservative Protestants by the David C. Cook Foundation tells the first-grade child:

> We have all done wrong things. . . . Jesus, God's son, took the punishment for our wrongdoings . . . because Jesus already took our punishment, God will forgive our sins. Now we trust in Jesus and are in God's family . . . we will live forever in Heaven. (West 1980, 5)

By grades five and six, children learn of "the universality of fear," and that the means of allaying fear is belief in the "central truth" of Jesus's resurrection (Grendahl 1982, 29, 42). Later, in middle school, young teenagers are taught:

> Because I sinned, God couldn't have any fellowship with me. . . . [God] doesn't force me to give my sins and my life to Him. but if I don't, I must accept the consequences. I must pay for my own sins. . . . Admit I've sinned and I can't get rid of that sin by myself. Ask Jesus to forgive me, and give me a new life. (Tomasik 1982, 41)

Fear over the fate of one's soul is not at all "universal." It is, as Darwin knew well from his own personal struggles with religious belief, inculcated by authorities.

Henry Morris borrowed a term favored by anthropologists when he wrote of

> . . . creation and evolution. Each model is essentially a complete world view, a philosophy of life and meaning, of origins and destiny. . . . There are only two possible ultimate world views—evolutionism or creationism. (Morris 1982, 9, 30)

Contrary to Morris's claim, of course, there are innumerable worldviews (Kehoe 1998b, 133–38). The antithesis that Morris holds is between basically agnostic worldviews premising human fallibility and worldviews premising sufficient and definitive knowledge of the cosmos. Henry Morris, being a reasonably modest man, places his faith in certain knowledge in Scripture revealed, he believes, by an omniscient God. Authority for certain knowledge is removed from humans—whom we all can see are fallible—to God, who miraculously vouchsafed Revelation. Scientific creationists incorporate miracles into their world.

Any explanation of observational data depending upon miraculous intervention is, *ipso facto*, not science. Observation of a scientific fact must, in principle, be replicable. (Pragmatically, most scientific facts are "virtually witnessed" through published reports; peer review is relied upon to separate fraud and inadequate and inept attempts from replicable observations [Shapin and Schaffer 1985, 51–72].) Because (as the creationists remind us) no human observed any evolution of species before the modern era, historical sciences are constrained to use the principle of actualism. Eminent paleontologist George Gaylord Simpson expounded the use of this principle. Disentangling the principle of actualism from the outdated assumption of uniform rate of change—two concepts previously conflated in Lyell's principle of "uniformity"—Simpson explained that the scientist will

> observe present configurations and from them infer configurations that preceded them. . . .
>
> [T]here are three phases:
>
> (1) obtaining and studying the historical data . . .
>
> (2) determination of present processes . . .
>
> (3) confrontation of (1) and (2) with a view to ordering, filling in, and explaining the history. (Simpson 1970, 81, 84–85)

Each of Henry Morris's creation model's four critical biblical events—"creation of all things in six days," a "universal Flood," "sudden proliferation of languages and other cultural distinctives at Babel," and the "curse upon all things . . . the entire cosmos" (Morris 1974, 109)—contravenes accepted observations of present processes. Creationists' denial that evolution has been observed ignores both laboratory and field records of evolutionary speciation in populations of organisms with life spans shorter than humans—from fruit flies and bacteria to birds.

Scientific creationists' worldview hides the human authorities upon whom they rely, such as Henry Morris and Tim LaHaye. Gary Parker—who holds an Ed.D. but chaired the Institute for Creation Research's graduate-level biology department, confessed that he had come to realize that "people who write books don't know any more about the world than I do! Where, then, do I put my trust? . . . Science can't be trusted, but God can" (1978). His bald statement begs the question: How does Parker know this? Did God miraculously reveal this Truth to him or has he been inculcated since early childhood to put his trust in the God presented in church and family talk?

Looking at Morris's four events, the simplicity of scientific creationism stands out. Authorities whom one was indoctrinated to respectfully obey from earliest youth continue to provide guidance— earthly fathers mirroring the eternal Father. The world outside one's suburb may be discordant, dark, and harsh—Satan's doing—but one's congregation and pastors are a warm, bonded, supportive pod. Immutability of species intimates immutability of white suburban America's comfortable middle class. "Intelligent design" proponents are more cosmopolitan, more sophisticated—like their forebears, Robert Boyle and William Paley, they cast a much wider net of observations and build more detailed arguments. Their God is more an architect than an artisan, standing above and respecting the declared boundaries of science. "Intelligent design" and scientific creationism have in common an *a priori* assumption that an omniscient, omnipotent, manipulative God manages the world. If this is so, then the status quo is God's work and should not be disrupted.

Scientific creationists are a very small, although organized, minority of Americans; "intelligent design" proselytizers are an amorphous, probably larger, minority. God's "Great Commission" to "go into all the whole world and preach the gospel" (Mark 16:15) impels them to lobby legislators and persuade school boards to "be fair," to endorse teaching "two models" in public-school science classes. With most Americans poorly educated in current principles of scientific thinking, scientific creationists and "intelligent design" proponents readily gain audiences and have an impact disproportionate to their numbers. Gary Parker said, "The whole evolution–creation controversy boils down to: Whom are you going to trust?" (1978). Charles Darwin told his friends that what he believed, or did not believe, was "of no consequence to any one but myself." His friend the village vicar replied, "How nicely things would go on if other folk were like Darwin" (Desmond and Moore 1992, 635).

References

Appleby, J. 1992. *Liberalism and Republicanism in the historical imagination.* Cambridge, MA: Harvard University Press.

Bozeman, T. D. 1977. *Protestants in an age of science.* Chapel Hill: University of North Carolina Press.

Chambers, R. 1994 [1844] *Vestiges of the natural history of creation.* Facsimile reprint, with "Other Evolutionary Writings," including 1845 *Explanations,* ed. James A. Secord. Chicago: University of Chicago Press.

Desmond, A. 1989. *The politics of evolution: Morphology, medicine, and reform in radical London.* Chicago: University of Chicago Press.

Desmond, A., and Moore, J. 1992. *Darwin.* New York: Warner Books.

Fisch, M. 1991. *William Whewell, philosopher of science.* Oxford: Clarendon Press.

Fletcher, A. 1995. *Gender, sex and subordination in England 1500–1800.* New Haven, CT: Yale University Press.

Gilson, E. 1984. *From Aristotle to Darwin and back again: A journey in final causality, species, and evolution,* trans. John Lyon. Notre Dame, IN: University of Notre Dame Press.

Grendahl, T., ed. 1982. *Junior teacher's guide.* Elgin, IL: David C. Cook Publishing.

Hobsbawm, E. J. 1962. *The age of revolution 1789–1848.* New York: New American Library.

Jewett, R. 1973. *The Captain America complex.* Philadelphia: Westminster Press.

Kehoe, A. B. 1983. *The word of God.* In *Scientists confront creationism,* ed. L. R. Godfrey, 1–12. New York: W. W. Norton.

———. 1985 Modern anti-evolutionism: The scientific creationists. In *What Darwin began: Modern Darwinian and non-Darwinian perspectives on evolution,* ed. L. R. Godfrey, 165–85. Boston: Allyn and Bacon.

———. 1986. Christianity and war. In *Peace and War,* ed. M. L. Foster and R. A. Rubinstein, 153–73. New Brunswick, NJ: TransAction Books.

———. 1992. Conflict is a Western worldview. In *The anthropology of peace,* ed. V. J. Rohrl, M. E. R. Nicholson, and M. D. Zamora, 55–65. Studies in Third World Societies, no. 47. Williamsburg, VA: College of William and Mary, Department of Anthropology.

———. 1998a. *The land of prehistory: A critical history of American archaeology*. New York: Routledge.

———. 1998b. *Humans: An introduction to four-field anthropology*. New York: Routledge.

LaHaye, Tim. 1975. Introduction. In *The troubled waters of evolution,* by Henry M. Morris. San Diego: Creation-Life Publishers.

Marrin, A., ed. 1971. *War and the Christian conscience.* Chicago: Henry Regnery.

Moore, J. R. 1979. *The post-Darwinian controversies: A study of the Protestant struggle to come to terms with Darwin in Great Britain and America 1870–1900.* Cambridge: Cambridge University Press.

———. 1989. Of love and death: Why Darwin gave up Christianity. In *History, humanity and evolution,* ed. J. R. Moore, 195–229. Cambridge: Cambridge University Press.

Morris, H. M. 1974. *Scientific creationism.* San Diego: Creation-Life Publishers.

———. 1982. *Creation and its critics.* San Diego: Creation-Life Publishers.

Murray, M. 1995. *The law of the father?* London: Routledge.

Noble, D. F. 1979. *America by design.* New York: Alfred A. Knopf.

Numbers, R. L. 1992. *The creationists.* New York: Alfred A. Knopf.

Parker, G. 1978. Evolution: My religion. *Today's Student* 2 (11) (Ames, IA). n.p.

Richards, E. 1989. Huxley and woman's place in science: The "woman question" and the control of Victorian anthropology. In *History, humanity and evolution,* ed. J. R. Moore, 253–84. Cambridge: Cambridge University Press.

Richards, J. L. 1997. The probable and the possible in early Victorian England. In *Victorian science in context,* ed. B. Lightman, 51–71. Chicago: University of Chicago Press.

Roberts, D. 1973. Tory paternalism and social reform in early Victorian England. In *The Victorian revolution: Government and society in Victoria's Britain,* ed. P. Stansky, 147–64. New York: Franklin Watts.

Schaeffer, F. A. 1968. *The God who is there.* Downers Grove, IL: InterVarsity Press.

———. 1976. *How shall we then live?* Old Tappan, NJ: Fleming H. Revell.

Secord, J. A. 2001. *Victorian sensation: The extraordinary publication, reception, and secret authorship of* Vestiges of the natural history of creation. Chicago: University of Chicago Press.

Shapin, S. 1996. *The scientific revolution.* Chicago: University of Chicago Press.

Shapin, S., and S. Schaffer. 1985. *Leviathan and the air-pump.* Princeton, NJ: Princeton University Press.

Shuttleworth, S. 1990. Female circulation: Medical discourse and popular advertising in the mid-Victorian era. In *Body/Politics: Women and the discourses of science,* ed. M. Jacobus, E. F. Keller, and S. Shuttleworth, 47–68. New York: Routledge.

Simpson, G. G. 1970. Uniformitarianism: An inquiry into principle, theory, and method in geohistory and biohistory. In *Essays in evolution and genetics in honor of Theodosius Dobzhansky,* ed. M. K. Hecht and W. C. Steere, 43–96. New York: Appleton-Century-Crofts.

Smiles, S. 1859. *Self-help with illustrations of conduct and perseverance.* London: John Murray.

Smith, C. 1998. *The science of energy: A cultural history of energy physics in Victorian Britain.* Chicago: University of Chicago Press.

Tomasik, K. M. 1982. *Junior High Teacher's Guide.* Elgin, IL: David C. Cook Publishing.

Torrey, R. A., ed. 2003. *The fundamentals.* Grand Rapids, MI: Baker Books.

Toumey, C. P. 1994. *God's own scientists: Creationists in a secular world.* New Brunswick, NJ: Rutgers University Press.

West, R., ed. 1980. *Primary-Junior Teacher's Guide.* Elgin, IL: David C. Cook Publishing.

White, L. A. 1959. *The evolution of culture.* New York: McGraw-Hill.

Young, R. M. 1989. Persons, organisms and . . . primary qualities. In *History, humanity and evolution,* ed. J. R. Moore, 375–401. Cambridge: Cambridge University Press.

Why Teach Evolution?

Andrew J. Petto & Laurie R. Godfrey

FROM THE EARLY TWENTIETH CENTURY ONWARD, teaching evolution in the public schools has been a contentious issue. When the first edition of *Scientists Confront Creationism* was published in 1983, legislators in several states had introduced resolutions and bills demanding "equal time," "balanced treatment," or "fair treatment" of "creation science" and "evolution science" (Edwords 1983). Although a series of federal court decisions derailed those approaches, the struggle over evolution in the curriculum did not disappear (Matsumura 2001). Recent conflicts over the content of science education standards in many states have arisen precisely because evolution was awarded its proper status as the fundamental theoretical construct underpinning modern biology (Cunningham 1999; Petto 2000; Evans 2001a). If, as previous chapters have demonstrated, each of the claims for the scientific validity of "alternative theories" about the history of life and the universe fails on its merits, how could their proponents justify the demand that the public schools teach these ideas in the science classroom?

Modern opponents of evolution base their claims on two generally well-regarded principles of modern educational practice: that students should be taught to think critically, and that they should understand the basis of scientific disciplinary controversies. Critical thinking means that students should learn to pose questions, gather and analyze data, synthesize and apply conclusions to the problem at hand, and evaluate the outcome. Furthermore, they argue, for students to understand completely the nature of scientific inquiry, they need to appreciate how scientific theories are formed, critiqued, and changed. This means that students should be exposed to contemporary scientific controversies and learn to evaluate them.

According to the definition proposed by the National Council for Excellence in Critical Thinking (NCECT):

> Critical thinking is the intellectually disciplined process of actively and skillfully conceptualizing, applying, analyzing, synthesizing, and/or evaluating information gathered from, or generated by, observation, experience, reflection, reasoning, or communication, as a guide to belief and action. (www.criticalthinking.org/ aboutCT/definingCT.shtml)

The NCECT also suggests a model description of a critical-thinking curriculum in higher education that is applicable to K–12 public education with only a few modifications. This model serves as a template for many disciplines, and the items in italics are inserted into the template to specify a critical-thinking foundation in the sciences.

> Students successfully completing a [course] in *scientific foundations* will demonstrate a range of *scientific* thinking skills and abilities which they use in the acquisition of knowledge. Their work at the end of the program will be clear, precise, and well-reasoned. They will demonstrate in their thinking command of the key *scientific* terms and distinctions, the ability to identify and solve fundamental *scientific* problems. Their work will demonstrate a mind in charge of its own *scientific* ideas, assumptions, inferences, and intellectual processes. They will demonstrate the ability to analyze *scientific* questions and issues clearly and precisely, formulate *scientific* information accurately, distinguish the relevant from irrelevant, recognize key questionable *scientific* assumptions, use key *scientific* concepts effectively, use *scientific* language in keeping with established professional usage, identify relevant competing *scientific*

> points of view, and reason carefully from clearly stated *scientific* premises, as well as sensitivity to important *scientific* implications and consequences. They will demonstrate excellent *scientific* reasoning and problem-solving. (www.criticalthinking.org/ resources /recomendations-for-self-evaluation.shtml)

What is wrong with the notion that students should acquire the skills of critical thinking by grappling with the controversies surrounding evolutionary theory? This chapter focuses on that essential question. We argue, first, that materials offered by opponents of evolution to promote critical thinking are actually designed to misinform and obfuscate the real controversies in evolutionary biology. Educational materials offered to encourage "critical thinking" consist mainly of "evidence against" evolution—a "catalogue" of specific cases for which evolutionary theory has presumably failed to provide a complete or "convincing" explanation (Kenyon and Davis 1989, 1993; Behe 1996; Dembski 1998; Dembski and Behe 1999; Wells 2000; Dembski 2001a). Second, we argue that opponents of evolution confuse the source of evolutionary controversies in two important ways. The first is semantic; Darwinism is incorrectly equated with neo-Darwinism, which is in turn incorrectly equated with all modern evolutionary theory. Materials from the Discovery Institute and its fellows, for example, typically trumpet the shortcomings of *Darwinian* theory—conveniently ignoring more than a century of research that has expanded, modified, and, in some cases, replaced strictly "Darwinian" ideas about evolution (see, for example, the critique of "intelligent design" models by Van Till [2002]). Furthermore, the Discovery Institute relies on its own unique definition of neo-Darwinian theory, which most modern scientists would not accept (Branch 2002b). According to the Discovery Institute, neo-Darwinism is characterized by

> the sufficiency of small-scale random variation and natural selection to explain major changes in organismal form and function;
>
> the equivalence, given enough time, of the processes of micro- and macroevolution;

the usefulness of "molecular clocks" to determine historical branching points between species;

the existence of a single Tree of Life, with its roots in a Last Universal Common Ancestor (LUCA);

the congruence or matching of evolutionary trees (that is, phylogenies) derived from morphological and molecular evidence;

the appearance, in embryology, of a conserved stage revealing the common ancestry of all vertebrates.

This interpretation puts most contemporary evolutionary biologists at odds with "neo-Darwinism" because most would not describe "neo-Darwinism" (and certainly not evolutionary theory) in such a particularistic and limited manner (see, for example, Evans 2001b).

The second way is more insidious, as it involves the meaning of the word *controversy* itself. If "controversies" are to be taught, they should be real matters of contention within the disciplines they are said to represent. But there is no controversy in biology over the occurrence of evolution (descent with modification, common ancestry, natural selection as a powerful mechanism for evolutionary change). The only controversy is one imposed from outside the discipline: It is about *teaching* evolution in public schools. This controversy is not *scientific* but socio-culturo-political (Petto 2003). There are, to be sure, scientific controversies about the relative contributions that different processes of biologic change make to the pattern that we recognize as evolution; however, none of these concerns *whether* evolution occurred (for example, see Sterelny 2001; Anonymous 2002). Nonetheless, textbook disclaimers (see Numbers, in this volume) as well as the "teach the controversy" appeal aimed at school boards and media outlets (Santorum 2002) continue the misinformation inherent in the "two-models" proposals—that is, that there are competing scientific models for the history and diversity of life on earth that challenge evolution's role as the theoretical foundation of the biological sciences.

We also argue in this chapter that, even when there *are* competing *scientific* models within a discipline, it is not always necessary or appropriate to include all alternatives in a K–12 science education. Models should be evaluated for curriculum inclusion (or exclusion) on pedagogical, scientific, and constitutional grounds. In the past three decades, constitutional issues have been paramount (Matsumura 2001; Reule 2001), but challenges to evolution education—although essentially unchanged in their fundamental arguments against evolution—have progressively been relabeled and reworded in more secular language in order to avoid obvious constitutional entanglements (Scott 1996). Thus far, the courts have rejected these repackaged versions of creationism for what they are, but Reule (2001) points out that the nonsecular purpose of these newer formulations is becoming more difficult for courts to discern.

Since the constitutional considerations are well covered elsewhere in this volume, we will concentrate here on the other two considerations for public education—scientific and pedagogical issues. The scientific issues relate to active debates within the sciences, carried on by scientific researchers and exchanged in professional scientific publications. There is no shortage of these both within the evolutionary sciences and elsewhere. A number of such "controversial" proposals and models are candidates for inclusion in science education.

The pedagogical issues relate to classroom practice—what we should teach and how we should teach it. Although the science education curriculum can be enriched by a discussion of scientific controversies (for example, Kipnis 2001), the goal of most science education is to provide students with a solid understanding of the current consensus in various scientific disciplines, not to present untested and speculative ideas, even if they are "cutting edge" research.

Scientific Considerations

If we consider only the scientific controversies related to "Darwinian" evolutionary models in the last half of the twentieth century, it is clear that such challenges are a common occurrence in the scientific literature. Table 1 shows challenges that have suc-

Non-Darwinian Construct	Author	Date	Outcome
Developmental Evolution	Waddington	1942	Area of active research; testing and application awaited technological innovations in molecular and cell biology; generally accepted as a contributor to significant evolutionary change by the late 1980s.
Cladistics	Hennig	1950	Gained acceptance as computer-assisted analysis became more powerful; began to dominate phylogenetic studies by the early 1990s.
Transposons	McClintock	1951	Examples accumulated throughout the mid-20th century; generally accepted as a contributor to significant evolutionary change by the early 1980s; model now also supported by applied agricultural research.
Molecular Clock Hypothesis	Sarich and Wilson	1967	Area of active research as scientists work to understand the correspondence between morphological and molecular change; generally accepted as an indicator of divergence and differentiation times since the mid-1980s, even though questions still remain about calibration of the "clock" and correspondence among rates of change in different regions of the genome.
Neutral Mutation	Kimura	1968	Settled by 1980s; generally considered a significant source of variation in DNA sequences, even in the absence of significant morphological change
Endosymbiosis	Margulis	1970	Still an area of active research with a number of unanswered questions, but generally accepted since the late-1980s as the most likely explanation for a number of cellular organelles, especially mitochondria and chloroplasts.
Punctuated Equilibria	Eldredge and Gould	1973	Model actively applied; generally accepted as a contributor to significant evolutionary change by the mid-1980s.
Epigenetics	Løvtrup	1974	Area of active research, especially in plant sciences; generally accepted as contributing to significant inheritable morphological change by the late 1990s; model also supported by a number of applied studies.
Horizontal Gene Transfer	Woese	1977	Well known as a means of DNA exchange in bacteria and other single-cell organisms, the impact of this form of genetic recombination on evolutionary inferences is still unsettled.
Somatic Hypermutation	Steele	1979	Area of active research; testing and application awaited technological innovations in molecular and cell biology; generally accepted as a contributor to significant evolutionary change by the mid-1990s; model also supported by a number of applied clinical studies.

Table 1. Some Late–20th-Century Scientific Controversies

Sources for Table 1

Eldredge, N., and S. J. Gould. 1973. Punctuated equilibria: An alternative to phyletic gradualism. In *Models in Paleobiology*, ed. T J. M. Schopf, 82–115. San Francisco: Freeman, Cooper and Co.

Hennig, W. 1950. *Grundzuge einer theorie der phylogenetischen Systematik*. Berlin: Deutscher Zentralverlag.

Kimura, M. 1968. Evolutionary rate at the molecular level. *Nature* 217: 624–626.

Løvtrup, S. 1974. *Epigenetics: A treatise on theoretical biology*. London: Wiley.

Margulis, L. 1970. *Origin of Eukaryotic Cells*. New Haven, CT: Yale University Press.

McClintock, B. 1951. Mutable loci in maize. *Carnegie Institution of Washington Year Book* 50: 174–81.

Sarich, V. M. and A. C. Wilson. 1967. Immunological time-scale for human evolution. *Science* 158: 1200–1203.

Steele, E. J. 1979. *Somatic selection and adaptive evolution: On the inheritance of acquired characters*. Toronto: Williams-Wallace.

Waddington, C. 1942. Canalization of development and the inheritance of acquired characters. *Nature* 150: 563–65.

Woese, C. R, and G. E. Fox. 1977. Phylogenetic structure of the prokaryotic domain: The primary kingdoms. *Proceedings of the National Academy of Sciences USA* 74: 5088–90.

ceeded in expanding our ideas of how biological change is generated and how it produces evolution. We present these as examples of scientific controversies from which we may draw some general principles. In particular, the chief questions are: "What is a *scientific* controversy?" and "What is the process by which controversial ideas are evaluated by scientists?"

In general, a scientific controversy exists whenever competing scientific explanations for any phenomenon or set of observations are used to support research into unanswered questions or to resolve conflicting interpretations in scientific research. The examples in table 1 illustrate proposals from biologists to account for observed biological change that did not correspond to the expectations of neo-Darwinian models derived from the New Synthesis (for example, Fisher 1930; Haldane 1932; Dobzhansky 1937; Mayr 1942). Some of these, such as Waddington's speculations on the evolutionary implications of developmental biology (Waddington 1942, 1956), had to wait decades for further testing and refinement because of technological limitations (Wilkins 2003). Others, such as Steele's model of somatic hypermutation (1979, et seq.), are still in the process of being fully tested and understood.

Stage	Characteristics
New Proposal	Propose new model to resolve difficult or unanswered questions in a field
Retrospective Research	Review existing research and apply new model to unresolved questions
Prospective Research	Propose new research that explores limits and tests hypotheses based on new model
Scientific Exchange	Presentation of research results in scientific forums—ultimately in the peer-reviewed scientific research literature
Revision and Recycling	Revision of models and re-examination of research results in light of scientific exchange or self-evaluation of research results
Incorporation	Contribution of new models to scientific understanding confirmed and models incorporated into contemporary understanding in the relevant scientific field(s)

Table 2. The Natural History of a Scientific Controversy

What all the examples in table 1 have in common, however, is a similar pattern of scientific activity from original proposal through to ultimate disposition. We may understand this process, in a sense, as a "natural history" of a scientific controversy. Controversial ideas undergo a maturation process in which their predictions and formulation are refined and, ultimately, find their proper place within contemporary scientific theory. Of course, there is no specific formula that new scientific ideas can follow to guarantee their ultimate acceptance, but all controversial scientific ideas, models, and theories do have one thing in common: They have been thoroughly tested in the scientific research literature, even (or especially) by critics who want nothing more than to prove them wrong!

In the earliest stages, a new idea or interpretation is proposed based on a reexamination of existing data or to accommodate a new type of data. When this new idea appears to perform well in this limited application, researchers will often carry out retrospective research, applying the new idea in turn to data derived from previous research. The next step is prospective research—defining and carrying out new research based on problems and questions suggested by the new model. Then, the proponents of these ideas engage in scientific exchange—participating in scientific meetings and publishing their findings in the peer-reviewed scientific research literature. As a result of these exchanges, the new model is often modified to some degree, reflecting its performance in contemporary research—a process called "recycling" by Wynn and Wiggins (2001). Finally, some or all of the ideas in the new model are incorporated into existing scientific models—or else they are revised and recycled for further investigation.

Contemporary Scientific Controversies

Not all scientific controversies are equally dramatic. Some—such as the debate in the 1920s between Darwinian naturalists and Mendelian geneticists that resulted in the "New Synthesis," or the more recent debate between Richard Dawkins and Stephen Jay Gould over the importance of natural selection as an agent of evolutionary change (see Sterelny 2001, for a recent review; see also Dawkins 1996; Gould 2002)—have had such profound scientific impact that they have been aired in the popular press. Other historically prodigious and contentious ideas, such as Darwin's notion of "pangenesis" as a vehicle for the generation of novel traits (see Endersby 2003), have vanished into obscurity, with no lasting impact on the way we do science. Still other debates are quite focused and relevant only within a limited subdisciplinary specialty. For example, few laypeople will care about the controversy over the relative importance of "ontogenetic scaling" (extensions or truncations of ancestral growth allometries) versus "size/shape dissociation" (normally resulting in noncoincident ancestral and descendant growth allometries) in models of evolutionary change, or the relative importance of various "heterochronic" or nonheterochronic models of evolutionary change (Gould 1966, 1977; Shea 1988, 2000, 2002; Godfrey and Sutherland 1995; Rice 1997; Godfrey et al. 1998; Klingenberg 1998; Vrba 1998; Vinicius and Lahr 2003). Few might even care to understand what "allometry," "heterochrony," and "size/shape dissociation" mean. However, this is an area of both active investigation and contentious disagreement among evolutionary biologists, and one with a long history of relevance to at least part of evolutionary biology (e.g., Huxley 1932, Gould 1977; McKinney and McNamara 1991).

Similarly, few laypeople will understand debates over the utility of "coordinate-free" landmark data versus traditional, multivariate methods of capturing biological shape (e.g., Bookstein 1986; Lele and Richtsmeier 1991), or of "common principal components" versus "matrix correlations" as tools for understanding "morphological integration" (the relationships among metric traits in organisms; see Olson and Miller 1958; Flury 1988; Chernoff and Magwene, 1999; Ackermann and Cheverud 2000, 2004; Magwene 2001). They might have little interest in understanding how selection operates on correlated traits (e.g., Lande and Arnold 1983), but this issue is vitally important in developing models that take into account that evolutionary change usually involves several traits and only rarely a single one at a time.

Even narrowly constructed debates that do not speak to the basic framework of "descent with modification" may ultimately have profound impacts on the field of evolutionary biology. They are invisible to the general public first and foremost because their resolutions will not significantly alter the general outline of evolutionary theory that we teach to K–12 students (or even most students in higher education). They are also invisible because the general outline of evolutionary theory provided to these students does not equip them with the training and background to be able to comprehend the issues involved. This second point is a double-edged sword, of course, because it also permits the general public to misinterpret controversies (such as those mentioned above) to suggest that these studies are controversies about evolution itself—not simply disagreements about the relative contributions of different processes to the outcome on which all the researchers agree.

Less obtuse to the layperson are debates about interpretations of particular fossils or the proximate causes of changes in species composition at particular times in the history of life on earth (see, for example, Padian and Angielczyk, in this volume). Yet even these debates often turn on geochemical or anatomical minutiae that would require years of training for students to understand completely. For example, Brian Richmond and David Strait put forth a hypothesis about the evolution of locomotion in australopithecines that immediately became a center of controversy in biological anthropology. Richmond and Strait (2000; Richmond et al. 2001) claimed that *Australopithecus afarensis* ("Lucy" in popular parlance) and another extinct member of the human family, *Australopithecus anamensis*, show relics of an earlier, knuckle-walking ancestry. The debate that ensued (e.g., Corruccini and McHenry 2001, Richmond and Strait 2001) was not over whether Lucy moved on two legs when on the ground. Indeed, evidence for that is incontrovertible (see Brace, in this volume). The knuckle-walking debate centered on exactly *when* specialized knuckle-walking appeared in the hominoid evolutionary tree—whether on a branch leading *only* to modern African apes, *after* the divergence of the human lineage, or on an earlier branch leading to African apes *and* humans). Nowhere in the controversy is there any disagreement over *whether* humans and African apes share a recent common ancestor, or that the African apes are our closest evolutionary relatives—or even that it was upright walking that marked the divergence of early humans from their nonhuman cousins. The resolution of this controversy would not change the evolutionary tree of human history to any significant degree. However, it would affect the interpretation by experts of the sequence and impact of certain anatomical, locomotor, and ecological specializations on the evolutionary heritage of humans (see Brace, in this volume).

Does "intelligent design" (ID) theory qualify as a scientific controversy? The chief characteristic of a *scientific* controversy is an active debate in the scientific literature. This is clearly absent in the case of ID (for example, Gilchrist 1997). Some ID proponents argue that it is the "Darwinian" bias that keeps it out of the research literature (for example, Johnson 1997; Behe 2000), but if that were true, then why has that same bias not prevented a host of other non-Darwinian models from appearing in the peer-reviewed research literature in biology?

Topic	N retrieved	Examined	Peer-Reviewed	Other	%Peer Reviewed
Cladistic Analysis	2058	200	198	2	99
Developmental Evolution	35	35	31	4	88.6
Endosymbiosis	283	200	192	8	96
Epigenetics	117	117	117	20	82.9
Horizontal Gene Transfer	418	200	183	17	91.5
Molecular Clocks	83	83	80	3	96.4
Neutral Mutations	2	2	2	0	100
Punctuated Equilibria	17	17	17	0	100
Somatic Hypermutation	325	200	190	10	95
Transposons	2862	200	192	8	96

Table 3a. Models of Biological Change in the Scientific Literature

Table 3a shows the results of a seven-year review of the *Biological Abstracts* bibliographic database for a number of non-Darwinian models of organic change. Although the number of peer-reviewed research and review articles varies by model, all except "neutral mutation" return several research articles per year over that seven-year period. The dearth of research articles on neutral mutation is similar to that on the "modern synthesis." Both of these can be considered "settled" issues in modern biology to the extent that they seldom need to be studied in and of themselves.

Now, well more than a decade after it was first proposed (Kenyon and Davis 1989), "intelligent design" theory has failed to move beyond the very first stage in the natural history of new scientific ideas—that of an interesting idea. Philosopher Kelly Smith (2000) has proposed that the problem is a lack of clarity and focus within the ID community itself—that is, ID proponents cannot seem to agree on what it is that their "alternative" explanatory model really means in terms of practical scientific research (Scott 1999). Even if it should prove productive some time in the future, at this stage ID remains unsettled even—or perhaps especially—among its few proponents, and its potential impact on how we view the history and diversity of life is vague and marginal. It appears to be nothing more than a twenty-first-century god-of-the-gaps stratagem that will be forced to retreat as

Topic	N retrieved	Examined	Peer-Reviewed	Other	%Peer Reviewed
Biological Complexity	45	45	42	3	95.6
Irreducible Complexity	3	3	1	2	33
"Intelligent Design" Theory	1	1	0	1	0
Information Theory	202	200	198	2	99
Biological Complexity and Irreducible Complexity	0	0	0	0	0
Biological Complexity and "Intelligent Design"	0	0	0	0	0
Biological Complexity and Information theory	0	0	0	0	0
Irreducible Complexity and "Intelligent Design"	0	0	0	0	0
Information Theory and "Intelligent Design"	0	0	0	0	0
Irreducible Complexity and Information Theory	0	0	0	0	0

Note: Tables 3a and 3b represent results from a literature search in Biological Abstracts *for the period January 1995 through December 2002.* Biological Abstracts *includes contemporary research literature from biological and related sciences. "Peer-reviewed" refers to research articles and professional review articles. "Other" refers to news reports, letters to the editor, commentaries, book reviews, and reports of conferences. In searches returning more than 200 citations, we examined the first 200.*

Table 3b. "Alternative" Models of Biological Change in the Scientific Literature

new evidence of self-organizing systems or macromutational pathways is described in the research literature (see Pennock, in this volume; Petto 2003).

Pedagogical Considerations

Even when a controversy is active and widespread in a discipline with far-reaching implications for our understanding of science, its introduction into the curriculum is not always appropriate. For example, current research into models of the origin of life (see Lazcano, in this volume) or the early evolutionary history of and relationships among the Bacteria, Archaea, and Eucarya (Woese 2000, 2002) are inappropriate for at least two reasons.

First, many of these studies are exploratory and very early in their "natural history" as new scientific ideas. Many of the items contained in its *Bibliography of Supplementary Resources for Ohio Science Instruction,* which the Discovery Institute sent to the Ohio Board of Education (OBE) during its development of science education standards, were of just such an exploratory nature. Since the OBE is concerned *only* with K–12 science education, Branch (2002b) solicited comments from contemporary researchers whose work was included in the DI's *Supplemental Bibliography* about the suitability of these articles for use at this level of science education. Responses from Leslie E. Orgel and Günther P. Wagner were typical:

> Orgel . . . remarked, "I work at the frontiers of present-day knowledge. I doubt that the time is ripe for a detailed and correct interpretation of my work at the high-school level." And Günther P. Wagner . . . explains, "This is cutting-edge research, and we cannot yet know whether it will stand up under the scrutiny of our colleagues. There is too much work to be done to determine whether our ideas and results turn out to be correct and useful for further research." Orgel's and Wagner's attitude instructively contrasts with that of the promoters of "intelligent design" who wish for their views to be taught at the high-school level before they have been accepted by the scientific community. (Branch 2002b, 23)

Brian Alters, director of the Evolution Education Research Centre at McGill University, added in an interview with Glenn Branch:

> When high school students read such relatively complex discussions written for scientists, they often believe that the authors are contending that evolution is a theory in crisis. But when such articles are read by those with the proper university training in science, those readers do not conclude that the authors are contending that evolution is a theory in crisis. This difference is very telling and probably explains why the Discovery Institute selected these particular papers. After all, the Institute gives no rationale

> for the selection. Of all the colleagues I know in North America, none of those university science educators with expertise in training high school teachers would have selected these papers for high school students. So again, why were these particular papers selected? Not only is this selection of papers inappropriate for the high-school level, it will likely engender numerous misconceptions among high school students about the science of evolution—something no science teacher would want. (Branch 2002b, 24)

Alters's comments also illustrate the second issue: Many of the scientific controversies cannot be fully and correctly appreciated without more advanced study and preparation than is usually available either to K–12 students or their teachers. After all, teacher training emphasizes the skills for understanding, presenting, and helping students learn consensus views in scientific disciplines. Rarely do K–12 teachers themselves have the advanced training to decide the impact of scientific controversies on scientific disciplines, much less to predict how these disciplines may resolve the controversies in the future. In some cases, controversies such as those listed in table 1 are resolved rather quickly (for example, Kimura and neutral mutation); in other cases, the jury is still out after two decades (for example, Steele and somatic hypermutation). The process of resolving scientific controversies takes time—which may be why Nobel Prizes in the sciences are often awarded for work that was originally performed decades earlier.

Even as controversial scientific ideas mature through their natural histories, not all aspects of the ideas should be equally represented in K–12 (or even university) curricula. For example, a typical grade 9 biology text (Biggs et al. 2000) includes Carl Woese's discovery of the Archaea as a distinct high-level taxon, but it does not explore Woese's more recent work on the evolutionary implications of horizontal gene transfer for the origin of modern biological domains. This is appropriate, as Alters argued above, because the details of Woese's analysis are based on detailed comparisons of patterns of variation in biochemical structures and pathways. The K–12 curriculum usually discusses these details, if at all, in so rudimentary a manner that students cannot be expected to understand the nature of the controversy. To make matters worse, Woese (2002) discusses how his work challenges a "Darwinian" explanation. Of course, his is still an *evolutionary* model; it is merely a *non-Darwinian* evolutionary model—and it seems to apply only to the Bacteria and Archaea for which horizontal gene transfer makes "descent" meaningless (Petto 2003).

Lynn Margulis's (1993) endosymbiosis model for the emergence of eukaryotes from prokaryotic ancestors—once an extremely controversial proposal—is included in the same textbook as the most likely explanation for the origin of the Eucarya. The punctuated equilibria model gets a brief mention (though the associated diagram resembles a gradualistic model), but McClintock's model of transposable genes is entirely absent, as are neutral mutations and developmental evolutionary models. Some of these are well-accepted evolutionary models and mechanisms, but simply are too sophisticated for the preparation and level of study that is generally found even among advanced high-school students. In general, models that can be easily explained in a few sentences or clearly illustrated in a chart or diagram are the ones that appear in K–12 texts. Table 4 provides a sampling of these non-Darwinian models that *already* appear in textbooks.

Should these "controversies" be a part of the K–12 curriculum? Except for the decidedly noneducational goal of calling into question evolutionary theory itself, what would be the point of telling K–12 students, for example, about Woese's (2002) claims that the roots of the evolutionary tree cannot be traced back farther than the "Darwinian threshold" at the emergence of the first nucleated cells? Since the scientific content is so advanced and relies on the interpretation of complex patterns of cellular biochemistry, since the matter is still unsettled scientifically, and since Woese's proposals call for a *modification* of Darwinian evolution models, not their elimination, there is no doubt that introducing Woese's "Darwinian threshold" model even into high-school

Text (Auth/Publ/Edition/year)	Cladistic Analysis	Develop-mental Evolution	Endo-symbi-osis	Epi-genetics	Horizon-tal Gene Transfer	Mole-cular Clocks	Neutral Mutation	Punctu-ated Equilibria	Somatic Hyper-mutation	Trans-posons	Total
Biology: Dynamics of Life (Biggs et al., Glencoe, 2000)	4	0	4	0	1	1	0	4	0	0	14
Biology: Living Systems. (Biggs et al., Glencoe, 1998)	0	0	0	0	0	2	0	3	0	3	8
BSCS Biology: An Ecological Approach, 8th ed. (Kendall/Hunt) 1998	0	0	4	0	0	1	0	2	0	0	7
Biology, 2nd ed. (Miller & Levine) Prentice-Hall, 1993	0.5	0	4	0	1	4	0	4	0	4	17.5
Biology: The Living Science, (Miller & Levine) Prentice-Hall, 1998	0	1	3	0	0	0	3	3	0	0	10
Biology, 2nd ed. (Addison-Wesley)	0	0	3	0	0	0	0	3	0	0	6
Biology: An Australian Perspective (Oxford, 1998)	0	0	0	0	0	0	0	2	0	0	2
The Nature of Biology, Book 2. 2nd ed. (Kinnear & Martin, Jacaranda, 2000)	0	3	2	0	0	0	0	3	0	3	11
Merrill Life Science (Glencoe, 1995)	.5	0	0	0	.5	2	0	2.5	0	0	5.5
Merrill Biology: An Everyday Experience. (Glencoe, 1995)	0	0	0	0	0	0	0	0	0	0	0

0 — Concept absent 1 — Concept named or described 2 — Concept defined or illustrated
3 — Concept explained in context of unifying themes and issues 4 — Concept presented with follow-up questions or other supporting activities

Table 4. Sample of K–12 Biology Instructional Materials

Sources for Table 4

Biggs, A., K. Gregg, W.C. Hagins, C. Kapicka, L. Lundgren, and P. Rillero. 2000. *Biology: The dynamics of life*. New York: Glencoe McGraw-Hill.

Biggs, A., A. Kaskel, L. Lundgren, and D. Mathieu. 1998. *Biology: Living systems*. New York: Glencoe McGraw-Hill.

Cairney, W.J., et al. 1998. *BSCS Biology: An ecological approach*, 8th ed. Dubuque, IA: Kendall/Hunt.

Daniel, L., E. P. Ortleb, and A. Biggs. 1995. *Merrill life science*. New York: Glencoe.

Essenfield, B., C. Gontang, and R. Moore. 1996. *Addison-Wesley Biology*, 2nd ed. Reading, MA: Addison-Wesley.

Huxley, X., and Y. Walter. 1998. *Biology: An Australian perspective*. South Melbourne, Australia: Oxford.

Kaskel, A., P. J. Humer, Jr., and L. Daniel. 1995. *Merrill biology: An everyday experience*. New York: Glencoe.

Kinnear, J., and M. Martin. 2000. *The nature of biology, Book 2*, 2nd ed. Victoria, Australia: Jacaranda.

Miller, K. R., and J. Levine. 1993. *Biology*, 2nd ed. Needham, MA: Prentice Hall.

———. 1998. *Biology: The living science*. Needham, MA: Prentice Hall.

science classes would do little more than "engender numerous misconceptions among high school students about the science of evolution—something no science teacher would want" (Alters, quoted in Branch 2002b).

The same is true for numerous other scientific controversies in evolutionary theory. Steele's somatic hypermutation (see Saunders 1994 for a review of the first decade of research into this model of evolutionary change) and competing models proposed for the origin of life (see Lazcano, in this volume; Deamer 1999) are still unsettled. Most scientists in these fields recognize the tentative nature of their work, as reflected in their comments to Branch (2002b; see above) about the incorporation of their ideas into high-school biology teaching. Most would argue that it is unclear at this point which of the ideas, models, and hypotheses will survive professional scrutiny and form the basis for future scientific discovery.

If the scientific research community has not yet decided which of these ideas are tenable, can K–12 students (or their teachers) resolve them in the curriculum? Do they have the knowledge and the intellectual sophistication to decipher the complexities of these models and their implications? What can the introduction of such untested ideas produce, other than deep confusion? The National Research Council's National Science

Education Standards make it clear that the prior academic preparation and intellectual maturity of K–12 students are important criteria for deciding what to include in the curriculum:

> The gradual development of understanding and ability will be realized only if the concepts and capabilities designated for each grade level are congruent with the students' mental, affective, and physical abilities. . . . [I]t is inappropriate to require students to learn terms and perform activities that are far beyond their cognitive and physical developmental level. (NRC 1995, 214)

It is the exceptional high-school student who will be able to appreciate both the technical aspects of this research and its proper relationship to existing evolutionary theory—that the challenge is to some of the details of how evolution works, *not* to whether evolution occurred or whether organisms are related by common descent. It is also the exceptional high-school teacher whose training has prepared him or her to guide students through this material. On the one hand, the training of high-school science teachers focuses on the consensus view––well-established scientific models and theories. On the other hand, high-school teachers are faced with the formidable task of "delivering" a comprehensive curriculum that seldom allows for the type of in-depth exploration and detailed planning that an adequate treatment of these models would require.

It is not that these teachers are incapable of performing this task; it is more that they are well prepared to perform a *different* task—one that is already very challenging and powerful: to produce a developmentally appropriate curriculum that will provide students with a fundamental knowledge of core scientific concepts and theories in order to help them understand and apply scientific concepts in the future. This pedagogical objective results in a "survey" course—one based on an overview of a discipline and its key concepts and facts. For the most part, this curriculum is untouched by the contemporaneous scientific controversies that rage on the pages of peer-reviewed scientific research journals. The Discovery Institute claimed that the research articles in its

Bibliography of Supplementary Resources for Ohio Science Instruction represented "dissenting viewpoints that challenge one or another aspect of neo-Darwinism (the prevailing theory of evolution taught in biology textbooks), discuss problems that evolutionary theory faces, or suggest important new lines of evidence that biology must consider when explaining origins" (Meyer and Wells 2002). However, Branch summed up the import of these articles for K–12 education:

> What must be understood is that, although these debates about the details of [evolutionary biology] and the mechanisms of macro-evolution are legitimate, they in no way affect the presentation of evolution at the high school level, which is simply not presented in enough detail for these highly technical debates to be relevant. (2002b, 14)

In other words, the material focused on evolution typically included in the K–12 curriculum does not generally contain the level of detail that must be invoked in order to engage these *scientific* controversies about the way that evolution works. To introduce them would be neither developmentally appropriate nor pedagogically productive in promoting a clear understanding of how contemporary science frames questions and tests possible answers.

What About That Other Controversy?

Proponents of "intelligent design" theory and their allies insist that the controversy is a *scientific* dispute that should have a place in the K–12 curriculum (for example, Dembski 2001b). Recent legislative actions indicate that ID proponents have convinced a number of state legislators, at least, of the *scientific* validity of their claims (Hudson 2003; Nevers 2003). However, since its introduction in the book *Of Pandas and People* (Kenyon and Davis 1989, 1993), ID has made no discernible progress from the very earliest stages of the natural history of a scientific controversy—that of an interesting proposal (Gilchrist 1997; Bennett 2000; Sonleitner 2000). Attempts to cast evolution as a *scientifically* controversial theory have failed repeatedly (for example, Matsumura 2001). Still, a number of states have introduced the "Alabama" disclaimer—a statement to be pasted into biology textbooks declaring that evolution is a controversial theory that "some scientists believe" explains the history and diversity of life on earth (see Numbers, in this volume). In an assessment of the scientific content of the disclaimer introduced in Oklahoma, Brown University biologist Kenneth R. Miller assigned it a failing grade of 42 percent: of the twelve "scientific" statements included in the disclaimer, "only 5 were free of major errors. Three are seriously misleading, and 4 are downright false" (Miller 2000, 33).

So, the claim often made by those in the public sphere—elected officials, bureaucrats, and other policymakers—that there is a *scientific controversy* over evolution, is based on a grossly inaccurate view of current scientific knowledge. It has prompted a listener to a radio program featuring Pennsylvania State Representative Sam Rohrer to comment that such inaccurate arguments against teaching evolution are the best rationale for insisting that students receive *more* science education, and particularly more *evolution* education (Moss-Coane 2000).

There is no doubt that evolution is a controversial topic, but it is clear from a review of the scientific literature that the source of the controversy is not scientific but sociopolitical, as others have pointed out in this volume (Evans 2001a; and see Cole, Kehoe, Numbers, and Scott, in this volume). So, if we are to "teach the controversy" about evolutionary theory, then we really are addressing a sociocultural controversy. Where—if anywhere—in the science education curriculum does this issue belong?

Based on the National Science Education Standards (NRC 1995), sociopolitical and sociocultural aspects of scientific theories and research fit into two sets of standards. The first is in the History and Nature of Science. These standards direct students to explore how contemporary scientific ideas and practices came to be realized in their current form, and how (and why) these ideas were embraced by the scientific community when they were generated. The standard explanation from ID proponents for the exclusion of "intelligent design" from the curriculum is that the "elite priesthood" of the "Darwinian establishment" excludes all challenges to the prevailing materialistic paradigm (Brooke 1998; Johnson 2001; DeMar 2004). However, a competent review of the history of scientific alternatives to Darwinian evolutionary models (see table 1) would refute that superficial reading of the situation in two ways.

First, during the last half of the twentieth century, several significant challenges to Darwinian models did enter both the scientific research literature and, once they had established themselves scientifically, the K–12 curriculum. Second, the "scientific" program of "intelligent design" theory has failed to produce scientific results that test the expectations and predictions of the models, despite the fact that ID was first proposed as an alternative scientific theory by Kenyon and Davis in 1989 (Gilchrist 1997; Smith 2000). Since these results are necessary to demonstrate that a model or theory is scientifically viable and valuable, the failure of ID to produce such results ensures that ID will not make any significant impact on scientific theory and practice—and thus should not have any presence in the science curriculum as an alternative *scientific* theory or model. Furthermore, as philosopher of science Kelly Smith (2000) pointed out, ID currently allows too many contradictory positions to be accommodated within the same "big tent" (Scott 2001). The fact that ID comfortably accepts proponents who claim the universe is less than 10,000 years old at the same time as other ID proponents who claim it is much older, perhaps billions of years old, weakens its claim to the status of a productive scientific theory (Smith 2000).

The second place for teaching this controversy is in the Science in Personal and Social Perspectives standards. These standards direct students to explore how contemporary society has responded to scientific theories and discoveries—how (or whether) the general public has come to accept and apply new scientific knowledge. Addressing sociocultural and political aspects of scientific theory is relatively new as an integral part of the science curriculum; however, such practice is a common part of K–12 social studies curricula (Gallo 1996). In a special issue of its journal *Social Education*, the National Council for the Social Studies (NCSS) explored classroom and curricular strategies for teaching about controversial issues—many no less contentious than creation/evolution, such as abortion, Holocaust denial, and affirmative-action programs (Gallo 1996; Simpson 1996). The NCSS recommendations represent the essence of a critical-thinking approach. They emphasize an examination of starting premises, available data, social setting, and sociocultural values and mores that guide judgments, legal and civic traditions, and so on (see, for example, Gallo's [1996] model lessons on the First Amendment).

The strategy proposed by the educators in this special issue is not to present all the evidence and let students decide but rather to guide students through an exploration of the empirical and philosophical issues. This contrasts pointedly with the "critical thinking" justification used by proponents of "equal time" in the 1980s. Instead, Lockwood's contribution to this special issue (1996, 28) suggests that educators first separate the controversy into its two major components: "(1) empirical issues associated with disciplined inquiry in [the relevant field of scholarship], and (2) issues associated with the exploration of values as they arise in public policy discussions." Lockwood (1996: 29) also provides four different roles that teachers could play in the classroom discussion of controversial issues. In the case of evolution education, where the scientific consensus is so strong, the most appropriate role for the teacher should be that of "determined advocate." The determined advocate would be certain to challenge factually incorrect statements—such as those in the textbook disclaimers—if students used them to make a case for a scientific controversy about evolution. However, Lockwood (1996, 31) emphasizes that the goal here is not simply content mastery but rather the ability to assess factual claims in order to apply them to a broader social issue—in our case, the relative public acceptance of evolution and evolution education. This approach fulfills Soley's admonition that we should develop classroom practice that provides students with "the kind of substantive knowledge that will promote a deeper understanding of their social world" (Soley 1996, 9).

Applying Lockwood's (1996) two major components of teaching controversial issues to the choice of curriculum materials to a recent article arguing a constitutional basis for the Discovery Institute's call to "teach the origins controversy" (DeWolf et al. 2000) shows why the DI's argument fails to meet the minimum standards for inclusion in the curriculum. DeWolf et al. make several serious errors in their presentation of the "empirical issues associated with disciplined inquiry" in evolutionary biology. First, the authors claim that the debate is "over how biological origins should be taught" (2000, 40). But evolution is not a theory of biological origins; it is a theory explaining the history and diversity of life on earth. DeWolf et al. have redefined the subject and scope of evolutionary theory in the way that makes it most at odds with common culture. Indeed, much of the recent controversy over science education standards in Ohio focused on how to teach "origins science"—a term that is not recognized by any scientific discipline but is used to focus public concern about the curriculum.

Second, despite their obvious awareness of all the ways that evolutionary theory has been modified by "new developments in paleontology, systematics, molecular biology, genetics, and developmental biology" (2000, 49), DeWolf et al. persist in referring to "Darwinism" and "Darwinian" evolution—defined as above in terms strictly of gradual cumulative change that results from the action of natural selection on random mutation (Branch 2002b). They point out correctly that scientific challenges to the neo-Darwinian models of the "new synthesis" of natural selection and population genetics have changed how we think about biological change (DeWolf et al. 2000; 50), and they suggest that openness to these models and controversies should also extend to "intelligent design" theory. However, the reasoning here is convoluted, and it is important to deconstruct it in order to expose its weaknesses, because the long chain of inference on which it depends is typical of the "evidence against" evolution materials proposed for classroom and curricular use.

In brief, DeWolf et al. argue that peer-reviewed scientific research has changed our view of the neo-Darwinian "new synthesis" as a comprehensive evolutionary theory. This research, as in the case with research cited in the *Supplemental Bibliography*, presents legitimate critique of and supplements to evolutionary models. On at least some of these theoretical models, there does not yet exist a scientific consensus (2000, 50). DeWolf et al. appear to conclude—as do many who cite the Discovery Institute's materials to oppose evolution education—that, since ID publications cite peer-reviewed scientific research literature as their "evidence against" a "Darwinian" theory, therefore ID should be admitted to the curriculum, because it is based on the "anomalous" data in evolutionary biology.

It is important to note, however, that the conclusions and logic suffer from two serious defects. The first is that "intelligent design" itself is not contained within that literature. Not one of the peer-reviewed scientific research articles in the *Supplemental Bibliography* is based on or recognizes "intelligent design" as a theoretical model for its research, and not one of the resources cited by DeWolf et al. (2000) that *does use* "intelligent design" in this way is in the peer-reviewed scientific research literature. The second is that the authors of the items included in the *Supplemental Bibliography* specifically deny that there is anything in their work that can be construed as arguing for anything other than evolutionary change as a result of naturalistic processes known or suspected to occur (Branch 2002b)—even if the field has not yet reached a consensus on which of the models should be preferred.

On the basis of the first guideline provided by Lockwood (1996)—"the empirical issues associated with disciplined inquiry in" evolutionary biology—proposals for the inclusion of "intelligent design" fail because their arguments depend on nonstandard definitions of terms; overgeneralization of narrow, technical studies as challenges to the whole structure of evolutionary theory; selective citation and quotation from the scientific literature ("quote-mining": Branch 2002b); and conflation of popular and professional sources as equally valid in providing "evidence against" so-called Darwinian theory.

However, even if the argument in favor of opening up the curriculum to this controversy fails on the basis of its handling of empirical issues, Lockwood (1996, 28) would support its inclusion in the curriculum under a second rubric: "issues associated with the exploration of values as they arise in public policy discussions." This approach recognizes that even if a scientific theory is considered well supported and well accepted within the relevant scientific discipline, and thus does not constitute a scientific controversy, there may be some aspects of the theory that are perceived by the general public as socially controversial.

All of the articles in the NCSS special issue recognize that students from different backgrounds may bring different values, goals, or meaning to the disciplinary content that forms the factual basis of the sociopolitical controversy (for example, Simpson 1996). Alters and Alters (2001) also detail many of the reasons—both religious and nonreligious—why students reject evolution and why teaching the subject is educationally controversial. Soley writes that these values are themselves fundamental to any social controversy:

> Subjects are controversial, in part, because they address basic questions of identity and worth—who am I (or who are we), how should we judge others, and how should we judge ourselves? These questions are highly subjective and depend, to a great extent, on one's view of the world and one's values. (1996, 10)

In a similar vein, Nelson (2000) explores how these values may raise the level of factual support needed to accept evolution because of the dire consequences that students or their parents perceive or believe will ensue were they to accept evolution. The greater the perceived consequence—in this case, eternal damnation or at least a life of immorality—the stronger the evidence for evolution must be in order for the students to accept it. Nelson (2000) used the metaphor of the rusty hand grenade to explain this: Even though a rusty hand grenade has only a 1:10,000 chance of exploding when someone pulls the pin, the consequences of the occurrence of that rare event are so devastating that most rational people would not take the chance. The anti-evolution literature is full of similarly dire warnings about the moral and spiritual dangers of learning evolution (see, for example, Colson and Pearcey 2001). Both Nelson (2000) and Skehan (2000) strongly urge classroom teachers to acknowledge the sociocultural values and traditions that may affect the students' acceptance of evolution. (Nelson also points out that the scientific evidence in support of evolution is stronger than for a number of other socially noncontroversial scientific theories; this is because critical evidence from numerous nonbiological research disciplines that might invalidate evolutionary theory is entirely absent—despite claims to the contrary by Denton [1986], Behe [1996], and Wells [2000].)

Applying Lockwood's (1996) approach, we see clearly that "intelligent design" does not depart significantly from the anti-evolutionary tradition that spawned "creation science" in the mid-twentieth century—either in its core objections to evolutionary mechanisms or in its expressed concern over the social, cultural, and moral implications of so-called Darwinian science (Brook 1998). The terminology is different, and decidedly more secular, but the concerns about a cultural slide into a strictly "materialist" worldview are unchanged, despite the new labels (Scott, in this volume). This concern was a key motivation for the establishment of the Discovery Institute's Center for the Renewal of Science and Culture (renamed the Center for Science and Culture in 2002; see Cole, in this volume; Branch 2002c). The "Renewal" in the original title referred to preventing the further deterioration of cultural values caused by a commitment to naturalism or materialism in scientific theory and practice (Still 1999) and to removing the commitment to naturalism from scientific theory and practice (Johnson 2001).

Once this anti-naturalism position was established by means of publications, public presentations, and online resources, the DI fellows and associates with a background in the law began pressing for the inclusion of "intelligent design" on the grounds of "viewpoint discrimination"—a civil-rights strategy (based on First Amendment issues) that was often pressed in parallel to the claims of an ongoing scientific controversy. By the end of the twentieth century, law-review papers and public discussions by fellows of the Discovery Institute began to argue that the resistance of the scientific "establishment" to "intelligent design" was an example of "viewpoint discrimination" (DeWolf et al. 2000). This approach appeared under a different label in the protracted struggle over state science education standards in Kansas (Cunningham 1999; Scott 1999). In Kansas, opponents of teaching evolution argued for "local control" of the science education curriculum—based on the idea that "community standards" ought to control the content of the K–12 curriculum. The argument for local control is laid out by DeWolf et al:

> The law provides no guidelines for determining how long a scientific theory must have existed in order to warrant teaching students about it. Further, good teachers know that exposing students to new (and even controversial) ideas can stimulate student interest and engagement and lead to greater subject mastery. Nor does science itself have a governing body that can issue binding rules about such matters. Instead, this constitutes a matter for local teachers and school boards to decide. (2000, 75)

DeWolf et al. go on to complain that teachers are required to teach that only "majority opinions constitute 'the scientific perspective.' " However, this claim only serves to confuse the scientific and curricular issues further. First of all, many scientific ideas are not presented at the K–12 level at all, despite being majority opinions. More important, scientific theories and models do not take their place in either the research or the education community simply because they have a "right" to be represented. Minority views in the sciences must earn their place by successful demonstration of their productivity and usefulness in addressing key research issues in the discipline; these are the criteria by which they are accepted. In contrast, public-school administrators—especially elected school boards—are likely to be persuaded not by the scientific evidence (which may be too specialized or complex for them to understand completely) but by the arguments based on our cultural value of fairness and equality—a perspective derived from First Amendment and other "civil rights."

Applying Lockwood's (1996) two-part categorization of controversial issues has one inadvertent benefit: It demonstrates how a "civil-rights" approach to alternative scientific theories undermines the nature of scientific inquiry and scientific progress. Kenneth R. Miller has argued repeatedly that proponents of the "viewpoint discrimination" are badly misinformed—or misled—about the nature of scientific inquiry (Milner and Maestro 2002). If "viewpoint discrimination" becomes a basis for inclusion of ideas in the science—or any—curriculum of ideas rejected by the current practitioners of a discipline, then, as Miller argued, there would be no basis for rejecting claims that the earth is flat or that it is the center of the universe. (See Bob Holmes's [2000] interview of Tom Willis in *New Scientist,* in which this anti-evolutionist who worked to remove evolution from the Kansas science education standards remarked that he was not convinced that scientists were sure that the earth revolved around the sun.) By putting this issue in the context of other sociocultural controversies, the NCSS approach helps students to see that the "intelligent design" controversy is one of conflicting cultural values and that the scientific arguments are secondary to the sociopolitical issues that evolution opponents regard as paramount. This is not to deny that evolution is controversial, but that the controversy is social, political, and cultural rather than scientific. Therefore, the solutions to the controversy are not to be found in making a change in the science curriculum, but rather in helping students to distinguish between the sociocultural and political *reactions* to evolutionary theory—only about half of the general public accepts evolutionary models—and the fundamental role that evolution serves within the life sciences, in which even "dissenting" scientists are proposing evolutionary models.

Conclusions

Scientists would be ill-advised to argue that evolution is not controversial. In North America, at least, we "know" it is controversial as well as we know anything (for example, Alters and Alters 2001). However, the controversy does not lie within the scientific disciplines that are grappling with and resolving difficult questions relating to processes of evolutionary change. The locus of the controversy is within the public domain. It is perhaps most remarkable that the "controversy" over teaching evolution has not changed its character, even as evolutionary theory has discovered and incorporated many modes of biological change previously "undreamt of in [our] philosophy." Indeed, even as the scientific community accepts new ideas and models that strengthen and invigorate evolutionary theory, critics of evolution continue to argue that these same models weaken the scientific position of evolutionary theory.

In science education, we stress the importance to scientific progress of researchers whose alternative or minority views have added significantly to our knowledge (see, for example, Hagen et al. 1996; Kipnis 2001), so it is no surprise that anti-evolutionists would try to cast their ideas as "scientific" alternatives to evolutionary theory. In the late 1990s, anti-evolutionists from the Discovery Institute began calling for public schools to "teach the controversy"—by including what they called "Darwinism" and their alternative "intelligent design" theory side by side. They proposed that students must learn why "growing numbers of scientists" are rejecting evolution in favor of a new scientific theory based on the idea that complex structures in nature are the product of a purposeful, intelligent designer.

Even though the terminology is recent, the tactics and the rationale behind "intelligent design" creationism (IDC) are not new. Opponents of evolution education call for students to hear "all the evidence" in the guise of "good science education."

Proposed revisions to state science education standards in Pennsylvania (Petto 2000) and Ohio (Evans 2001a) included language that required teachers to provide students with scientific "evidence against" evolution. Although based on language in the dissenting opinion by U.S. Supreme Court Justice Antonin Scalia (1987) in *Edwards v. Aguillard,* this tactic has its roots in the "fair treatment," "balanced treatment," and "equal time" arguments made in the 1980s—the notion that any anti-evolutionary idea should be presented in the public-school curriculum in equal measure with evolutionary theory. The stated goal of these arguments is to promote "critical thinking"—a phrase borrowed from contemporary educational research and objectives found in recent science education standards (NRC 1995)—but the real goal is merely to weaken evolution education, since the only consistent association among the proposed "alternatives" is their perceived antithesis to evolution.

Then, as now, two questions face science educators. The first is whether excluding these so-called alternatives results in students' receiving inadequate science education. The second is how and when to introduce scientific controversies into the classroom. It is not enough merely to make students aware of minority views; educators must also provide a framework within which students may evaluate them. The old call to present "both sides of the evolution debate" and let the students decide for themselves does not promote critical thinking. The goal should be for teachers to help students understand the foundations and frameworks of science and to avoid producing "numerous misconceptions among high school students about the science of evolution" (Branch 2002b, 24). This can perhaps best be accomplished by devoting more time and resources in the curriculum to the study of evolutionary biology as it is currently understood and practiced by research scientists.

Acknowledgments

We thank Kim Bilica and Bob Cooper for comments on earlier drafts of this chapter. We thank Joanna Day, Alan Gishlick,

Tony Hiatt, Harry Kanasa, and Jeff Witters for information on concept coverage in high-school biology texts.

References

Ackermann, R. R, and J. M. Cheverud. 2000. Phenotypic covariance structure in tamarins (Genus *Saguinus*): A comparison of variation patterns using matrix correlation and common principal component analysis. *American Journal of Physical Anthropology* 111: 489–501.

———. 2004. Morphological integration in primate evolution. In *Phenotypic integration: Studying the ecology and evolution of complex phenotypes*, ed. M. Pigliucci and K. Preston, 302–19. Oxford: Oxford University Press.

Alters, B., and S. M. Alters. 2001. *Defending evolution: A guide to the creation/evolution controversy.* Sudbury, MA: Jones and Bartlett.

Anonymous. 2002. Buzzing up the wrong tree: Discovery Institute caught out on a limb. *Reports of the National Center for Science Education* 22 (4): 25–27.

Behe, M. J. 1996. *Darwin's black box: The biochemical challenge to evolution.* New York: Free Press.

———. 2000. Correspondence with science journals: Response to critics concerning peer-review. Available at www.arn.org/docs/behe/mb_correspondencewithsciencejournals.htm (last accessed September 4, 2003).

Bennett, G. L. 2000. A review of *Of pandas and people* as a textbook supplement. *Reports of the National Center for Science Education* 20 (1–2): 31–34, 39.

Biggs, A., K. Gregg, W. C. Hagins, C. Kapiçka, L. Lundgren, and P. Rillero. 2000. *Biology: The Dynamics of Life.* New York: Glencoe McGraw-Hill.

Bookstein, F. L. 1986. Size and shape spaces for landmark data in two dimensions. *Statistical Science.* 1: 181–242.

Branch, G. 2002a. Analysis of the Discovery Institute's "Bibliography of Supplementary Resources for Ohio Science Instruction." *Reports of the National Center for Science Education* 22 4: 12–18; 23–24.

———. 2002b. Quote-mining comes to Ohio. *Reports of the National Center for Science Education* 22 (4): 11–13.

———. 2002c. Evolving banners at the Discovery Institute. *Reports of the National Center for Science Education* 22 (5): 12.

Brook, T. 1998. The terrible strength and weakness of naturalism: An interview of Phil Johnson. *SCP Journal* 21 (4)–22 (1). Available at www.scp-inc.org/publications/journals/J2104/j2104_1.htm (last accessed October 8, 2004).

Chernoff, B., and P. Magwene. 1999. Morphological integration: Forty years later. In *Morphological integration*, ed. E. C. Olson and R. Miller, 319–48. Chicago: University of Chicago Press.

Colson, C., and N. Pearcey. 2001. *Developing a Christian worldview of science and evolution.* Wheaton, IL: Tyndale House Publishers.

Corruccini, R. S., and H. M. McHenry. 2001. Knuckle-walking hominid ancestors. *Journal of Human Evolution* 40 (6): 507–11.

Cunningham, D. L. 1999. Creationist tornado rips evolution out of the Kansas science standards. *Reports of the National Center for Science Education* 19 (4): 10–15.

Dawkins, R. 1996. *Climbing Mount Improbable.* New York: W. W. Norton.

Deamer, D. W. 1999. Self-assembly of organic molecules and the origin of cellular life. In *Evolution: Investigating the evidence,* ed. D. Springer and J. Scotchmoor. College Park, PA: Paleontological Society Press.

DeMar, G. 2004. *The religion of evolution.* Available at www.americanvision.org/articlearchive/religion_of_evolution.asp (last accessed October 8, 2004).

Dembski, W. A. 1998. *The design inference: Eliminating chance through small probabilities.* Cambridge: Cambridge University Press.

———. 2001a. *No free lunch: Why specified complexity cannot be purchased without intelligence.* Lanham, MD: Rowman and Littlefield.

———. 2001b. Teaching intelligent design. Metaviews 010 (February 2). www.metanexus.net/archives/message_fs.asp?ARCHIVEID=2675 (last accessed, March 31, 2003).

Dembski W. A., and M. J. Behe. 1999. *Intelligent Design: The bridge between science and theology.* Downers Grove, IL: InterVarsity Press.

Denton, M. 1986. *Evolution: A theory in crisis.* Bethesda, MD: Adler and Adler.

DeWolf, D. K., S. C. Meyer, and M. E. DeForest. 2000. Teaching the origins controversy: Science, or religion, or speech? *Utah Law Review* 39: 39–110.

Dobzhansky, T. 1937. *Genetics and the Origin of Species.* New York: Columbia University Press.

Edwords, F. 1983. Is it really fair to give creationism equal time? *Scientists confront creationism,* ed. L. R. Godfrey. New York: W. W. Norton.

Endersby, J. 2003. Darwin on generation, pangenesis and sexual selection. In *The Cambridge Companion to Darwin,* ed. J. Hodge and G. Radick, 69–91. Cambridge: Cambridge University Press.

Evans, S. 2001a. Ohio: The next Kansas? *Reports of the National Center for Science Education* 22 (1–2): 4–5.

———. 2001b. Doubting Darwinism through creative license. *Reports of the National Center for Science Education* 21 (5–6): 22–23.

Fisher, R. A. 1930. *The genetical theory of natural selection.* Oxford: Clarendon Press.

Flury, B. 1988. *Common principal components and related multivariate models.* New York: John Wiley.

Gallo, M. 1996. Classroom focus: Controversial issues in practice. *Social Education* 60 (1): C1–C4.

Gilchrist, G. 1997. The elusive scientific basis of intelligent design theory. *Reports of the National Center for Science Education* 17 (3): 14–15.

Godfrey, L. R., and M. R. Sutherland. 1995. What's growth got to do with it? Process and product in the evolution of ontogeny. *Journal of Human Evolution* 29: 405–31.

Godfrey, L. R., S. J. King, and M. R. Sutherland. 1998. Heterochronic approaches to the study of locomotion. In *Primate locomotion: Recent advances,* ed. E. Strasser, J. Fleagle, A. Rosenberger, and H. McHenry, 277–307. New York: Plenum Press.

Gould, S. J. 1966. Allometry and size in ontogeny and phylogeny. *Biological Reviews* 41: 587–640.

———. 1977. *Ontogeny and phylogeny.* Cambridge, MA: Harvard University Press.

———. 2002. *The structure of evolutionary theory.* Cambridge, MA: Belknap Press of Harvard University.

Hagen, J., D. Allchin, and F. Singer. 1996. *Doing biology.* New York: HarperCollins.

Haldane, J. B. S. 1932. *The causes of evolution.* London: Longmans Green.

Holmes, B. 2000. Take me to your leader. *New Scientist* 166 (2235): 40–43.

Hudson, E. 2003. Creationism vs. evolution central debate behind rejection of textbooks. Maryville, (TN) *Daily Times* (April 5): 1. Available at www.thedaily times.com/sited/story/html/127192 (last accessed April 7, 2003).

Huxley, J. S. 1932. *Problems of relative growth.* Baltimore: Johns Hopkins University Press.

Johnson, P. E. 1997. The unraveling of materialism. *First Things* 77: 22–25.

———. 2001. *The wedge of truth: Splitting the foundations of naturalism.* Downers Grove, IL: InterVarsity Press.

Kenyon, D. H., and P. W. Davis. 1989. *Of pandas and people: The central question of biological origins.* Dallas: Haughton Publishing Co.

———. 1993. *Of pandas and people: The central question of biological origins.* 2nd ed. Dallas: Haughton Publishing Co.

Kipnis, N. 2001. Scientific controversies in teaching science: The case of Volta. *Science and Education* 10 (1–2): 33–49.

Klingenberg, C. P. 1998. Heterochrony and allometry: The analysis of evolutionary change in ontogeny. *Biological Reviews of the Cambridge Philosophical Society* 73: 79–123.

Lande, R., and S. J. Arnold. 1983. The measurement of selection on correlated characters. *Evolution* 37: 1210-26.

Lele, S., and J. R. Richtsmeier. 1991. Euclidean distance matrix analysis: A coordinate-free approach for comparing biological shapes using landmark data. *American Journal of Physical Anthropology* 86: 415–27.

Lockwood, A. L. 1996. Controversial issues: The teacher's crucial role. *Social Education* 60 (1): 28–31.

Magwene, P. M. 2001. New tools for studying integration and modularity. *Evolution* 55: 1734-45.

Margulis, L. 1993. *Symbiosis in cell evolution,* 2nd ed. New York: W. H. Freeman.

Matsumura, M. 2001. Ten significant court decisions. Available from The National Center for Science Education. www.ncseweb.org/articles/5690_10_significant_court_decisi_2_15_2001.asp (updated April 18, 2001; last accessed August 27, 2007).

Mayr, E. 1942. *Systematics and the Origin of Species.* New York: Columbia University Press.

McKinney, M. L., and K. J. McNamara. 1991. *Heterochrony: The evolution of ontogeny.* New York: Plenum Press.

Meyer, S. C., and J. Wells. 2002. Bibliography of supplemental resources for Ohio science instruction. Available from the Discovery Institute's Center for Science and Culture www.discovery.org/viewDB/index.php3/program=CRSC&command=view&id=1127 (created March 11, 2002; last accessed April 9, 2003).

Miller, K. R. 2000. Dissecting the disclaimer. *Reports of the National Center for Science Education* 20 (3): 30–33.

Milner, R. M., and V. Maestro, eds. 2002. Intelligent design? *Natural History* 111 (3): 73–80.

Moss-Coane, M. 2000. Pennsylvania science education standards. *Radio Times.* Philadelphia: WHYY. Available at www.whyy.org/rameta/RT/RT20001221_20.ram (last accessed May 11, 2003).

National Research Council (NRC). 1995. *National science education standards.* Washington, DC: National Academy Press.

Nelson, C. E. 2000. Effective strategies for teaching evolution and other controversial topics. In *The creation controversy and the science classroom,* by J. W. Skehan and C. E. Nelson, 19–50. Arlington, VA: National Science Teachers Association (NSTA) Press.

Nevers, B. 2003. HLS 03-537. House Concurrent Resolution Number 50. Available at www.legis.state.la.us/leg_docs/03RS/CVT1/OUT/0000K4Q7.PDF (last accessed April 7, 2003).

Olson, E., and R, Miller. 1958. *Morphological integration.* Cambridge: Cambridge University Press.

Petto, A. J. 2000. Creeping creationism in Pennsylvania's science education standards. *Reports of the National Center for Science Education* 20 (4): 13–15.

———. 2003. Modification without descent? Non-Darwinian . . . but not anti-evolutionary . . . cellular theory. *Reports of the National Center for Science Education* [forthcoming].

Reule, D. A. 2001. The new face of creationism: The Establishment Clause and the latest efforts to suppress evolution in public schools. *Vanderbilt Law Review* 54 (2455): 2555–2610.

Rice, S. H. 1997. The analysis of ontogenetic trajectories: When a change in size or shape is not heterochrony. *Proceedings of the National Academy of Sciences USA* 94: 907–12.

Richmond, B. G., and D. S. Strait. 2000. Evidence that humans evolved from a knuckle-walking ancestor. *Nature* 404 (6776): 382–85.

———. 2001. Knuckle-walking hominid ancestor: A reply to Corruccini and McHenry. *Journal of Human Evolution* 40: 513–20.

Richmond, B. G., D. R. Begun, and D. S. Strait. 2001. Origin of human bipedalism: The knuckle-walking hypothesis revisited. *Yearbook of Physical Anthropology* 44: 70–105.

Rossi, J. A. 1996. Creating strategies and conditions of civil discourse about controversial issues. *Social Education* 60 (1): 15–21.

Santorum, R. 2002. Illiberal education in Ohio schools. *Washington Times* (March 14): A14.

Saunders, S. 1994. The enduring tension: Darwinian and Lamarckian models of inheritance. In *Strength in diversity: A reader in physical anthropology*, ed. A. Herring and L. Chan, 1–20. Toronto: Canadian Scholars Press.

Scalia, A. 1987. Dissenting opinion. *Edwards v. Aguillard.* 482 U.S. 578.

Scott, E. C. 1996. Evolving euphemisms. *NCSE Reports* 16 (2): 5, 18.

———. 1999. Bleeding Kansas: What happened? What's next? *Reports of the National Center for Science Education* 19 (4): 7–9.

———. 2001. The big tent and the camel's nose. *Reports of the National Center for Science Education* 21 (2): 39–41.

Shea, B. T. 1988. Heterochrony in primates. In *Heterochrony in evolution: A multidisciplinary approach*, ed. M. L. McKinney, 237–66. New York: Plenum Press.

———. 2000. Current issues in the investigation of evolution by heterochrony, with emphasis on the debate over human neoteny. In *Biology, brains, and behavior: The evolution of human development*, ed. S. T. Parker, J. Langer, and M. L. McKinney, 181–213. Sante Fe, NM: SAR Press.

———. 2002. Are some heterochronic transformations likelier than others? In *Human evolution through developmental change*, ed. N. Minugh-Purvis and K .J. McNamara, 79–101. Baltimore: Johns Hopkins University Press.

Simpson, M. 1996. Teaching controversial issues. *Social Education* 60 (1): 5.

Skehan, J. W. 2000. Modern science and the book of Genesis. In *The creation controversy and the science classroom*, by J. W. Skehan and C. E. Nelson, 1–18. Arlington, VA: National Science Teachers Association (NSTA) Press.

Smith, K. C. 2002. Can intelligent design become respectable? *Reports of the National Center for Science Education* 20 (4): 40–43.

Soley, M. 1966. If it's controversial, why teach it? *Social Education* 60 (1): 9–14.

Sonleitner, F. J. 2000. Pandas update. *Reports of the National Center for Science Education* 20 (1–2): 40–46.

Steele, E. J. 1979. *Somatic selection and adaptive evolution: On the inheritance of acquired characteristis.* Chicago: University of Chicago Press.

Sterelny, K. 2001. *Dawkins vs. Gould: Survival of the fittest.* Cambridge, UK: Icon Books Ltd.

Still, J. 1999. Discovery Institute's "Wedge Project" circulates online. Available at www.infidels.org/secular_web/feature/1999/wedge .html (last accessed October 8, 2004).

Van Till, H. 2002. *E. coli* at the no-free-lunchroom: Bacterial flagella and Dembski's case for intelligent design. Washington, DC: American Association for the Advancement of Science. Available at www.aaas.org/spp/dser/evolution/perspectives/vantillecoli.pdf (last accessed September 4, 2003).

Vinicius, L., and M. M. Lahr. 2003. Morphometric heterochrony and the evolution of growth. *Evolution* 57 (11): 2459–68.

Vrba, E. S. 1998. Multiphasic growth and the evolution of prolonged growth exemplified by human brain evolution. *Journal of Theoretical Biology* 190: 227–39.

Waddington, C. H. 1942. Canalization of development and the inheritance of acquired characters. *Nature* 150: 563–65.

———. 1956. *Principles of embryology.* London: Allen and Unwin.

Wells, J. 2000. *Icons of evolution: Science or myth?* Washington, DC: Regnery.

Wilkins, A. 2003. Canalization and genetic assimilation. In *Keywords and concepts in evolutionary developmental biology*, ed. B. K. Hall and W. M. Olson, 23–30. Cambridge, MA: Harvard University Press.

Woese, C. R. 2000. Interpreting the universal phylogenetic tree. *Proceedings of the National Academy of Sciences USA* 97 (15): 8392–96.

———. 2002. On the evolution of cells. *Proceedings of the National Academy of Sciences USA* 99 (13): 8742–47.

Wynn, C.M., and A. W. Wiggins. 2001. *Quantum leaps in the wrong direction: Where real science ends and pseudoscience begins.* Washington, DC: Joseph Henry Press.

AUTHORS

Kenneth Angielczyk is a postdoctoral researcher at the California Academy of Sciences. Most of his work focuses on reconstructing evolutionary relationships among nonmammalian synapsids. His interests also include mass extinctions, phylogenetics theory, and the morphology of the turtle shell. Dr. Angielczyk received a bachelor's degree in biology and geology from the University of Michigan and a PhD in integrative biology from the University of California–Berkeley. He also worked as a researcher at the University of Bristol.

C. Loring Brace is curator for physical anthropology at the University of Michigan Museum of Anthropology in Ann Arbor. He is the author of numerous books and articles on human variation, the concept of race, and human evolution, including *Man in Evolutionary Perspective*; *Human Evolution: An Introduction to Biological Anthropology*; *Atlas of Human Evolution*; and *The Stages of Human Evolution: Human and Cultural Origins.* Since 1967, Dr. Brace has been involved in efforts to test the relationships of the skeletal remains in the museum's collections with pre-Columbian inhabitants both from the Western Hemisphere and from the Old World.

John R. Cole is an anthropologist. He is past president and currently a member of the board of directors of the National Center for Science Education. He is the author of hundreds of publications, including scholarly research and journalistic reports. He is a fellow of the American Anthropological Association and the Committee for the Scientific Investigation of Claims of the Paranormal (CSICOP). He is particularly interested in how the public understands and misunderstands science, and how scientists misunderstand the public. John received his PhD from Columbia University.

G. Brent Dalrymple has a PhD in geology from U.C. Berkeley and is a professor emeritus at Oregon State University. Dr. Dalrymple's primary research interests involve the development and application of isotopic dating techniques to a broad range of geological and geophysical problems. An early research effort to test the hypothesis of geomagnetic field reversal and to determine the time scale for the reversals led to the theory of plate tectonics. He is widely considered an authority on the age of the Earth and has written two books on that subject. Dr. Dalrymple is a member of the national Academy of Sciences and the American Academy of Arts and Sciences, and a Fellow of the American Geophysical Union, where he served as president and a member of the Board of Directors. He received the 2001 Public Service Award from the Geological Society of America and the 2003 National Medal of Science.

Robert Dorit is associate professor of biological sciences at Smith College. His research involves experimental, retrospective, and computational approaches to molecular evolution, including evolutionary changes in weed plants and antibiotic-resistant microbes. He has published several articles on this research in journals focused on molecular biology and biochemical evolution.

Wesley R. Elsberry is information project director at the National Center for Science Education. His research interests include biosonar signal production and physiology in bottlenose dolphins; the effects of sound on marine mammals; cognition and behavior; artificial neural systems; and evolutionary computation. Dr. Elsberry is an expert on the history and rhetoric of anti-evolutionism and president of the TalkOrigins Archive Foundation (www.talk origins.org).

Laurie R. Godfrey is professor of anthropology at the University of Massachusetts–Amherst and editor of the original edition of *Scientists Confront Creationism.* Her research interests are in anatomy and evolution of nonhuman primates. She studies how individual development provides clues to behavior and to the "life history strategies" of extinct species. She uses a variety of techniques to reconstruct the behavior and "lifeways" of extinct animals, with the ultimate goal of being able to reconstruct whole communities of primates in the past, and their transitions to the present. Her particular area of expertise is the lemurs of Madagascar, where she has worked in the field with colleagues for several decades to better understand the recent extinctions of the "megafauna" (including the giant lemurs).

Norman A. Johnson, an evolutionary geneticist, received his PhD from the University of Rochester in 1992. Johnson's primary research interests are in the genetics and evolution of reproductive isolation between nascent species, and the evolution of development. Johnson has taught at the University of Chicago, University of Texas at Arlington, and the University of Massachusetts, where he is currently an adjunct research assistant professor in the Department of Plant, Soil, and Insect Sciences. He is also the author of the forthcoming book *Darwinian Detectives: Revealing the Natural Histories of Genes and Genomes* (Oxford University Press).

Alice Beck Kehoe is professor emerita in anthropology at Marquette University and adjunct professor of anthropology at the University of Wisconsin–Milwaukee. Her books include *Humans: An Introduction to Four-Field Anthropology*; *America Before the European Invasions*; *The Ghost Dance: Ethnohistory and Revitalization*; *North American Indians: A Comprehensive Account*; and *The Kensington Runestone: Approaching a Research Question Holistically.*

Antonio Lazcano is distinguished professor of biology at the National University in Mexico City. He has received several major awards for his contributions to science, scientific journalism, and teaching, including the Gold Medal for Biological Research granted by the University of Puebla. He is the author of several books in Spanish, including *The Miraculous Bacteria,* a collection of scientific essays; *The Spark of Life,* an exploration of the heterotrophic theory of the emergence of life, written for a general audience; and *The Origin of Life,* which has become a best seller with more than a half-million copies sold. He focuses his research on the study of the deepest branches of the tree of life, with a special interest in the last common ancestor of extant life forms and the origin and development of metabolic pathways.

Ronald L. Numbers is Hilldale and William Coleman Professor of the History of Science and Medicine at the University of Wisconsin–Madison. He has written or edited more than two dozen books, including, most recently, *Darwinism Comes to America* (Harvard University Press, 1998); *Disseminating Darwinism: The Role of Place, Race, Religion, and Gender* (Cambridge University Press, 1999), coedited with John Stenhouse; *When Science and Christianity Meet* (University of Chicago Press, 2003), coedited with David Lindberg; and *The Creationists* (expanded edition, Harvard University Press, 2006). He is currently completing a history of science in America (for Basic Books) and coediting the eight-volume *Cambridge History of Science.* He is a past president of both the History of Science Society and the American Society of Church History and the current president of the International Union of History and Philosophy of Science/Division of History of Science and Technology.

Kevin Padian is professor of integrative biology at the University of California–Berkeley and curator at the University of California Museum of Paleontology. He currently serves as the president of the Board of Directors of the National Center for Science Education. Dr. Padian is well known as an international expert on the evolution of major adaptations in the history of vertebrates and as a tireless advocate of educational outreach. His major research interests include the study of evolutionary transitions, especially in the evolution of flight, and in the changes in structure and function in land animals across the Triassic-Jurassic boundary. He has written extensively on the evolution of flight and on the meaning and importance of transitional forms in the fossil record. He is a recipient of the Carl Sagan Prize for the Popularization of Science.

Robert T. Pennock is professor of history and philosophy of science at Michigan State University, where he is on the faculty of the Lyman Briggs School of Science, the Philosophy Department, the Department of Computer Science, and the ecology, evolutionary biology, and behavior graduate program. He has published more than a dozen articles about the "intelligent design" creationist movement. His book *Tower of Babel: The Evidence against the New Creationism* has been reviewed in more than fifty publications; the *New York Review of Books* called it "the best book on creationism in all its guises." He edited *Intelligent Design Creationism and Its Critics: Philosophical, Theological, and Scientific Perspectives*, which is the most comprehensive sourcebook on the topic. Dr. Pennock's scientific research on experimental evolution and evolutionary computation has been featured in many magazines, including a cover story in *Discover.* He was named a national distinguished lecturer by Sigma Xi, the Scientific Research Society, and speaks regularly around the country on issues relating to science and values.

Andrew J. Petto is lecturer in anatomy and physiology in the Department of Biological Sciences at the University of Wisconsin–Milwaukee and editor of *Reports of the National Center for Science Education.* He has been active in public education—especially in evolution education—for more than thirty years, conducting professional development programs for teachers, special summer research programs for students in grades 6–12, and undergraduates. He has served as a consultant on science education standards in life sciences and evolution in several U. S. states and for the District of the Western Cape in South Africa. His current research is focused on how students learn underlying principles, such as evolution, in the context of general education courses in colleges.

J. Michael Plavcan is an associate professor in anthropology at the University of Arkansas. He received his PhD in biological anthropology and anatomy from Duke University in 1990, and he specializes in primate and human evolution. He has closely followed and studied creation science and creation scientists for more than two decades.

Eugenie C. Scott is executive director of the National Center for Science Education. She has been both a researcher and an activist in the creationism/evolution controversy for more than twenty years and has been involved in many aspects of this controversy—including educational, legal, scientific, religious, and social issues. She has received national recognition for her NCSE activities, including awards from the National Science Board, the American Society for Cell Biology, the American Institute of Biological Sciences, the Geological Society of America, and the American Humanist Association. Her most recent book is *Evolution vs. Creationism.*

Victor J. Stenger is emeritus professor of physics at the University of Hawaii and adjunct professor of philosophy at the University of Colorado. He has held visiting positions at the Universities of Heidelberg, Oxford, and Florence. Professor Stenger's research career spanned the period of great progress in elementary particle physics that ultimately led to the current standard model. Professor Stenger is the author of *Not by Design*; *Physics and Psychics*; *The Unconscious Quantum*; *Timeless Reality*; *Has Science Found God?*; *The Comprehensible Cosmos*; and the forthcoming *God: The Failed Hypothesis.*

INDEX

Page numbers in *italics* refer to illustrations and tables.